Lithium Niobate-Based Heterostructures (Second Edition)

Synthesis, properties, and electron phenomena

Online at: https://doi.org/10.1088/978-0-7503-6305-1

Lithium Niobate-Based Heterostructures (Second Edition)

Synthesis, properties, and electron phenomena

Maxim Sumets

Department of Science, Grayson College, Denison, TX, USA

IOP Publishing, Bristol, UK

ISBN   978-0-7503-6305-1 (ebook)
ISBN   978-0-7503-6303-7 (print)
ISBN   978-0-7503-6306-8 (myPrint)
ISBN   978-0-7503-6304-4 (mobi)

DOI   10.1088/978-0-7503-6305-1

Version: 20241001

IOP ebooks

British Library Cataloguing-in-Publication Data: A catalogue record for this book is available from the British Library.

Published by IOP Publishing, wholly owned by The Institute of Physics, London

IOP Publishing, No.2 The Distillery, Glassfields, Avon Street, Bristol, BS2 0GR, UK

US Office: IOP Publishing, Inc., 190 North Independence Mall West, Suite 601, Philadelphia, PA 19106, USA

*To my sons Maxim and Ustin, whose boundless curiosity and love inspire
me to reach beyond what I once thought possible.*

Contents

Preface		**x**
Acknowledgments		**xiv**
Author biography		**xv**

1   **Lithium niobate thin films: potential applications, synthesis methods, structure, and properties**   **1-1**

1.1	The structure and main properties of bulk lithium niobate	1-1
1.2	Applications of thin $LiNbO_3$ films	1-4
	1.2.1 Waveguide-based optical devices	1-4
	1.2.2 Surface acoustic wave devices	1-5
	1.2.3 Memory units and neuromorphic systems	1-6
1.3	Fabrication methods of thin $LiNbO_3$ films	1-7
	1.3.1 Liquid-phase epitaxy	1-8
	1.3.2 Chemical vapor deposition	1-8
	1.3.3 Sol–gel process	1-8
	1.3.4 Pulsed laser deposition	1-8
	1.3.5 Radio-frequency magnetron sputtering	1-9
1.4	Fundamentals and critical parameters of the RFMS method and the ion-beam sputtering method	1-11
1.5	Electrical properties and charge transport phenomena in $LiNbO_3$-based heterostructures	1-22
	1.5.1 Charge carriers in $LiNbO_3$	1-22
	1.5.2 Electrical conductivity limited by contact phenomena	1-24
	1.5.3 Electrical conductivity limited by bulk properties	1-28
1.6	Major application-focused challenges in the synthesis of thin $LiNbO_3$ films	1-35
	1.6.1 Synthesis of epitaxial LN films with low mechanical stress	1-36
	1.6.2 Ferroelectric properties and polarization reversal	1-39
	1.6.3 Defect concentration and performance of thin LN films	1-42
1.7	Summary and discussion	1-44
	References	1-45

2   **Synthesis, structure, and surface morphology of $LiNbO_3$ films**   **2-1**

2.1	Technological regimes for the synthesis of thin $LiNbO_3$ films by radio-frequency magnetron sputtering and ion-beam sputtering methods	2-1

2.2   Direct growth of polycrystalline LiNbO$_3$ films   2-3

     2.2.1   Films deposited by RFMS   2-4

     2.2.2   Films deposited by ion-beam sputtering   2-10

2.3   Influence of the synthesis regime and subsequent annealing   2-14
on the composition and structural properties of LiNbO$_3$ films

     2.3.1   Plasma effect   2-14

     2.3.2   Effect of reactive gas pressure   2-17

     2.3.3   Effect of substrate type   2-21

     2.3.4   Effect of oxygen presence in the gas reaction environment   2-21

     2.3.5   Effect of thermal annealing   2-25

     2.3.6   Early stages of lithium niobate film growth   2-27

     2.3.7   Crystallization of amorphous Li–Nb–O films   2-37

2.4   Summary and discussion   2-43

     References   2-44

3   **Electron phenomena in LiNbO$_3$-based heterostructures**   **3-1**

3.1   Basic electrical properties of LiNbO$_3$ thin films in Si–LiNbO$_3$ 3-1
heterostructures

     3.1.1   Capacitance–voltage and current–voltage characteristics of   3-2
LiNbO$_3$-based heterostructures

     3.1.2   Ferroelectric properties of LiNbO$_3$ thin films   3-13

3.2   Conduction mechanisms in (001)Si–LiNbO$_3$ heterostructures   3-20

     3.2.1   Region of average electric fields in the temperature interval   3-23
of $T = 90$–140 K

     3.2.2   Region of average electric fields in the temperature range   3-23
of $T = 140$–300 K

     3.2.3   Region of high electric fields   3-24

3.3   Band diagram of the Si–LiNbO$_3$ heterostructures   3-27

3.4   Impedance spectroscopy and AC conductivity of thin LiNbO$_3$ films   3-38

3.5   Summary and discussion   3-48

     References   3-51

4   **Effect of sputtering conditions and post-growth treatment on**   **4-1**
electron phenomena in Si–LiNbO$_3$ heterostructures

4.1   Effects of spatial plasma inhomogeneity and composition and the   4-2
relative target–substrate position

     4.1.1   Capacitance–voltage characteristics and ferroelectric properties   4-2
of Si–LiNbO$_3$–Al heterostructures

4.1.2 Current–voltage characteristics of Si–LiNbO$_3$–Al heterostructures   4-11

4.1.3 Impedance spectroscopy and AC conductivity of Si–LiNbO$_3$–Al heterostructures   4-17

4.2 Thermal annealing effect on electrical properties of Si–LiNbO$_3$ heterostructures   4-31

4.2.1 Capacitance–voltage characteristics and ferroelectric properties of Si–LiNbO$_3$–Al heterostructures   4-34

4.2.2 Current–voltage characteristics of Si–LiNbO$_3$–Al heterostructures   4-42

4.3 Impedance spectroscopy of Si–LiNbO$_3$–Al heterostructures after thermal annealing   4-47

4.4 Optical bandgap of thin LiNbO$_3$ films produced by different fabrication regimes   4-50

4.4.1 Dependence of the optical bandgap shift in thin LiNbO$_3$ films on RFMS conditions and subsequent thermal annealing   4-50

4.4.2 Effect of pulsed photon treatment on the optical bandgap of LiNbO$_3$ films   4-55

4.5 Temperature-induced transition from p-type to n-type conduction in LiNbO$_3$/Nb$_2$O$_5$ polycrystalline films   4-65

4.6 Summary and discussion   4-74

References   4-77

5 Oxide charge: localization, evolution, and related phenomena at heterojunctions   **5-1**

5.1 Electrical properties and crystallization of amorphous Li–Nb–O films on silicon   5-2

5.2 Evolution of oxide charge during the crystallization of amorphous Li–Nb–O films   5-11

5.3 Transport properties and crystallization of Li–Nb–O films   5-20

5.4 Charge phenomena at Si–LiNbO$_3$ heterojunctions   5-29

5.5 Charge phenomena in NiSi$_2$–LiNbO$_3$ heterostructures   5-40

5.6 Summary and discussion   5-51

References   5-52

6 Bonus chapter: multifunctional Si–LiNbO$_3$ heterostructures for nonvolatile memory units   **6-1**

References   6-11

Appendix A   **A-1**

Preface

The use of ferroelectric materials in memory devices and high-speed integrated optics is of significant interest due to their exceptional properties, including energy nonvolatility, fast switching, radiative stability, and unique optoacoustic and electro-optic characteristics. Thin-film forms of these materials are particularly attractive because of their potential for integration with conventional semiconductor technology and optoelectronic devices. In addition, their excellent optical localization in waveguide geometries allows for the utilization of their nonlinear optical properties.

Lithium niobate (LN, $LiNbO_3$), a ferroelectric material, is distinguished by a unique combination of physical properties, such as a high Curie temperature ($T_c = 1210$ °C), a wide bandgap, and high piezoelectric and electro-optic coefficients. These attributes make it one of the most promising materials for ferroelectric and nonlinear optical applications.

In the last two decades, the potential integration of LN with conventional silicon technology has spurred the development and fabrication of versatile integrated optical devices. These devices include nonvolatile memory devices, Mach–Zehnder modulators, surface acoustic wave (SAW) filter modulators, frequency filters, microring resonators, and optical amplitude modulators, among others.

In integrated optics, thin-film technology is particularly advantageous. It allows materials to be deposited onto inexpensive, wide-area substrates, a process that overcomes numerous technological challenges associated with bulk material growth. Consequently, LN-based photonic elements, such as heterostructures and photonic wires, can significantly advance the development of integrated devices with active elements, including optical modulators, tunnel filters, nonlinear wavelength converters, and various tunnel lasers. The fabrication of thin LN films with a high degree of crystallinity is a critical goal in this field. Therefore, the synthesis of high-quality LN films with excellent substrate interfaces is of paramount interest.

Radio-frequency magnetron sputtering (RFMS) is one of the most effective vacuum deposition techniques for producing thin films of complex oxides. This method preserves the initial elemental composition and ensures better adhesion to the substrate. However, the high sensitivity of film properties to RFMS parameters offers both opportunities and challenges. It enables the fabrication of films with the desired characteristics but also poses reproducibility issues. Addressing these issues requires the development and maintenance of precise RFMS regimes to produce LN films with the desired structural, electrical, ferroelectric, and optical properties. The complexity of achieving stable fabrication of LN films with these properties is due to the wide range of critical parameters involved.

Solving this multifaceted problem is essential for the practical application of LN-based heterostructures. In numerous studies of the synthesis of LN films over the last two decades, the greatest focus has been given to the impact of technological parameters on the structure, composition, and morphology of these films. The electrical properties of LN-based heterostructures, which depend on the fabrication regime, are less extensively covered in the research literature due to the uncertain nature of the factors that influence them.

The electrical properties of LN films reported in the literature are often complementary to structural studies, largely due to the lack of systematic analysis of electronic phenomena in complex structures such as LN-based heterostructures.

Considering the above points, a comprehensive characterization of the substructure, composition, optical, and ferroelectric properties of LN-based heterostructures fabricated by various methods, along with an in-depth analysis of their internal electronic phenomena, is of significant practical and fundamental interest.

This book delves into the fundamental principles of fabricating LN-based heterostructures, exploring their properties and the electronic phenomena that underpin their efficient application in various devices. The majority of the experimental results presented in this monograph are original, stemming from fifteen years of dedicated research on LN thin films, their fundamental properties, and their applications. The first edition has garnered very positive feedback from numerous leading experts worldwide and has attracted significant attention from readers.

This amazing book covers the majority of fundamental and applied aspects associated with LiNbO₃-based heterostructures as a basis for a new generation of multifunctional devices.
—Dr Laurent Bellaiche, University of Arkansas (USA)

There is no doubt that Dr Sumets's results, reflected in his comprehensive monograph, are extremely valuable not only for our group but also for many researchers working in the Materials Science field.
—Prof. Chennupati Jagadish, Australian National University (Australia)

This is a comprehensive book covering the original optimal synthesis methods and basic properties of LN-based heterostructures, with their potential applications for nonvolatile memory units, optoelectronic devices, and many other prospective devices.
—Dr Housei Akazawa, Nippon Telegraph and Telephone Corporation (Japan)

Compared to the first edition, which had four chapters, the second edition has been expanded to include new material on the growth and crystallization of amorphous Li–Nb–O films, along with their electrical and optical properties. In addition, the new edition presents a comparative analysis of conventional thermal annealing and a novel post-deposition treatment, pulsed photon treatment, and their influence on the properties of LN-based heterostructures. A new fifth chapter, written specifically for the second edition, discusses the formation and evolution of oxide charge and other electronic phenomena in LN-based heterostructures. Furthermore, a sixth chapter on the application of the fabricated heterostructures for nonvolatile memory units has been added to the second edition of the monograph.

The second edition of this book contains five chapters.

The first chapter of the monograph presents both the basic physical properties of bulk LN and advancements in using LN in thin-film form. It includes a comparative analysis of the most efficient fabrication techniques, focusing on radio-frequency magnetron sputtering, and discusses their advantages and disadvantages. This chapter also provides a contemporary review of some existing and unresolved technological problems related to the formation of epitaxial LN films, their quality, and the combination of physical properties required for their successful application in new device generations.

In the experimental section (chapter 2), two alternative approaches for synthesizing LN films on various substrates are described and discussed: (1) the direct growth of polycrystalline LN films during RFMS, and (2) the deposition of amorphous Li–Nb–O films followed by subsequent crystallization. All steps and regimes are investigated, and the optimal sputtering conditions are proposed for forming high-quality c-oriented LN films on various substrates. Special attention is paid to the evolution of the structure, composition, and morphology of amorphous LN during crystallization. The early stages of LN film growth are described in detail, which is vital for the controllable fabrication of LN-based heterostructures. For the first time, based on the results obtained, the structure of deposited Li–Nb–O films previously thought to be amorphous is defined as 'pseudo amorphous'—it is extremely nanostructured, consisting of appropriate nuclei of crystalline lithium oxides, niobium oxides, and LN produced by the reaction between them.

Following this, based on experimental current–voltage (I–V) and capacitance–voltage characteristics, many of the electrical properties of as-grown LN-based heterostructures are discussed in chapter 3. Specifically, the charge transport mechanisms are identified in various temperature and electric field ranges. An energy band diagram is proposed for the studied LN–Si heterostructures. The essential parameters of the electronic states in LN films and at the LN/substrate interface are determined. It is revealed that the RFMS and IBS regimes significantly affect these parameters as well as the ferroelectric properties of LN films. Using the impedance spectroscopy technique, the relaxation phenomena in LN–Si heterostructures are investigated and thoroughly described.

Since sputtering parameters affect the properties of LN-based heterostructures, a significant part of the monograph (chapter 4) focuses on studying how the reactive gas composition and pressure, the substrate temperature and position, and the reactive plasma impact these properties. Considering the requirements for the successful application of thin LN films in multifunctional optoelectronic devices, optimal RFMS regimes are developed for LN–Si heterostructures. In addition, the original results presented in the monograph reflect how post-deposition treatments affect the electrical, ferroelectric, and optical properties of LN-based thin-film heterostructures. Specifically, the positive effect of thermal annealing (TA) at temperatures of up to 700 °C on minimizing positive oxide charge is discussed. Although conventional TA demonstrates good performance, a new energy-saving and fast-working alternative technique called pulsed photon treatment has been successfully applied.

Currently, there is little information regarding the positive charge inevitably present in LN–Si heterostructures, which limits their applications. Therefore, our research group made considerable efforts to investigate this phenomenon. The nature, localization, and evolution of the positive oxide charge during the crystallization of amorphous LN films are studied and presented in a separate fifth chapter of the monograph. In addition, the electronic phenomena at several heterojunctions, which play an important role in applications, are thoroughly examined.

As a bonus chapter, this monograph presents well-structured results on the potential applications of synthesized LN-based heterostructures as foundational elements for nonvolatile memory units.

All scientific results presented in this monograph have been reported at various conferences and symposiums, and this is reflected in the cited publications.

Acknowledgments

This work would not have been possible without the outstanding talent, collaboration, and support of my research colleagues. The formulation of objectives, research planning, and discussions were carried out jointly with Professor Valentin Ievlev (Voronezh State University, Russian Academy of Sciences). I would like to acknowledge their valuable help of Dr Alexander Kostuchenko (Voronezh State Technical University, Russia), Dr Evgeny Belonogov (Voronezh State Technical University, Russia), Dr Gennadiy Kotov (Voronezh State University of Engineering Technologies, Russia), Dr Oleg Ovchinnikov (Voronezh State University, Russia), Professor Vladimir Shur (Ural Federal University, Russia), Dr Vladislav Dybov (Voronezh State University, Russia), and Dr Dmitri Serikov (Voronezh State Technical University, Russia) with the synthesis and measurements associated with LN-based heterostructures.

I would also like to thank my brother, Dr Pavel Sumets, for assisting me with editing this monograph.

As was the case when I was writing the first edition of this book, my special and sincere thanks go to Natalia, my beloved wife. Supporting my work, she still expected me to spend much more time with her, while I spent most of my free time conducting research, discussing results, and preparing the second edition of this monograph.

Author biography

Maxim Sumets

Maxim Sumets is a physics professor in the Science Department at Grayson College, USA, specializing in Materials Science with a focus on thin films, semiconductor heterostructures, ferroelectrics, and their applications. He earned his master's and PhD degrees from Voronezh State University, Russia and has been actively engaged in research and education for over 25 years.

As an esteemed expert in his field, Dr Sumets serves as a reviewer for numerous reputable scientific journals. His research interests encompass the electrical and structural properties of materials.

IOP Publishing

Lithium Niobate-Based Heterostructures (Second Edition)
Synthesis, properties, and electron phenomena
Maxim Sumets

Chapter 1

Lithium niobate thin films: potential applications, synthesis methods, structure, and properties

This chapter provides a contemporary overview of the main properties of bulk lithium niobate (LN, LiNbO₃), highlighting the advantages of thin LN films and their various applications in integrated electronics and optoelectronics. The chapter discusses different deposition techniques, each with its specific advantages and disadvantages. The structural and electrical properties of thin LN films are significantly affected by the deposition regime used. Charge transport in LN-based heterostructures is influenced by both contact and bulk electron phenomena. In addition, this chapter addresses the major application-focused challenges in the synthesis of thin LN films.

1.1 The structure and main properties of bulk lithium niobate

In 1965, bulk lithium niobate was grown by the Czochralski method for the first time [1], and Abrahams investigated its structure in a series of works [2, 3]. LN belongs to an R3c space group, and its oxygen atoms are arranged in nearly hexagonal close-packed planar sheets, as shown in figure 1.1. The unit cell is either rhombohedral (trigonal, $a = 5.4944$ Å, $\alpha = 55°52'$) or hexagonal ($a_H = 5.1483$ Å, $c_H = 13.8631$ Å, $c/a = 2.693$) with six formula units per unit cell. The stacking sequence of cations in these octahedral sites is Nb, Li, vacancy, Nb, Li, vacancy, and so on [3]. The octahedral interstices formed in this structure are one-third filled by lithium atoms, one-third filled by niobium atoms, and one-third vacant.

Thus, bulk crystals with a stoichiometric composition ($R = $ Li/Nb $= 1$) have a nearly ideal structure. In both lithium niobate crystals and congruent crystals ($R = 0.946$), a cation sublattice is considerably disordered. As demonstrated, excess niobium ions replace lithium ions at the lithium positions; the resulting fairly loose cation sublattice allows various ions to be introduced into it [4].

doi:10.1088/978-0-7503-6305-1ch1

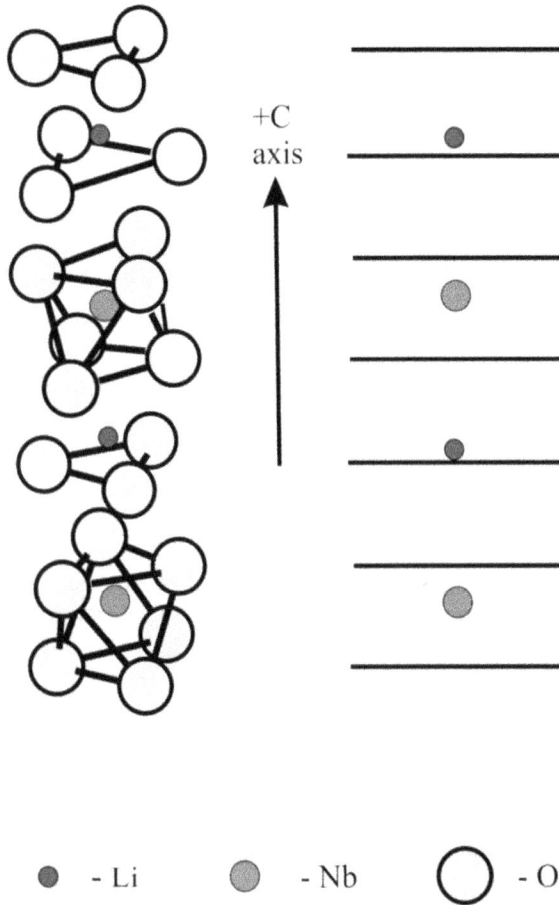

Figure 1.1. Positions of the lithium atoms and niobium atoms with respect to oxygen octahedra in the ferroelectric phase of lithium niobate. Horizontal lines in the diagram on the right represent the oxygen layers.

The distance from the Li^+ ion to the nearest oxygen plane is 0.37 Å. Its lattice asymmetry makes lithium niobate a polar material, and oxygen atoms reside in the oxygen triangles shown in figure 1.1. To reverse the polarization, the Li^+ ion must be replaced on the other side of the nearest oxygen triangle, whereas the Nb^{5+} ion moves only within the oxygen cages. Two stable positions (before and after reversal) determine two possible spontaneous polarization directions.

In stoichiometric crystals, defects are affected by the cation sublattice ordering, and they play a crucial role in the formation of the material's optical characteristics. Even insignificant changes in the Li/Nb ratio or the presence of impurities—for example, the replacement of Li and Nb ions and the occupation of vacant octahedral voids—lead to local changes in the cation order along the polar axis. Impurity concentrations in the order of hundredths of a percent can greatly influence the dielectric and optical properties of these crystals, leading to negative effects such as laser damage.

The observed inhomogeneity in the cation sublattice of lithium niobate changes the translational invariance of a structure and does not disrupt the unit cell symmetry. The uniform composition along the growth axis does not mean that the crystal corresponds to the congruent melting conditions. The degree of crystallinity can be greatly improved if the growth process is performed under an electric field or through thermal annealing near the melting point accompanied by a weak electric field. Single crystals of lithium niobate grown under congruent melting conditions have a disordered cation sublattice. They are very sensitive to laser damage, limiting their application in optical devices. There is no principal difference between congruent and stoichiometric crystals from the physical and chemical points of view—they differ only in the degree of defectiveness.

At the Curie temperature of 1210 °C [5], LN manifests a phase transition from the ferroelectric to a paraelectric state, corresponding to the $R\bar{3}$ space group. The density of LN single crystals depends on the Li content, and it equals 4.648 g cm^{-3} for congruent LN [4].

A schematic phase diagram of the Li_2O–Nb_2O_5 system is shown in figure 1.2.

As shown in figure 1.2 and demonstrated in [6], LN is a material with a variable composition, demonstrating a wide range of solid solutions. Specifically, there are areas in the phase diagram consisting of both LN oxide and either Li_3NbO_4 or $LiNb_3O_8$ that are centrosymmetric (non-ferroelectric) phases. Therefore, LN single crystals must be grown with extra care and at low temperatures because a slight deviation results in the formation of Li_3NbO_4 or $LiNb_3O_8$ oxides, provoking degradation of ferroelectric properties.

With the aim, for example, of observing better grain boundary or forming the required topology for thin film devices, chemical etching sometimes attracts interest. Although lithium niobate is relatively inert at room temperature, it has been

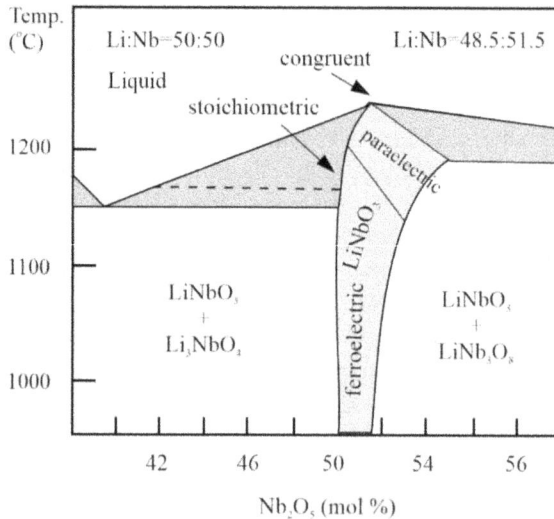

Figure 1.2. Schematic phase diagram of the Li_2O–Nb_2O_5 pseudo binary system near the congruent and stoichiometric compositions of LN.

reported that LN single crystals can be etched by boiling (110 °C) in an HF:HNO$_3$ (1:2) mixture [7]. Nevertheless, the etching of thin LN films by this method leads to their exfoliation from the substrate after a few minutes. Reactive ion etching (RIE) ('dry') using various reactive gas recipes is among the most popular methods in contemporary thin film technologies. Regarding lithium niobate, there are a wide range of etching approaches that utilize a variety of fluorine-based and chlorine-based plasmas, such as CF_4, CCl_2F_2, and CHF_3 [8]. For instance, using a $CF_4/Ar/H_2$ reactive gas environment, it was revealed [9] that the etch rate and degree of surface crystallinity depend on the etching duration and H_2 flow rate. Other authors have reported that depending on the compound ratio in their Ar/C_3H_8 reactive gas mixture, various degrees of etching anisotropy are attainable [10].

It is generally accepted that the coefficient of thermal expansion is a critical parameter for understanding lattice distortion. Due to symmetry, there are two linear coefficients of thermal expansion for lithium niobate (longitudinal and transverse) with magnitudes of $\alpha_{11} = 15.0 \times 10^{-6}$ and $\alpha_{33} = \alpha_{11} = 7.5 \times 10^{-6}$, respectively.

The design of memory units requires controllable and tuned dielectric permittivity. According to axial symmetry, the dielectric permittivity tensor for lithium niobate has only two independent elements, whereas the low-frequency dielectric permittivity of bulk LN is $\varepsilon_{11} = 30$, $\varepsilon_{33} = 80$ [7].

Both optical properties and electro-optic effects are the most attractive features of lithium niobate. Bulk lithium niobate is an optically transparent material in the wavelength range from 0.35 to 5 µm where lattice absorption is observed. Even though the optical properties of lithium niobate are well established, the optical bandgap reported in the research literature varies significantly. For example, theoretical and experimental works have shown that the optical bandgap ranges between 3.57 eV [11] and 4.7 eV [12]. Some authors argue that this phenomenon occurs due to the presence of defects in LN, generating localized states (even the band tails) and influencing absorption in single crystals. Lithium niobate manifests birefringence, exhibiting principal ordinary and extraordinary refractive indexes of $n_o = 2.289$ and $n_e = 2.201$, respectively ($\lambda = 633$ nm). The refractive indexes depend on the Li/Nb ratio and vary significantly depending on the composition [13].

The results reported for the electrical properties of lithium niobate in the research literature are controversial; they are affected by the structure and composition of the studied material. Thus, this issue will be discussed in the following chapters.

1.2 Applications of thin LiNbO$_3$ films

The idea of ferroelectric–semiconductor integration was realized in the 1960s. Since then, a wide range of multifunctional devices utilizing the unique properties of ferroelectrics (nonvolatile memory, electro-optic modulation, ferroelectric sensitivity) has been created. In this section, we briefly review some of the most popular devices.

1.2.1 Waveguide-based optical devices

Lithium niobate shows promise as a key component of electro-optic devices, particularly high-speed (>20 GHz) modulators. In contrast to bulk material, thin

LN films provide higher intensity per unit power in waveguides and hence stronger nonlinear optical effects and shorter interaction lengths. The ability to fabricate such films on various substrates extends the spectrum of their possible applications.

One of the key issues associated with the application of thin LN films in waveguides is an understanding of the mechanisms that allow us to exclude or minimize optical losses (absorption, leakage, bulk scattering, surface scattering, and scattering triggered by grain boundaries) and to satisfy the requirements of the film grain boundary structure.

Optical leakage takes place, for instance, when a thin film waveguide is fabricated on a substrate with a high refractive index (for example, Si with $n = 3.8$ for $\lambda = 633\,nm$), and an optical mode then inevitably 'leaks' into the substrate. The degree of leakage depends on the film thickness. To achieve the maximum allowed leakage of $1\,dB\,cm^{-1}$, LN films must have a thickness of about 400 nm [14].

Bulk scattering is an essential process for optical mode coupling in a waveguide when the optical mode interacts with the radiative one. The change in the refractive index inevitably leads to scattering in the waveguide. This can be caused by any inhomogeneity in the waveguide material, for example, the presence of amorphous inclusions, parasitic phases ($LiNb_3O_8$, Li_3NbO_4), or grain disorder in a polycrystalline film. Even in epitaxial films, significant bulk scattering can be observed due to the presence of subgrains and mechanical stress. The surface scattering and scattering at the grain boundaries depend greatly on surface roughness and can overwhelm the bulk scattering. Roughness of the order of 1 nm is critical, posing a complex technological problem that requires a solution in the growth stage of LN film production. It has been shown that regardless of the fabrication method of thin ferroelectric films on sapphire substrates, there is a stable correlation between film thickness and surface roughness, which is 1% of the thickness [15]. The authors explained that this tendency was due to the effect of grain boundary groove formation caused by competition between the surface energy of grain boundaries and the energy of the free surface in the process of island coalescence. Total energy minimization originates from groove formation.

As a result, the main way to decrease the optical losses is to fabricate high-quality, two-axis textured or epitaxial LN films, which is a challenging problem.

1.2.2 Surface acoustic wave devices

Surface acoustic waves (SAWs) transfer the major part of their energy in the surface layer of a material, i.e. within a thickness of about one or two acoustic wavelengths. This allows them to interact with matter on such surfaces. Moreover, SAWs can have an electric field associated with them if they are generated in a piezoelectric material. This interaction leads to changes in the wave characteristics (amplitude, phase, speed, etc.) that underlie the operation of chemical sensors. Here, we do not discuss the operational principle of such sensors (it has been described in detail in some comprehensive reviews, for example, [16]); instead, we focus only on requirements relevant to the fabrication of these materials.

Materials for SAW devices should primarily possess the following properties: the piezoelectric effect, a high Curie temperature, significant resistivity, and high piezoresponse. Lithium niobate has all the listed properties, so it is ranked as the most popular piezoelectric material, surpassing competitors such as $Pb(ZrTi)O_3$, $BaTiO_3$, and $Bi_4Ti_3O_{12}$. Specifically, lithium niobate is characterized by extremely high values for both its Curie temperature and its piezoelectric constant, making it a unique material for high-temperature acoustic sensors.

It has been emphasized [17–19] that the parameters of SAW filters are affected to a great extent by the structure and surface morphology of LN. It was demonstrated [18] that SAW filters had poor characteristics due to the high surface roughness and porosity of LN films. The authors of [19] associated the presence of preferable orientation and low surface roughness with the basic film characteristics for SAW filters, which depend on the fabrication technique. Emphasizing the challenge of forming c-oriented LN films using the chemical vapor deposition (CVD) process, the authors of [20] demonstrated that their fabricated SAW device operated at a frequency of 6.3 GHz, which corresponds to an SAW velocity of the order of $12\,500\,\text{m s}^{-1}$, i.e. a value that exceeds the analogous characteristic for bulk LN.

The authors of [21] investigated the fabrication of an SAW filter based on LN films deposited on diamond substrates by the radio-frequency magnetron sputtering (RFMS) method. They reported a strong dependence of the structure on the fabrication regime, and the optimal parameters (substrate temperature, RF magnetron power, reactive gas composition) were given. Single-phase films with low surface roughness ($\sim 10\,\text{nm}$) allowed SAW filters to operate at a frequency of 2 GHz and an SAW speed of $8200\,\text{m s}^{-1}$.

1.2.3 Memory units and neuromorphic systems

Thin LN films can be the basis for the creation of various types of RAM[1], such as FRAM (FeRAM)[2] and RRAM (ReRAM)[3].

A wide range of works has focused on the study of the possible application of thin LN films in FRAM units (for example [22–24]). In doing so, the following issues arise:

- the fabrication of single-phase, c-oriented, and defect-free LN films with ferroelectric properties close to those of single-crystal lithium niobate;
- the creation of an ideal interface at the film–substrate heterostructure.

Provided these issues are solved, thin LN films can replace traditional ferroelectric materials ($PbZr_xTi_{1-x}O_3$, $BaTiO_3$, and $SrBi_2Ta_2O_9$) as a functional element of contemporary FRAM units. The use of metal, LN, and Si in metal–oxide–semiconductor (MOS)-like heterostructures potentially allows the introduction of FRAM units, which adds two masking steps to the process of fabricating regular

---

[1] RAM—*random-access memory.*
[2] FRAM—*ferroelectric random-access memory.*
[3] RRAM—*resistive random-access memory.*

complementary MOS (CMOS) structures. Such an approach makes it possible to integrate FRAM into microcontrollers. Simplifying the technological stages required for the creation of FRAM units would allow FRAM to compete with the currently dominant flash memory due to its significant advantages [25].

The possibility of utilizing LN films in RRAM units and neuromorphic systems was demonstrated in [26–29]. The main aim of this research was the development of a fundamental model describing the behavior of memristors. For example, [30, 31] reported the memristive effect in TiO_2 films, where the change of resistance was associated with the motion of oxygen vacancies activated by electrical current (the electroforming effect). In [32], the memristive effect was revealed in VO_2 films, which was caused by the insulator–metal transition in these structures.

The memristive effect in LN-based heterostructures is explained in [26–29] by the motion of oxygen vacancies. However, the resulting theoretical predictions do not correlate with experimental data (see [27]). Moreover, some evidence pointing to the role of Li ions in the process of memristive switching was presented in [28].

Currently, LN seems to be an alternative material for RRAM, although solutions based on this material are more technologically complex, making it hard for them to compete with RRAM technology based on amorphous SiO_2. Nevertheless, LN is more promising in the context of neuromorphic systems [26, 28].

As a result of the creation of RAM units based on FRAM and RRAM, it will be possible to produce more reliable, high-speed, durable, and protected storage memory compared to flash memory units or HDDs.[4] At the same time, neuromorphic systems can become the basis for creating self-learning artificial intelligence.

The potential application of LN films has been greatly expanded due to the discovery of their ferromagnetic property. The authors of [33, 34] proposed the doping of films by Co ion implantation such that the Co ions replaced the Nb ions in lithium niobate. This mechanism is supported by the fact that oxygen vacancies are responsible for the ferromagnetic properties of LN nanoparticles [35, 36]. As stated in [35], the mechanical–chemical synthesis of LN with subsequent annealing leads to the formation of a high concentration of oxygen vacancies and allows for the fabrication of samples manifesting ferromagnetic properties.

Although the practical application of the multiferroic properties of lithium niobate is at the development stage, active and promising work on LN film synthesis is a precursor to the integration of LN films into optoelectronic devices.

1.3 Fabrication methods of thin LiNbO$_3$ films

When selecting an appropriate synthesis method for thin LN films, it is crucial to consider the method's ability to preserve the elemental composition and structure of the bulk material, as well as its reproducibility. Various fabrication techniques have been proposed to meet these requirements, including the sol–gel method, pulsed laser deposition (PLD), discrete thermal evaporation in vacuum, RFMS, CVD, liquid-phase epitaxy (LPE), ion-beam sputtering (IBS), and the Pechini method [37].

---

[4] HDD—*hard (magnetic) disk drive.*

1.3.1 Liquid-phase epitaxy

In 1975, the possibility of synthesizing LN films by LPE on the surface of single-crystal $LiTaO_3$ was demonstrated [38, 39]. Although the LPE method is now rarely used for fabricating LN films, some researchers have reported the successful fabrication of optical waveguides on single crystals of LN [40, 41]. It has been noted that doping the films with Zn atoms increases surface roughness as the dopant concentration rises. Under these conditions, a low degree of optical loss was achieved (0.154 ± 0.002 dB mm^{-1} at $\lambda = 1$ μm), and the surface roughness was minimized to 0.2–0.3 nm. During the process of pseudomorphic epitaxial growth, an intermediate Zn:LN layer forms, in which the Zn content varies according to the substrate temperature. In addition, the refractive index difference at the film–substrate interface depends on the Zn concentration in the LN film [40].

1.3.2 Chemical vapor deposition

In the CVD process, the reactive gas mixture used in the reaction chamber significantly impacts the synthesis of LN films. At low reactive gas pressures (approximately 266.6 Pa), polycrystalline films comprising two phases, namely LN and $LiNb_3O_8$, are formed on sapphire substrates. When the reactive gas pressure is increased to around 2666 Pa, single-phase LN films with a $\langle 0001 \rangle$ texture are produced [42]. In addition, the mechanical stress in the films decreases from 1.3 to 0.9 GPa as the pressure increases from 666.5 Pa to 101.3 kPa.

1.3.3 Sol–gel process

The first instance of the successful fabrication of complex oxides was demonstrated in [43]. Since the 1980s, the sol–gel method has been extensively employed for depositing LN films [44]. This method facilitates film formation at significantly lower temperatures compared to alternative techniques due to the high reactivity of the gel. However, it is noteworthy that the sol–gel method is both time- and resource-intensive, as it necessitates the use of expensive precursors for gel fabrication. In addition, the synthesized LN films often contain additional non-ferroelectric phases, such as $LiNb_3O_8$ and Nb_2O_5.

The emergence of the $LiNb_3O_8$ phase can be attributed to the phase reaction between LN and the inherent oxide present on a Si substrate [45]. The quantity of the $LiNb_3O_8$ phase present in synthesized films is influenced by the annealing temperature.

The surface roughness varies depending on deposition conditions and the molar concentration of compounds in the solution. Notably, as the concentration increases from 0.25 to 1 mol l^{-1}, surface roughness ranges from 10.6 to 16 nm [46].

1.3.4 Pulsed laser deposition

The initial demonstration of the feasibility of depositing ferroelectric films using the PLD method was first illustrated in [47]. In the 1990s, this method became widely adopted for synthesizing LN films [48–50].

The PLD method operates on the principle of ablating a target using short pulses (20–30 s) of focused laser radiation. The substrate holder is positioned in an atmosphere with high oxygen pressure, which is necessary for oxide formation. Film parameters are contingent upon numerous factors, including the distribution of compounds during ablation, the reactive gas pressure, the substrate–target distance, and the energy and frequency of laser radiation.

LN films fabricated via the PLD technique exhibit nonuniform composition and structure. Initial studies [48, 49] disclosed the formation of a Li-deficient phase, $LiNb_3O_8$. To circumvent this issue, the use of Li-rich targets was suggested [48]. In addition, there is evidence linking the appearance of the $LiNb_3O_8$ phase to the relative positioning of the substrate and target, as well as the presence of oxygen in the reaction chamber [51].

Research findings [52–54] have demonstrated that the elemental composition, structure, and surface morphology are correlated with the target structure and PLD parameters. Uniform single-phase, crack-free LN films with grain sizes ranging from 50 to 200 nm were deposited onto sapphire substrates and SiO_2/Si heterostructures at a substrate temperature (T_{sub}) of 700 °C [52]. Interestingly, films deposited onto sapphire exhibited greater transparency compared to those on SiO_2/Si heterostructures.

In [53], a nonlinear relationship between average roughness and partial oxygen pressure for LN films deposited onto diamond/Si heterostructures ($T_{sub} = 650$ °C) was unveiled. This relationship was elucidated in terms of the competition between two processes: scattering by atoms in a reactive environment (at high pressure) and re-sputtering of the growth front (at low pressure). Moreover, the successful synthesis of highly oriented LN films with $\langle 0001 \rangle$ texture onto SiO_2/Si substrates at $T_{sub} = 650$ °C has been reported [54]. It was noted that the degree of preferred orientation produced during film formation was contingent upon substrate temperature, as an increase in T_{sub} resulted in a decrease in the orientation degree, ultimately leading to the formation of films with arbitrary grain orientation at 750 °C. Notably, single-phase c-oriented LN films were exclusively formed at an oxygen pressure of 40 Pa in the reaction chamber, whereas further increases in oxygen pressure suppressed the formation of the $LiNb_3O_8$ phase.

Minimal roughness (4.3 nm) was achieved at a substrate temperature of 600 °C; however, at $T_{sub} = 500$ °C, roughness increased to 5.1 nm, and at $T_{sub} = 700$ °C, it further rose to 5.8 nm. The minimum optical losses corresponded to minimal roughness. The authors attributed this roughness–T_{sub} dependency to an alteration in the mobility of adatoms.

1.3.5 Radio-frequency magnetron sputtering

The first sputtering triode setup to produce LN films utilized a powdered LN target [55]. Subsequently, the synthesis of LN films progressed through radio-frequency diode sputtering (RF sputtering) [56–59], which was associated with a low deposition rate and compositional irregularities in the deposited films.

The advent of magnetron sputtering and RFMS supplanted RF sputtering entirely, offering several advantages:

- reduced reactive gas pressure, facilitating optimal free path length for ion motion within the space charge area;
- capacity to modulate ion energy across a broad spectrum via magnetic field manipulation while maintaining constant source power;
- elevated sputtering rates;
- insensitivity of sputtering coefficients to material melting points;
- consistency in the elemental composition of sputtered materials.

The initial exploration of the synthesis of LN films through the RFMS method commenced in the latter half of the 1980s [60]. Nevertheless, ongoing research is still elucidating the influence of technological parameters and plasma effects (specifically, the ion-assist effect) on the structural characteristics of deposited LN films.

Table 1.1 delineates the advantages and drawbacks associated with each of the aforementioned methods.

As can be seen in table 1.1, the RFMS process is one of the most effective deposition methods for LN films, satisfying the basic requirements dictated by the applications.

Table 1.1. Comparative analysis of advantages and disadvantages associated with the most popular deposition techniques.

Method	Advantages	Disadvantages
PLD	• Elemental composition is conserved • Deposition at relatively high oxygen pressure • Relatively high deposition rate	• Nonuniform sputtering due to the small size of the laser impact area • Deviation from the elemental composition of a target due to long-lasting ablation • Careful selection of many technological parameters (many of them are cross-dependent) • Difficult fabrication of layered structures due to the inevitable presence of drops in the flow • Relatively high process temperature
Sol–gel process	• Homogeneity of deposited films • It is possible to deposit films onto large substrate areas • The simplicity of integration with existing semiconductor technology	• Process conditions influencing the structure and electrical properties of deposited films • High probability of film cracking due to solvent evaporation • Particle agglomeration and inhomogeneous composition • Subsequent annealing required
CVD process	• The capability of synthesizing film on stepped surfaces	• Careful precursor selection (usually extremely toxic and corrosive)

	• Deposition at relatively high oxygen pressure • High deposition rate • The possibility of depositing high-purity films onto large substrate areas	• Relatively high process temperature (up to 900 °C) • The complexity of the process and the strong dependence of the film properties on the regime used.
LPE	• The simplicity of the equipment • The ability to dope the film • Low compound toxicity	• Inhomogeneity of the film structure • High surface roughness • The presence of the solvent compounds and a catalyst in deposited films • Low reproducibility of stoichiometry and structural characteristics
RFMS	• Controllable sputtering process • Suitable for the synthesis of a wide range of thin film materials: metals, semiconductors, dielectrics, and complex oxides • Relatively high deposition rate • The capability of aligning the elemental composition of the film with that of the target • The possibility of depositing thin films onto large substrate areas	• Target erosion causes its degradation • The complexity of the plasma effect on the growing process

1.4 Fundamentals and critical parameters of the RFMS method and the ion-beam sputtering method

Sputtering may be defined as the expulsion of atoms or molecules from a target material owing to its bombardment by highly energetic ions of reactive plasma. When a *direct voltage* not exceeding the breakdown voltage is applied to two electrodes positioned in a gas, the current between these electrodes is infinitesimal. Traditionally, argon serves as a reaction gas, which is ionized by electrons between an electrode conjugated with a target (cathode) and a substrate (anode). The current arises due to emission from the cathode and ceases upon the cessation of emission. This current escalates with an increase in the distance between electrodes, since electrons undergo more collisions and ionize reactive gas atoms in the chamber. Each collision act engenders an electron as well as an ion through ionization and acceleration by the field. At relatively high voltage, certain ions reach and strike the target surface, inducing sputtering of its surface and the generation of secondary electrons. When the ionic current reaches a significant fraction of the total current, the ions commence accumulating near the cathode, engendering localized space charge and amplifying the electric field. When breakdown is induced, the number of secondary electrons is adequate to sustain autonomous plasma discharge. Owing to their low mass, electrons traverse the space charge region swiftly and reach the neutral region, where the count of electrons approximates that of ions. This region is known as a plasma. Since ions are shielded by electrons, they remain unaware of the electrode's presence and diffuse accordingly. When an ion reaches the space charge area, it begins to perceive the potential of the cathode and promptly rushes toward it. Atoms ejected from the target due to bombardment may be deposited onto a substrate placed in the reaction chamber, culminating in the steady growth of a thin film on the substrate surface. This process forms the basis for the ion sputtering of materials (IBS method).

A primary drawback of this method is the relatively low film deposition rate and the high cathode voltage. To augment the efficiency of reactive gas ionization by electrons, diminish the reactive gas pressure, and lower the cathode voltage, the RFMS technique was proposed. RFMS has several advantages vis-à-vis traditional diode sputtering systems that lack a magnetic field:

- reduction in reactive gas pressure, facilitating collision-free motion of ions in the space charge area
- prevention of diffusion of the sputtered particles
- reduction in ion energy of up to hundreds of electron-volts
- the ability to regulate the ion energy over a broad range via a magnetic field at constant discharge power
- ease of establishing the optimal energy of bombarding ions for a given process
- increase in sputtering rate compared to sputtering systems that lack a magnetic field at the same power, attributable to more efficient utilization of energetic electrons in plasma
- virtually zero dependence of sputtering coefficients on the melting point of the material
- high reproducibility of sputtering material composition
- capability to sputter dielectric materials

The magnetron sputtering system resides within an evacuated reaction chamber maintained at a pressure not exceeding 10^{-4} Pa prior to the introduction of the reactive gas, which then increases to 10^{-1} Pa upon gas injection.

Figure 1.3(a) illustrates the fundamental layout of the RFMS system, while figure 1.3(b) depicts one ring magnetron configuration employed in RFMS systems.

In the ring planar magnetron (figure 1.3(b)), all components are housed within the hull 18, connected to the reaction chamber via an intermediate isolating ring 2 and flange 4, and sealed with vacuum gaskets 1 and 3 (refer to figure 1.3(b)). The disc-shaped target (cathode) 19 is cooled by water circulating through pipes 14 and 17, with voltage supplied through clamp 16. Below the cathode rests the magnet block, comprising the central magnet 15 and peripheral magnet 12, mounted on the base of block 13. This magnet block generates a magnetic field whose component runs parallel to the cathode plane. Anode 10, positioned above the cathode, is grounded, ensuring the generation of an electric field 9 with its component perpendicular to the cathode plane.

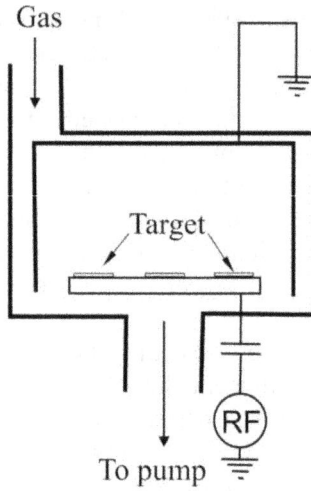

(a)

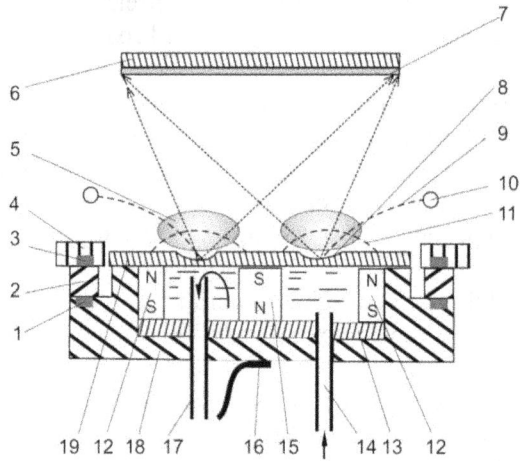

(b)

Figure 1.3. Schematic diagram of the RFMS system (a) and schematic of a ring magnetron used in RFMS systems (b). 1,3—vacuum gaskets, 2—isolating ring, 4—chamber flange, 5,8—plasma zone and erosion zone, 6—substrate, 7—thin film, 9,11—electric and magnetic fields, 10—anode, 12,15—peripheral and central magnets, 13—base of the magnetic block, 14,17—pipes, 16—clamp, 18—hull, 19—target.

The magnet system exerts a physical influence on plasma discharge properties. The magnetic field primarily alters electron motion within the glowing discharge, affecting electrons more than ions due to the relatively higher ionic mass. Magnetic field lines form closed loops between the system poles. The target surface, located where magnetic field lines enter and exit, experiences more intense sputtering, resulting in the formation of a closed groove. For magnetrons such as that depicted in figure 1.3(b), this groove adopts a ring shape. Consequently, the magnetic field's effect is akin to an increase in reactive gas pressure. Another significant effect is the 'magnetic trap effect,' which greatly reduces radial electron diffusion from the discharge zone and prevents electron loss.

A crucial characteristic of the RFMS process is the plasma potential—the electrical potential of the central part of the glowing discharge measured relative to the grounded electrode potential. When a high-frequency signal is applied to the cathode, the plasma potential tracks the applied signal frequency. Over time, the plasma potential transitions from the potential associated with the ground electrode to several hundred volts, depending on magnetron power (RF signal), reactive gas pressure, and electrode geometry. The asymmetric electrode configuration commonly utilized in RFMS systems enhances reactor efficiency. Electrons, as carriers with significantly higher mobility than ions, track the RF voltage on the cathode, which typically operates at a frequency of 13.5 MHz, while ions do not. Consequently, ions only follow the time-averaged voltage on the cathode, creating an intrinsic plasma field. This field directs the ion flux toward the target, causing bombardment. The intensity of this bombardment varies based on magnetron power, reactive gas pressure, target composition, intrinsic field, and other parameters. In addition to argon, gaseous oxygen is introduced into the reaction chamber to facilitate the formation of oxide compounds with the desired composition on a substrate. It is noteworthy that the relative positioning of the target and the substrate is a key parameter influencing film formation. For instance, when the substrate aligns precisely above the target erosion zone at a short distance, plasma ions reflected from the target during sputtering impact the growing film surface. This condition is termed *the plasma effect* regime. The degree of this effect depends on magnetron power, gas pressure in the reaction chamber, target–substrate distance, and their relative orientation.

Since sputtering is a physical rather than a chemical process, it enables the deposition of a wide array of thin film materials on diverse substrates. However, the outcome of the sputtering process may vary depending on target composition and other parameters, making it uncertain whether the resulting film will precisely match the target material's composition. Atoms from the target that are deposited onto the substrate surface are not statically positioned; they can diffuse (the Gibbs-Thomson effect) or migrate along the surface, given the absence of secondary evaporation. While migrating, these atoms may combine with others, forming clusters of the target material (nucleation). Further progression of this process follows various possible scenarios [61], leading to the creation of thin LN films (as illustrated in figure 1.4), provided that the sputtering parameters are meticulously controlled. Due to the differing adhesion energies of various materials on different substrate–film combinations, epitaxial growth can be categorized into three fundamental models: the Volmer–Weber (island) model, the Frank and van der Merwe (layer) model, and the Stranski–Krastanov (layer followed by island) model.

In the Volmer–Weber model (figure 1.4(a)), nucleation aggregates of a new phase initially form on a substrate, eventually culminating in the formation of a

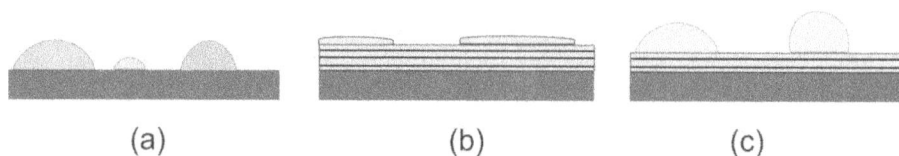

Figure 1.4. Schematic representation of three basic models of epitaxial film growth: the Volmer–Weber (island) model (a), the Frank and van der Merwe (layer) model (b), and the Stranski–Krastanov (layer followed by island) model (c).

continuous layer due to coalescence in the later stages. This mechanism occurs when there is poor adhesion between crystallites and the surface. On the other hand, the Frank and van der Merwe mechanism (figure 1.4(b)) occurs due to strong adhesion between crystallites and a substrate, resulting in layered film formation that occurs layer by layer. The Stranski–Krastanov model (figure 1.4(c)) represents an inter-mediate case between the aforementioned approaches and is commonly observed in metal–metal and metal–semiconductor systems.

For the deposition methods and regimes elucidated in this book, it is postulated that the Volmer–Weber mechanism prevails, as many researchers contend that this model accurately describes the synthesis of LN films by sputtering in a vacuum. For example, in a study by the authors of [62], c-oriented LN films were deposited onto Si-SiO$_2$ heterostructures using the PLD method. The authors observed the for-mation of 3D islands (nucleation) in the initial stages of deposition, consistent with the Volmer–Weber model. Similarly, another research group reported the nucleation of islands with a size of approximately 105 nm during the early stages of deposition [63, 64]. Subsequent deposition led to the induction of coalescence, resulting in the formation of thin LN films.

The key parameters of the RF magnetron sputtering process include:
- the type and temperature of the substrate,
- the composition of the reactive gas,
- the pressure of the reactive gas,
- the power reflected by the magnetron,
- the composition of the target and its position relative to the substrate.

Consequently, considerable effort has been directed toward elucidating the relation-ship between the aforementioned technological parameters of RFMS and the structure, composition, and electrical properties of the fabricated LN films.

The deviation of *the target material* from stoichiometry, target density, and the presence of impurities significantly impact the degree of crystallinity of synthesized films. For instance, another study [65] conducted a comparative analysis of film properties. Films were fabricated by sputtering using a single-crystal target grown via the Czochralski method and a target fabricated by sintering Li$_2$CO$_3$ and Nb$_2$O$_5$ powders at 1200 °C according to the reaction:

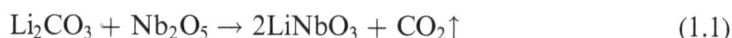

$$Li_2CO_3 + Nb_2O_5 \rightarrow 2LiNbO_3 + CO_2\uparrow \tag{1.1}$$

Compositional analysis revealed that the Li_2CO_3 oxide was not completely removed from the target, resulting in a surface layer exhibiting a preferable orientation along the LN polar axes due to segregation. Bombardment of such a target during the RFMS process inevitably incorporates Li_2CO_3 oxide into the synthesized film, leading to the formation of Li-poor LN layers. In addition, the presence of Li_2CO_3 in the film generates point defects, impeding the formation of a large-block LN polycrystalline structure. Some researchers have suggested [66] adding approximately 5% of Li_2O oxide to the target composition to ensure the formation of stoichiometric LN films. Another research group utilized a Li-rich target to achieve an optimal elemental composition in a reaction chamber through lithium evaporation [67].

The utilization of a single-crystal LN target is preferable for preserving the initial elemental composition and promoting the formation of films with a high degree of crystallinity.

The RF power significantly influences the properties of LN films. Figure 1.5 illustrates the x-ray diffraction (XRD) pattern of LN films fabricated under various RF power regimes at a substrate temperature of 580 °C [67].

As illustrated in figure 1.5, the RF power influences both the composition of the film and its degree of crystallinity. The formation of solely amorphous films at an RF power of 70 W has been explained in the literature by the inadequacy of this power level to facilitate stable nucleation on the substrate surface. Conversely, it has been noted that at an RF power of 100 W, polycrystalline films exhibiting a (012) preferred orientation are produced, which is attributed to the minimal surface energy

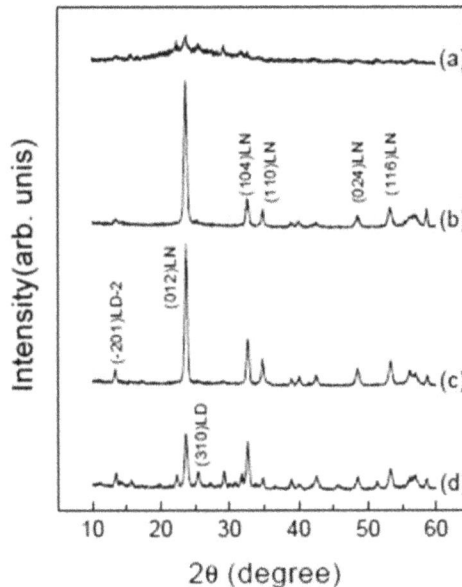

Figure 1.5. XRD patterns of thin films deposited at the substrate temperature of 580 °C and 1 mTorr working pressure. The RF power is 70 W (a), 100 W (b), 130 W (c), 170 W (c); LN, LD, and LD-2 stand for LN, $LiNb_3O_8$ and $Li_{1.9}Nb_2O_5$ respectively. Reproduced from [67]. Copyright © IOP Publishing Ltd 2018.

associated with the LN plane [67]. Subsequent escalation in RF power results in the emergence of additional phases such as $LiNb_3O_8$ and $Li_{1.9}Nb_2O_5$, concomitant with the cessation of oriented film growth. Moreover, studies [67, 68] have indicated a nearly linear correlation between film growth rate and RF power, ranging from 17 to 26 Å min^{-1} at 100 W. Notably, various researchers advocate sputtering at an RF power of 100 W to foster the formation of highly oriented (006) LN films [68–70]. As elucidated in [69], sputtering at this power level leads to the production of films with minimal surface roughness (refer to figure 1.6).

Simultaneously, the degree of crystallinity escalates with RF power, as evidenced by the heightened intensity of XRD peaks, as depicted in figure 1.7.

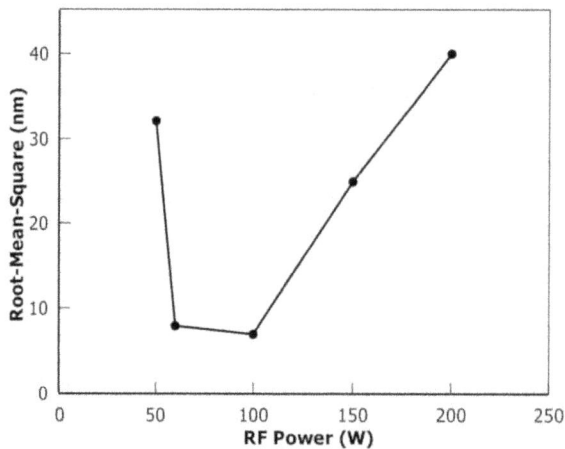

Figure 1.6. Dependence of the surface roughness of LN films on RF power. Reprinted from [69], Copyright (2007), with permission from Elsevier.

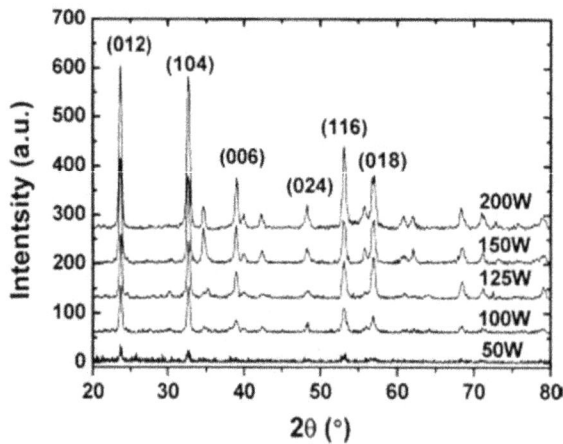

Figure 1.7. XRD pattern of LN films fabricated by RFMS as a function of RF power. Reprinted from [69], Copyright (2007), with permission from Elsevier.

Detailed analysis presented in [70] indicates that relatively high RF power can induce substantial heating of the target, leading to its cracking and the dissociation of lithium and oxygen, consequently resulting in lithium depletion from the target. Consequently, the films fabricated in this way comprise a Li-poor, non-ferroelectric phase of $LiNb_3O_8$. An RF power of 100 W emerges as the optimal parameter for synthesizing LN films with an index of refraction closely resembling that of the bulk material [58].

The temperature and type of substrate in the RFMS process significantly influence the growth rate, structure, and composition of LN films. While the substrate temperature has a less pronounced effect on the growth rate, it greatly affects the degree of crystallinity and film orientation [67, 69]. Evidence suggests that at a reactive gas pressure of 1 mTorr and RF power, only amorphous LN films are formed on unheated substrates, whereas 560 °C is the optimal temperature for oriented film growth [67].

At excessively high substrate temperatures, lithium evaporates from the surface of the deposited film, leading to the formation of an undesirable Li-poor $LiNb_3O_8$ phase [71]. Regarding substrate type, the ideal choice is one with lattice parameters close to those of LN to reduce surface strain in the film and minimize defect concentration at the film–substrate interface. Various intermediate layers, such as SiO_2, Si_3N_4/SiO_2, and ZnO, are used for this purpose. For example, some researchers have successfully fabricated a highly oriented ZnO film (as an intermediate layer with a thickness of 100 nm) on a silicon substrate, which facilitates the growth of c-oriented LN films with minimal lattice strain [70].

Among the most critical parameters in RFMS are *the reactive gas pressure and composition*. As mentioned earlier, the reactive gas pressure significantly impacts plasma properties, as well as the growth rate, degree of orientation, and surface roughness of synthesized films. Researchers have noted that the reactive gas pressure directly affects the concentration of Li in a film [67]. At high gas pressures, the mean free path of Li ions decreases, due to their relatively low atomic mass. This leads to strong scattering in the plasma. Consequently, the number of ions reaching the substrate decreases, resulting in the formation of the Li-poor $LiNb_3O_8$ phase.

Studies have shown that films deposited at low pressure in the reaction chamber (1–2 mTorr) exhibit a high degree of crystallinity and the preferred orientations (012) [71] and (006) [72]. Researchers recommend sputtering at an RF power of 100 W and a substrate temperature of 600 °C. The dependence of the growth rate on gas pressure follows a nonlinear relationship (see figure 1.8).

Initially, the deposition rate increases with working pressure but then sharply decreases. This can be explained by a decrease in the mean free path of the deposited compounds in the plasma. Since thin LN films fabricated by various methods (including RFMS) demonstrate a Li deficit, the most natural way to manage this process is through careful selection of the reactive gas composition. Specifically, it has been demonstrated that the presence of oxygen as a reactive gas component increases the partial pressure of lithium in the reaction chamber [73, 74]. However, using pure oxygen as a reactive environment does not allow for the fabrication of single-phase LN films [49].

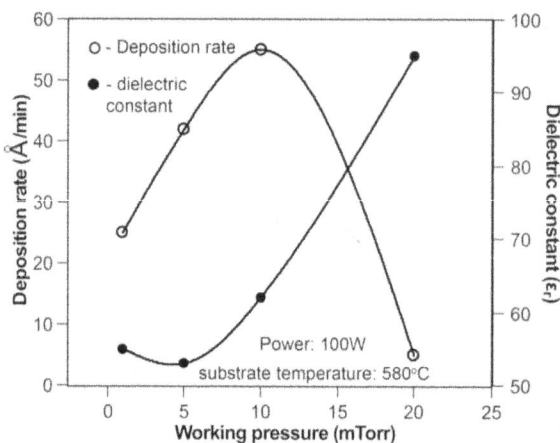

Figure 1.8. Deposition rate and dielectric constant as a function of working pressure. Reprinted from [71], Copyright (1999), with permission from Elsevier.

Many efforts have focused on finding the optimal Ar/O_2 ratio for the reaction chamber, taking into account the other RFMS parameters and the desired properties of the synthesized films. A study of the influence of oxygen presence on the structure of LN films deposited onto silicon substrates at a temperature of 575 °C and a working power of 100 W recommended using an Ar/O_2 ratio of 1:1 [69]. Under these conditions, highly oriented (006) LN films with good optical properties are formed.

Further research demonstrated that the concentration of oxygen in the reaction chamber significantly affects the lattice constant along the symmetry axis of synthesized c-oriented LN films and consequently their surface tension [70]. It was found that films with minimal strain and a lattice constant close to that of bulk material ($c = 1.3867$ nm) are formed only in an $Ar + O_2$ plasma with an 80% to 20% ratio. The study concluded that increasing the oxygen concentration in the reaction chamber leads to intense bombardment of the deposited film, encouraging the insertion of oxygen into the crystal lattice interstices, which causes deformation.

It was also shown that films fabricated at low oxygen pressure exhibit a lower degree of orientation and contain $LiNb_3O_8$, due to the evaporation of lithium from the film surface at low pressure [75]. With an increase in pressure to 30 Pa, only single-phase, highly oriented (006) LN films are formed, but this effect diminishes with further pressure increases. The dependence of preferred orientation and degree of crystallinity on oxygen pressure was investigated, revealing that the kinetic energy of plasma components and the size of the plasma area vary with oxygen pressure [53]. This variation, along with the substrate–target distance, provides optimal conditions for the synthesis of high-quality films.

The kinetic energy of particles reaching the substrate should be limited to an optimal value. At high oxygen pressure, the probability of collisions with plasma components increases, decreasing the energy of bombarding ions, which negatively affects the crystallinity of the deposited film. Conversely, at low oxygen pressure, low collision intensity allows bombarding ions to reach the substrate with excessive energy, generating many defects in the film and preventing oriented growth. In addition, numerous collisions at high reactive gas pressure prevent sufficient Li atoms from reaching the substrate surface, leading to the formation of Li-poor phases ($LiNb_3O_8$).

Despite these challenges, many authors recommend using an Ar/O_2 reactive plasma with a proportion of 60/40 to fabricate single-phase, highly oriented LN films [22, 71, 72, 76]. It has been demonstrated that an inappropriate choice of the Ar/O_2 ratio generally leads to the dissociation of LN, forming Li-poor phases ($LiNb_3O_8$) and oxides such as Nb_2O_5 [22]. This dissociation can occur at the target surface, in the plasma, or in the film after deposition. To prevent dissociation, the substrate temperature should not exceed 600 °C. Research aimed at optimizing technological regimes concluded that an Ar/O_2 ratio of 65/35 is required for the deposition of highly oriented (006) LN films. Nb_2O_5 oxide is a product of the dissociation of the $LiNb_3O_8$ phase according to the following reactions [22]:

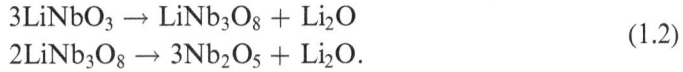

$$3LiNbO_3 \rightarrow LiNb_3O_8 + Li_2O$$
$$2LiNb_3O_8 \rightarrow 3Nb_2O_5 + Li_2O. \tag{1.2}$$

Nevertheless, it is important to stress that the Ar/O_2 ratio is not an independent parameter; rather, it should be considered in conjunction with other parameters influencing the properties of reactive plasma, such as working pressure and RF power.

In addition to the issues discussed above, it is important to mention the methods used for the post-deposition treatment of LN films. One of the most popular methods used to increase the degree of crystallinity of deposited films, decrease defect concentration, and reduce mechanical strain is thermal annealing (TA) in vacuum or a gas environment. Pure oxygen or air is the most frequently used gas environment for TA. The oxygen atmosphere affects the crystallization process of LN films [77, 78]. The role of oxygen is to prevent the evaporation of light compounds such as Li_2O, control stoichiometry, and improve electrical properties.

The majority of authors have reported the formation of amorphous LN films on unheated substrates [70, 77, 79]; such films exhibit ferroelectric properties after TA. The surface roughness depends on the annealing temperature in a nonlinear manner. For example, it was shown in [77] that the minimal roughness of 2.6 nm, corresponding to amorphous films, increased to 210 nm for films annealed at 550 °C and then declined to 80 nm for those annealed at 1000 °C. The minimal average surface roughness (15 nm) corresponded to an annealing temperature of 700 °C. TA of amorphous films leads to the formation of films with random grain orientation [77, 80]. When oriented films deposited onto heated substrates underwent TA, the final effect depended on the temperature and duration of TA. After the TA of LN films in air, the O/Nb ratio approached the stoichiometric value [81].

Studying the TA of LN films in static and dynamic oxygen atmospheres showed that films annealed in the dynamic regime demonstrated a more uniform size distribution of polycrystalline grains compared to films annealed in a static atmosphere [82]. The authors explained this phenomenon by noting that during oxygen flow, more intensive evaporation of components occurs from the film surface, causing a decline in its growth and consequently lower surface roughness. Nevertheless, most researchers report the formation of a Li-poor (parasitic) $LiNb_3O_8$ phase after TA of as-grown LN films in both vacuum and oxygen atmospheres [78, 83, 84].

Detailed studies of the structural properties of LN films after TA resulted in the development of an original model that describes the formation of the $LiNb_3O_8$ phase during TA. According to this model, the following reactions take place simultaneously during TA:

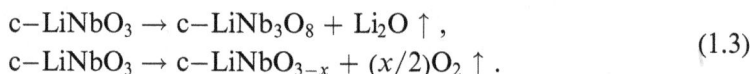

$$c-LiNbO_3 \rightarrow c-LiNb_3O_8 + Li_2O \uparrow ,$$
$$c-LiNbO_3 \rightarrow c-LiNbO_{3-x} + (x/2)O_2 \uparrow . \tag{1.3}$$

The first equation in (1.3) corresponds to the loss of Li, which predominates in the initial stage, whereas the second equation describes the loss of oxygen. The hexagonal atomic position in the (0001) crystallographic plane of LN is comparable to the atomic position in the $(\bar{6}02)$ plane of $LiNb_3O_8$ with a slight lattice mismatch. It is most likely that $LiNb_3O_8$ crystallites precipitate while maintaining their epitaxial relationship with the original LN crystallites. Figure 1.9 clarifies the mechanism of the reactions and the structural phase conversion taking place during TA of LN films.

At the beginning stages of annealing in *vacuum*, Li_2O is desorbed from the surface and grain boundaries, whereas the single grains merge, forming interphase boundaries (figure 1.9(a)). When the desorption of Li_2O slows down (reaches saturation), oxygen

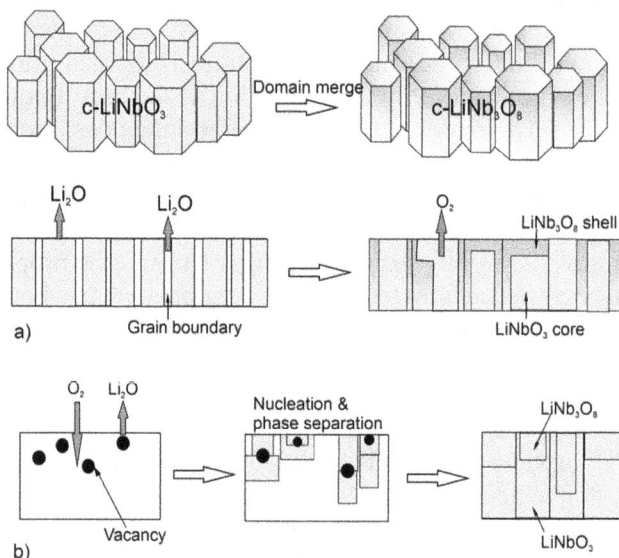

Figure 1.9. Structural model that illustrates the phase conversion that takes place during the high-temperature annealing of: (a) c-oriented LN textured film in a vacuum (upper and middle rows) and (b) a-LN film in ambient O_2 (lower row). The upper row shows perspective views, and the middle and low rows are cross-sectional views. Reproduced from [83], with permission from Springer Nature.

atoms continue evaporating from the inner areas of the film. Finally, the textured crystallites of LN become embedded in the crystalline 'shell' of the $LiNb_3O_8$ phase, forming a structure where each grain has a core (LN) and an outer layer ($LiNb_3O_8$).

During TA in an O_2 atmosphere, oxygen molecules penetrate the film, preventing the formation of vacancies that block the migration ability of adatoms. Within this model, during the incubation period, Li_2O evaporates according to the following reaction [83]

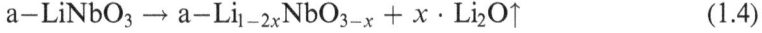

$$a-LiNbO_3 \rightarrow a-Li_{1-2x}NbO_{3-x} + x \cdot Li_2O\uparrow \tag{1.4}$$

This process creates vacancies, facilitating atomic motion and causing the nucleation of LN-$LiNb_3O_8$ pairs due to phase separation as follows:

$$a-Li_{1-2x}NbO_{3-x} \rightarrow (1 - 3x) \cdot c-LiNbO_3 + x \cdot c-LiNb_3O_8 \tag{1.5}$$

Once the nuclei of LN and $LiNb_3O_8$ are formed, the processes of crystallization and phase transition continue at a speed dependent on the annealing conditions. It is worth noting that similar transitions are possible not only for crystalline but also for amorphous films.

1.5 Electrical properties and charge transport phenomena in $LiNbO_3$-based heterostructures

The fact that the structural properties of LN films strongly depend on deposition parameters is widely documented in many papers. However, their electrical properties, such as ferroelectric properties or charge transport in LN-based heterostructures, are not systematically presented; instead, they are often reported as a complementary part of structural studies. Surprisingly, papers on the electrical properties of LN often report only general trends in current–voltage characteristics or DC or AC conductivity, rather than providing quantitative parameters that can be derived from the observed electrical phenomena. In addition, the information regarding conduction mechanisms and the type of charge carriers is sometimes controversial. Below, based on literature sources, we present a systematic overview of charge transport phenomena in LN-based heterostructures.

1.5.1 Charge carriers in $LiNbO_3$

According to the traditional view, charge carriers in single-crystal LN are triggered by point intrinsic or extrinsic defects [85, 86]. Sources of ionic defects in LN can be classified into three main types, as follows: impurity content, non-stoichiometry in Li/Nb, and intrinsic ionic disorder. Intrinsic ionic disorder is the predominant source of ionic defects. The shortage of Li during the growth process can be described by various possible processes of Li_2O effusion [85]

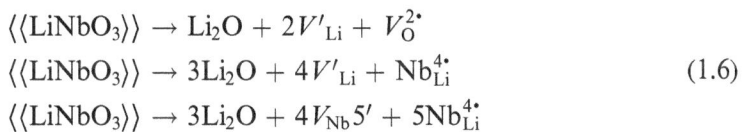

$$\langle\langle LiNbO_3 \rangle\rangle \rightarrow Li_2O + 2V'_{Li} + V_O^{2\bullet}$$
$$\langle\langle LiNbO_3 \rangle\rangle \rightarrow 3Li_2O + 4V'_{Li} + Nb_{Li}^{4\bullet} \tag{1.6}$$
$$\langle\langle LiNbO_3 \rangle\rangle \rightarrow 3Li_2O + 4V_{Nb}5' + 5Nb_{Li}^{4\bullet}$$

In these reactions, V_{Li}, V_{Nb}, and V_O represent lithium, niobium, and oxygen vacancies, respectively. Nb_{Li} is a Nb antisite defect at V_{Li} (where the bullet and the prime denote the positive and negative charge states of a defect with respect to the host, respectively). The second reaction in equation (1.6) describes the most popular model. However, the generation of antisite defects and the degree to which Li sites are filled by Nb ions greatly depend on the growth conditions [87, 88].

A variety of experimental results and different types of observed charge carriers are influenced by the wide range of growth conditions and fabrication methods of LN films. The polaronic model, which describes charge phenomena in lithium niobate, is one of the most widely used approaches [89]. A small polaron can be created when a sufficiently strong interaction between a carrier and the surrounding lattice induces self-trapping of the carrier at one site in condensed matter. Due to the vibration of surrounding ions, the equilibrium positions of this carrier are displaced. Even insignificant irregularities in the lattice cause complete localization at one site. The main charge transport mechanisms in this situation are thermally activated hopping or tunneling. A small bound polaron is created if electrons are captured at the $Nb_{Li}^{4\bullet}$ antisite defect (induced according to equation (1.6) by the concomitant lattice distortion). Furthermore, a bound bipolaron $Nb_{Li}^{4\bullet} - Nb_{Li}^{4\bullet}$ is formed when the nearest defect pair is capable of capturing two electrons. In addition, extrinsic ions inevitably present in LN (for instance, Fe_{Li}^{2+}) can capture an electron, generating small bound electron polarons. The polaronic model in LN, which underlies many nonlinear phenomena in this oxide, is supported by optical investigations [90–92].

Alternatively, experiments suggest that hole polarons are the major charge carriers in LN [92, 93]. In lithium niobate, like other oxide materials, valence band holes can be trapped at negatively charged acceptor defects. Despite the fact that the oxygen ions in a lattice are equivalent, hole–lattice coupling causes hole localization, forming an isolated O^- ion. Recall that non-stoichiometric Li deficit leads to the formation of intrinsic acceptors (Li vacancies, V_{Li}) that can capture an isolated O^- ion, generating the 'O^- next to V_{Li}' structure (see figure 1.10) [93].

Nonetheless, electrical measurements suggest that electrons are the major carriers in LN films [94]. This is because potentially free electrons can be generated in lithium niobate according to the following reactions [87]:

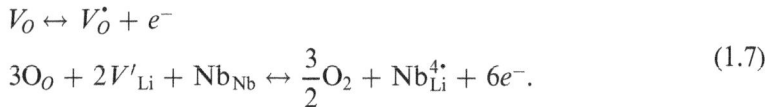

$$V_O \leftrightarrow V_O^\bullet + e^-$$
$$3O_O + 2V'_{Li} + Nb_{Nb} \leftrightarrow \frac{3}{2}O_2 + Nb_{Li}^{4\bullet} + 6e^-. \tag{1.7}$$

It is important to note that, regardless of the fabrication method, LN films are primarily polycrystalline. This results in a high density of interface states (defects) at the grain boundaries, which act as donor centers. Regardless of the type of major charge carrier, the polycrystalline structure of fabricated LN films leads to additional complications in interpreting experimental results related to charge transport phenomena. In the next section, we present a brief overview of the main charge transport mechanisms observed in LN-based heterostructures.

Generally, all conduction mechanisms in an insulator can be divided into two main groups: bulk limited (influenced by the properties of the dielectric itself) and

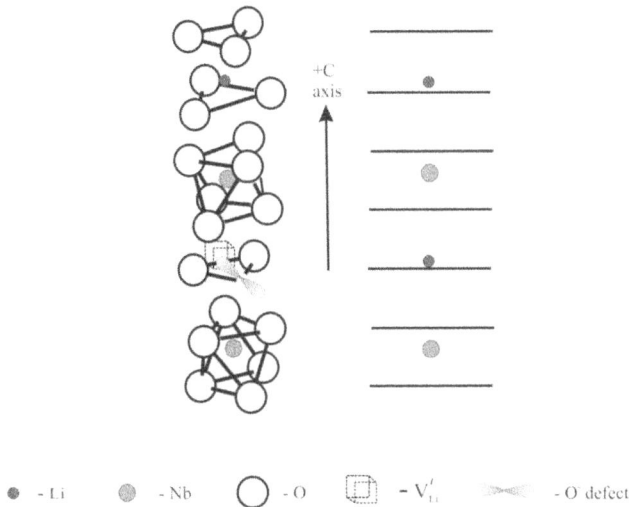

Figure 1.10. Model of an 'O$^-$ next to V_{Li}' structure in LN.

contact limited (affected by contact properties) [95]. However, two or more mechanisms can occur simultaneously, depending on experimental conditions (applied voltage, temperature, illumination, film thickness, etc). Because electrical conductivity is influenced by temperature and applied voltage in different ways, it is possible to isolate specific charge transport mechanisms.

1.5.2 Electrical conductivity limited by contact phenomena

Traditionally, the creation of a heterostructure, an interface between two different materials (e.g. dielectric–dielectric, dielectric–semiconductor, metal–dielectric, or metal–semiconductor), is an essential fabrication stage of any electronic or opto-electronic device. As a result, charge transport in such devices can be greatly affected by the contact properties. The following contact-limited conduction mechanisms have been identified: (1) thermionic or Richardson–Schottky emission, (2) Fowler–Nordheim tunneling, (3) thermally assisted tunneling, and (4) grain-boundary-limited conduction. It is important to note that contact-limited conductivity can be effectively analyzed based on an appropriate band diagram of a heterojunction.

Richardson–Schottky emission
Under an applied external electric field, the energy barrier at the metal–insulator interface (see figure 1.11) is lowered by the Schottky effect.

As a result, electrons gain sufficient energy through thermal activation and flow into the dielectric, overcoming the barrier. This effect, called Richardson–Schottky emission, influences the current–voltage (I–V) characteristics, which obey the following law [96]:

$$J = 2q\left(\frac{2\pi mkT}{h^2}\right)^{3/2}\mu E_0 \exp\left(-\frac{q\varphi_b}{kT}\right)\exp\left(\frac{\beta q E_0^{1/2}}{kT}\right)$$

(1.8)

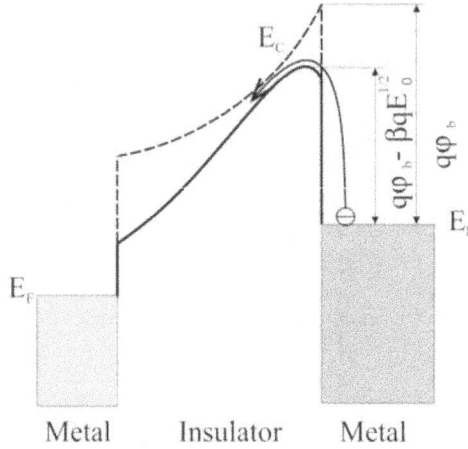

Figure 1.11. Band diagram of a metal–insulator–metal heterostructure undergoing Richardson–Schottky emission.

Here, J is the current density, q is the elementary charge, m is the electron mass, k is Boltzmann's constant, T is the temperature, h is Planck's constant, $q\varphi_b$ is the conduction band offset (the Schottky barrier), E_0 is the electric field at the metal–insulator interface, and μ is the carrier mobility. The parameter β depends only on the dielectric permittivity of the material, as follows:

$$\beta = \sqrt{\frac{q}{\pi\varepsilon\varepsilon_0}} \tag{1.9}$$

where ε_0 is the dielectric permittivity of free space. This relationship highlights how the electrical properties at the metal–insulator interface are influenced by temperature and the applied electric field. It follows from equation (1.8) that in the case of Richardson–Schottky emission, the current–voltage characteristic is linear in $\ln(J/E_0 T^{3/2})$ $-E_0^{1/2}$ coordinates (Simmons coordinates) with a slope of $q\beta/kT$. It is worth noting that the barrier height $q\varphi_b$ can be determined from the slope of a graph describing the temperature dependence (in Arrhenius coordinates) of the intercept of the I–V characteristic (in Simmons coordinates) with the vertical axis. At relatively high temperatures, thermionic emission often affects conduction mechanisms in semiconductor heterostructures. Thus, the barrier height $q\varphi_b$ at the heterojunction and the dielectric permittivity of an insulator can be derived from the parameter β (see equation (1.9)).

Thermionic emission has been reported in many studies of LN-based heterostructures [94, 97–99]. For instance, in a LN/GaN heterostructure fabricated by the PLD method, the potential barrier for holes with a height of 0.34 eV (for an applied electric field up to 4×10^6 V m^{-1}) is the limiting factor on charge transport within the framework of Richardson–Schottky emission [98]. In LN/Si heterostructures fabricated by the RFMS method, conductivity is also limited by thermionic emission with a barrier height of 0.2 eV for electrons, associated with a conduction band offset of 0.6 eV [99]. The dielectric constant of LN films derived experimentally from the parameter β is $\varepsilon = 28$, which is close to the value estimated from capacitance–voltage measurements [100, 101].

Fowler–Nordheim tunneling
In semiconductor heterostructures, when thermal emission is negligible (i.e. at relatively low temperatures), direct tunneling through a thin potential barrier ($\leqslant 50\,\text{Å}$) dominates. For sufficiently high applied electric fields, electrons can penetrate the triangular potential barrier in a metal–insulator–metal structure (see figure 1.12(a)).

The current–voltage characteristics in this case obey the Fowler–Nordheim formula [102]:

$$J_{F-N} = AE^2 \exp\left(-\frac{B}{E}\right). \tag{1.10}$$

Here, A and B are material-specific constants, and E is the electric field strength. The constant B is given by the following expression:

$$B = \frac{8\pi\sqrt{2m^*}\,\phi^{3/2}}{3qh} \tag{1.11}$$

where h is Planck's constant, m^* is the electron effective mass, and ϕ is the average potential barrier height. It follows from equation (1.10) that the current–voltage characteristics should be a straight line in $\ln(J/E^2)$ vs. $1/E$ coordinates (Fowler–Nordheim coordinates) with a slope of B. An average barrier height ϕ can be derived using equation (1.11). Although tunneling through thicker insulators ($d \gg 50\,\text{Å}$) appears improbable, this phenomenon occurs due to the presence of shallow traps in the bandgap. In this case, an electron tunnels to the trap level and can then flow through the insulator via hopping or tunneling between traps (see figure 1.12(b)). However, the barrier height ϕ given by equation (1.11) is actually the trap energy E_t in the bandgap of the insulator (relative to the Fermi level of the metal, see figure 1.12(b)).

Normally, polycrystalline films contain electron traps in their bandgap, so the tunneling process described above can be observed at liquid nitrogen temperatures. Indeed, in our recent paper [103], we reported Fowler–Nordheim tunneling at low temperatures ($T = 90\,\text{K}–140\,\text{K}$) in (001)Si–LN heterostructures fabricated by the IBS method. Although the thicknesses of the LN films were as much as $1\,\mu\text{m}$, tunneling was possible because of traps with an average energy of 1.7 eV in the bandgap.

Other research groups [104, 105] have also reported Fowler–Nordheim emission at low temperatures in LN-based heterostructures. Specifically, [106] demonstrated that the I–V characteristics of a LN/AlGaN/GaN heterostructure, fabricated by the pulse-laser deposition method, clearly showed Fowler–Nordheim tunneling due to the presence of deep traps (Li vacancies) with $E_t \sim 0.93\,\text{eV}$ in the conduction band of LN.

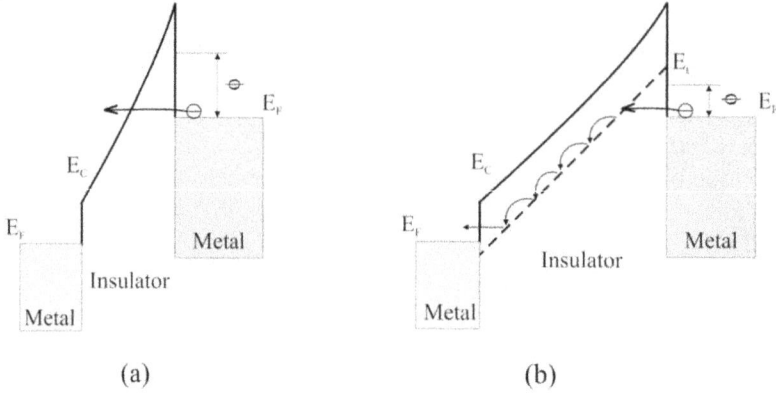

Figure 1.12. Band diagrams of metal–insulator–metal systems undergoing Fowler–Nordheim tunneling. (a) A thin insulator. (b) A thick insulator whose trap level E_t is below the conduction band edge E_c.

Thermally assisted tunneling

At temperatures between the Fowler–Nordheim emission temperature and the Schottky emission temperature, thermally assisted tunneling occurs. When a sufficiently strong electric field E is applied, the energy of thermally activated electrons lies between the Fermi level of a metal and the conduction band edge of an insulator. Thus, electrons can penetrate the triangular potential barrier of $q\varphi_b$ (see figure 1.13).

Under these conditions, the I–V characteristic can be described by the following expression [107]:

$$J = Js\left(\exp\left(\frac{E}{E_o}\right) - 1\right) \tag{1.12}$$

where

$$J_s = \frac{A \cdot T}{k}\sqrt{qE_{oo}\pi}\sqrt{\frac{q\varphi_b}{\cosh((E_{oo}/kT)^2)}}\ \exp\left(-q\varphi_b/E_o\right)$$

$$E_o = E_{oo}\coth\left(\frac{E_{oo}}{kT}\right) \tag{1.13}$$

and

$$E_{oo} = \frac{h}{4\pi}\sqrt{\frac{N_d}{m^*\varepsilon\varepsilon_0}}\ . \tag{1.14}$$

Here, N_d is the concentration of ionized donors in the dielectric layer, and A is Richardson's constant. Thermally assisted tunneling has been observed in (001)Si–LN heterostructures fabricated by the ion sputtering method in an Ar atmosphere [103] and by the RFMS method in an Ar/O$_2$ gas mixture [108]; the potential barriers were 0.7 and 1.25 eV, respectively, at the Si–LN interface. As mentioned above, various intrinsic and extrinsic defects are the main sources of donors in LN films.

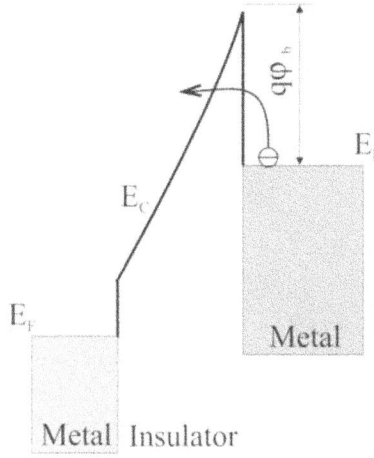

Figure 1.13. Band diagram of a metal–insulator–metal system undergoing thermally assisted tunneling.

1.5.3 Electrical conductivity limited by bulk properties

The bulk-limited conduction mechanisms are influenced by the intrinsic properties of an insulator or semiconductor, and the presence of traps distributed within its bandgap plays a crucial role. The following mechanisms are included in this group: (1) Poole–Frenkel emission, (2) space-charge-limited currents (SCLCs), (3) hopping conduction, (4) grain-boundary-limited conduction, and (5) ionic conduction. Several important parameters of LN films, such as trap density and energy distribution, drift mobility, intergrain barrier height, and dielectric relaxation time, can be derived by analyzing these bulk-limited conduction mechanisms.

Poole–Frenkel Emission
The Poole–Frenkel effect, also known as field-assisted thermal ionization, is the reduction of the Coulomb potential barrier under the influence of a relatively strong electric field [109]. This process is analogous to Richardson–Schottky emission at the interfacial barrier (see figure 1.14).

The current density in an insulator with shallow traps follows the law [109]

$$J = J_s \exp\left(\frac{\beta_{P-F}E^{1/2}}{kT}\right).$$ (1.15)

Here, J_s is influenced by the low-field conductivity, and β_{P-F} is the Poole–Frenkel coefficient, which can be defined as

$$\beta_{P-F} = (q^3/\pi\varepsilon\varepsilon_0)^{1/2}$$ (1.16)

where ε_0 is the permittivity of free space and ε is the dielectric constant of the insulator. As can be seen from equation (1.15), the I–V characteristics should be linear in $\ln(J) - E^{1/2}$ coordinates with a slope proportional to β_{P-F}. The trap concentration N_t can be determined through the Poole–Frenkel coefficient as follows [110]:

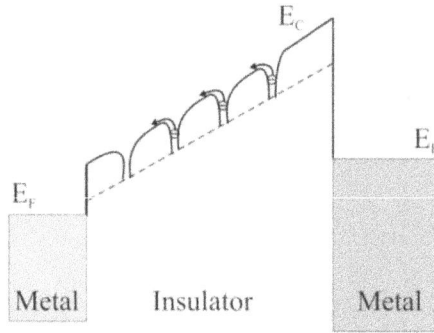

Figure 1.14. Schematic energy band diagram of Poole–Frenkel emission in a metal–insulator–metal structure.

$$N_t = \left(\frac{q\sqrt{E_k}}{\beta_{P-F}}\right)^3 \qquad (1.17)$$

where E_k is the field corresponding to the beginning of a linear section of the I–V characteristic in $\ln(J)$–$E^{1/2}$ coordinates. Poole–Frenkel emission has been identified as the dominant conduction mechanism in Si-LN heterostructures fabricated using the sol–gel technique [111], RFMS method [101], and IBS method [112]. I–V analysis within the framework of the Poole–Frenkel effect revealed shallow traps in lithium niobate with a concentration of $N_t = 2.4 \times 10^{17}\,\mathrm{cm}^{-3}$ [101] and an energy level of 0.1 eV below the conduction band [112].

Space-charge-limited currents
The SCLC mechanism occurs when ohmic contacts are present on an insulator with shallow traps. When space charge is injected into an insulator which has traps distributed in its bandgap, a significant portion of the charge accumulates within the insulator. SCLC demonstrates a specific temperature dependence because trap occupancy is strongly influenced by temperature. In this case, the I–V characteristic is affected by the trap distribution. In the simplest case of monoenergetic shallow traps with an energy E_t that lies below the conduction band, the I–V characteristic can be expressed as [109]

$$J = \frac{9}{8}\varepsilon\varepsilon_0\mu\theta\frac{V^2}{d^3}. \qquad (1.18)$$

Here, V is the applied voltage, μ is the carrier mobility, d is the film thickness, and θ is the ratio of free charge to trapped charges. It is important to note that when SCLC occurs, the I–V characteristics should be linear in double logarithm coordinates ($\ln(I)$ vs. $\ln(V)$) with a slope of two (the 'trap square law' beyond the trap-filled regime), which serves as a 'fingerprint' of this conduction mechanism. If a different trap distribution exists, then the I–V curve differs from that predicted by equation (1.18). The theory of this phenomenon is thoroughly reviewed in [109].

It should be noted that regardless of the fabrication method, LN films possess widely distributed traps in their bandgap. Therefore, it is not surprising that SCLC is

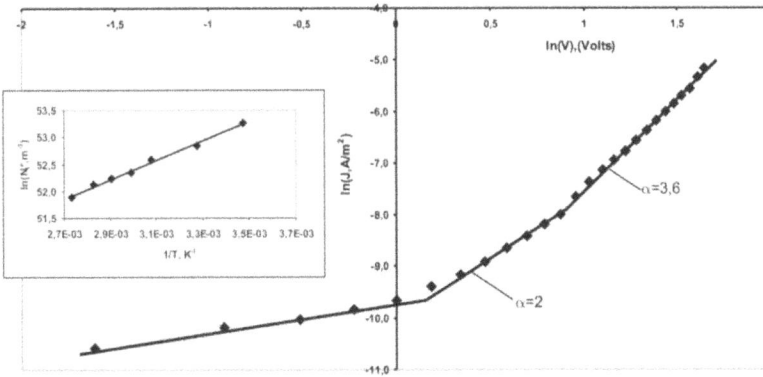

Figure 1.15. *I–V* characteristic of (001)Si–LN heterostructures in ln*J*–ln*V* coordinates. Temperature dependence of trap concentration N_t^* is shown in the insert. Reproduced from [113], with permission from Springer Nature.

a frequently observed charge transport mechanism [97, 98, 111, 113]. Figure 1.15 demonstrates the typical *I–V* characteristics of LN-Si heterostructures fabricated using the RFMS method. Two distinct linear sections with different slopes α (depending on specific SCLC regimes) are clearly visible.

The *I–V* analysis in [113] revealed an exponential trap distribution starting at 0.4 eV below the conduction band edge of LN. Another research group reported degenerate vacancies (e.g. V_{Li}) acting as traps in the bandgap of LN and gradually being filled by electrons injected from a metallic electrode [97]. The *I–V* characteristics displayed several linear regions in double logarithmic coordinates, similar to those shown in figure 1.15.

Hopping conduction
Hopping conduction is commonly observed in disordered materials, and LN is no exception. This type of conductivity is characterized by several attributes:
1. The current depends linearly on the applied voltage.
2. The activation energy required for conductivity is relatively low.
3. AC conductivity follows the 'universal power law' $\sigma = A\omega^s$, where A is a frequency-independent parameter and s is a power exponent dependent on the specific hopping mechanism and configuration.

The weak temperature dependence of hopping conductivity arises because sufficient energy must be transferred from phonons to charge carriers to overcome the energy difference between initial and final states under the applied electric field. According to Mott's model [114], if there is a spatial and energetic distribution of sites near the Fermi level in the bandgap, an electron primarily hops between sites with the smallest energy difference and the shortest distance between them (see figure 1.16).

In Mott's model, the temperature dependence of DC conductivity is described by the following expression [114]:

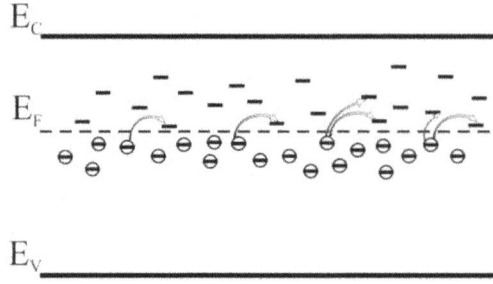

Figure 1.16. Energy diagram of a disordered semiconductor illustrating the level distribution within the bandgap and the possible hops of electrons (occupied states are marked by circles).

$$\sigma = \sigma_0(T)\exp\left(-\frac{A}{T^{1/4}}\right) \tag{1.19}$$

where $A = 2.1\ [\alpha^3/kN(E_F)]$. The pre-exponential factor $\sigma_0(T)$ is given by [114]

$$\sigma_0(T) = \frac{q^2\nu_{ph}}{2(8\pi)^{1/2}}\left[\frac{N(E_F)}{\alpha kT}\right]^{1/2}. \tag{1.20}$$

Here, $N(E_F)$ is the density of electronic states, $\alpha = 1/a$ (where a is the size of the localized site), and ν_{ph} is the phonon frequency. The parameters α and $N(E_F)$ can be estimated from the slope of a linear graph $\ln(\sigma T^{1/2})$ vs. $T^{-1/4}$ (Mott's coordinates) and from the intercept at $T^{-1/4} = 0$.

The increase in AC conductivity with frequency is due to carriers that hop back and forth over a certain number of hops; they follow the power law $\sigma = A\omega^s$ with a power exponent s ranging between 0.5 and 1.

In [115], it was reported that despite the significantly higher AC conductivity of highly oriented LN films (fabricated by the PLD method) compared to DC conductivity, both σ_{dc} and σ_{ac} are associated with the same hopping mechanism. Due to randomly distributed defects in the bandgap of LN films, electrons, as the major charge carriers, facilitate the hopping charge transport with an activation energy of 0.3 eV. Electronic hopping conductivity in thin LN films formed by thermal evaporation onto glass substrates was also reported in [116]. Two activation processes with activation energies of 0.067 and 0.57 eV were detected at temperatures below and above 423 K, respectively. In our work [101], the temperature dependence of the DC hopping conductivity of (001)Si–LN heterostructures fabricated by the RFMS method was linear in Mott's coordinates. Our research revealed the main charge transport mechanism in the studied heterostructures. We demonstrated that DC electronic hopping conductivity is the main process, characterized by an activation energy of 0.27 eV and an average hopping length $R = 126$ Å between defect centers with a concentration of $N_t = 2.5 \times 10^{17}\ cm^{-3}$. In the framework of various hopping mechanisms, the AC conductivity of Si-LN heterostructures fabricated by the RFMS method was analyzed in the frequency range of 30–10^4 Hz [117]. Activated hopping transport with an activation energy of 0.4 eV occurred due to the presence of defects at a concentration of $N_t = 7 \times 10^{18}\ cm^{-3}$.

It is important to note that if defect centers are distributed in an insulator with high density near the Fermi level, non-activated hopping conductivity can occur under sufficiently high applied electric fields. Its I–V characteristics obey the following law [118]

$$J = J_0 \exp(-(E_0/E)^{1/4}) \tag{1.21}$$

where E is the applied electric field, J_0 is a field-independent constant, and E_0 is the characteristic field, expressed by

$$E_0 = \frac{16}{D(E)a^4 q}. \tag{1.22}$$

Here, $D(E)$ is the energy density of localized states near the Fermi level and a is the localization length. According to equation (1.21), the I–V characteristics should be a straight line in $\ln(J)$ vs. $E^{-1/4}$ coordinates, with a slope giving E_0. In our previous work [95], we observed this conductivity mechanism in LN films, with linear I–V characteristics corresponding to equation (1.21) (see figure 1.17).

In LN films fabricated by the RFMS method, non-activated hopping conductivity under high electric fields ($E = 5 \times 10^5$–$3 \times 10^6 \, \text{V m}^{-1}$) plays a dominant role in the charge transport process. This occurs across defect centers (Li vacancies), which are distributed throughout the material's bandgap at a density of $D(E) = 1.5 \times 10^{27}$ $\text{eV}^{-1} \, \text{m}^{-3}$ [113, 119].

Grain-boundary-limited conduction
In polycrystalline materials, the resistivity of grain boundaries can exceed that of the grains themselves. It is influenced by factors such as temperature, grain size, and orientation. Consequently, charge transport in polycrystalline films is governed by the electrical properties of the grain boundaries rather than the bulk material, a phenomenon known as grain-boundary-limited conduction. Charge carriers moving from grain to grain are scattered by charged interfaces at intergrain boundaries.

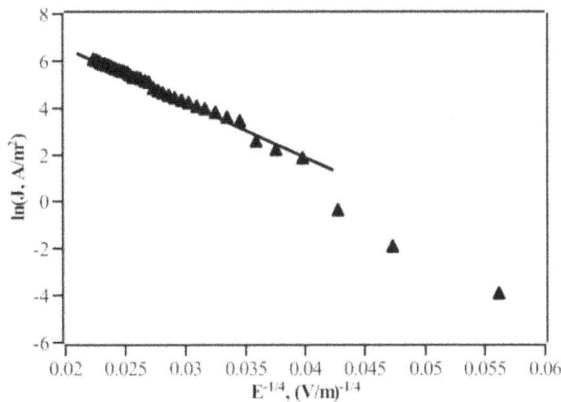

Figure 1.17. I–V characteristic of LN film in $\ln(J)$–$E^{-1/4}$ coordinates.

Only electrons with sufficient energy $E > \phi_b$, corresponding to their thermal velocity, can cross the barrier ϕ_b present at the grain boundaries (see figure 1.18).

Impedance spectroscopy can be used to split the total conductivity into the grain boundary contribution and the bulk contribution [120]. An appropriate equivalent circuit for the studied film can help distinguish bulk and grain boundary contributions due to different relaxation times associated with their response to an AC signal. Specifically, figure 1.19 shows the simplest equivalent circuit representing the conductive properties of LN films. This circuit depicts the resistivity and capacitance of grain boundaries (R_{gb}, C_{gb}), and the grain bulk (R_b, C_b).

The characteristic relaxation time or time constant corresponding to each parallel RC element is denoted by $\tau = RC$. A maximum in the impedance spectra occurs at a characteristic frequency of $\omega_m = 1/RC$. Thus, both the grain boundary and bulk parameters of a polycrystalline film can be estimated if experimental data is represented by a frequency dependence of the imaginary parts of the impedance ($Z''(\omega)$) and the dielectric modulus ($M''(\omega)$) with a peak at the frequency of ω_m. By analyzing the impedance spectra (see figure 1.20) of thin polycrystalline LN films grown by the RFMS method, we estimated the ratio between the average depletion area width within a single grain and the size of a neutral (bulk) area $d_{gb}/d_b = 0.4$ [121]. We also demonstrated that the grain-boundary-limited electrical conductivity of the studied films is caused by intergranular barriers of 0.4 eV. TA of as-grown films can increase the d_{gb}/d_b ratio to one, making LN films more homogeneous and increasing the role of bulk conductivity [121].

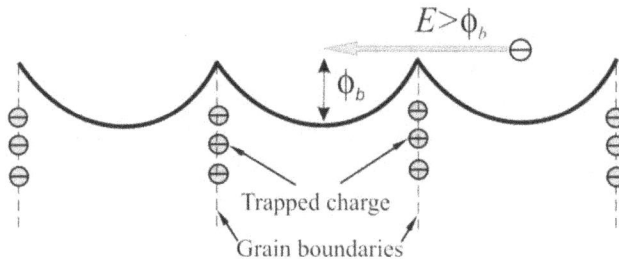

Figure 1.18. Schematic band diagram of an n-type polycrystalline semiconductor with grain boundaries and intergranular barriers of height ϕ_b.

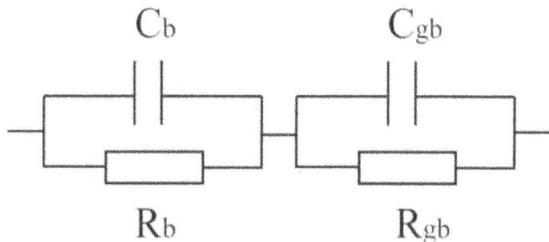

Figure 1.19. A possible equivalent circuit for an LN film.

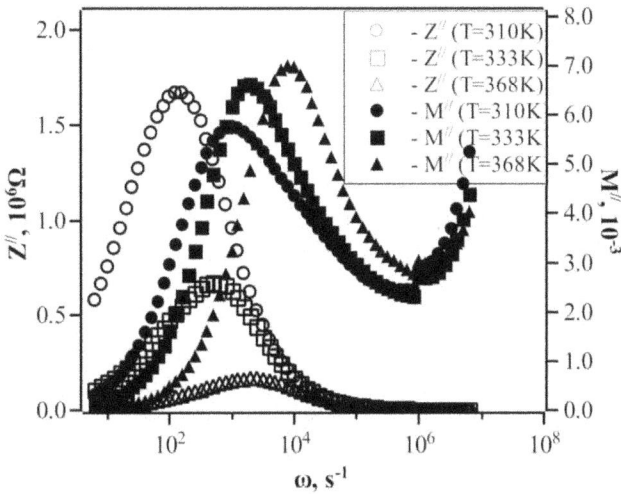

Figure 1.20. Imaginary parts of the complex impedance (Z'') and the dielectric modulus (M'') of as-grown (001)Si–LN heterostructures at different temperatures. Reproduced from [121], with permission from Springer Nature.

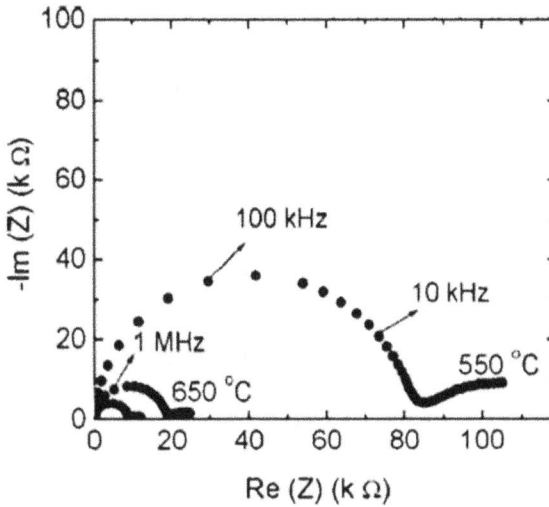

Figure 1.21. Complex impedance data at different temperatures of a LN ceramic sample. Reproduced with permission from [122]. © 1999 American Institute of Physics.

Another method of separating the grain boundary and bulk contributions to the total complex impedance Z^* of LN films deposited by a chemical evaporative technique was described in [122]. Figure 1.21 shows a Nyquist plot (Im(Z^*) vs. Re(Z^*) graph) with two semicircles attributed to grain boundaries (a wide semicircle at low frequencies) and the grain bulk (a smaller semicircle at higher frequencies). This study, which aimed to investigate the bulk dielectric properties of polycrystalline LN films, reported thermally activated bulk conductivity with an activation energy of $E_a = 1.17\,\text{eV}$, which apparently resulted from the thermal activation of electrons from deep donors.

Since frequency response is crucial in certain applications of LN films (such as selective filters, electro-optic modulators, and tunable microring optical resonators), grain boundaries should be considered. Studies [113, 117] have clearly demonstrated that TA of LN films reduces (or even suppresses) the contribution of grain boundaries to total conductivity, which is responsible for a low-frequency response ('slow process'), thereby enhancing the performance of LN-based devices.

Ionic conduction

When an insulator contains various defects, ions may jump from one defect site to another under an applied electric field. This process is known as ionic conduction. If the applied electric field satisfies the condition $E \approx kT/qd$, the current–voltage characteristics of ionic conduction follow the expression [102]

$$J = C \exp\left[-\left(\frac{\phi}{kT} - \frac{Eqd}{2kT}\right)\right]. \tag{1.23}$$

Here, C is a constant, ϕ is the potential barrier for ions between two sites separated by a distance d, and E is the applied electric field. Some researchers suggest that due to the relatively high mobility of lithium ions, ionic conductivity is the dominant charge transport mechanism in LN films [123, 124]. Initially, LN films deposited by an E-gun evaporation method exhibited electron-hopping AC conductivity due to lithium defects. However, when the surface of the LN film was coated with a Li–Nb–O layer followed by TA, the resulting film demonstrated ionic conductivity due to the diffusion of lithium ions into the film.

Another research group that analyzed dielectric relaxation and AC conductivity of congruent LN single crystals reported that thermally activated conductivity ($E_a = 0.9\,\text{eV}$) can be attributed to the long-range motion of Li^+ ions [125]. However, in our view, there is little evidence to support this opinion. It is challenging to distinguish between ionic and electronic conductivity in dielectric films. If a relatively high current flows through the film for a long period, ionic species are deposited upon the electrodes; in other words, material transport takes place, which is a strong criterion of ionic conduction. Polarization effects are the second crucial factor associated with ionic conductivity. In this case, film resistivity rises over time due to the formation of space charges near the electrodes. Since these effects were observed in the aforementioned studies, more detailed investigations are needed to determine whether electronic or ionic charge transport (or both) affects the conductivity of the studied films.

1.6 Major application-focused challenges in the synthesis of thin LiNbO$_3$ films

It is generally accepted that there is no 'universal' method of film synthesis, since each technique has both advantages and disadvantages [126]. Normally, the use of a particular deposition process is affected by the required heterostructure properties, which depend on the specification of the fabricated device. For example, in memory devices, the polar axis of the deposited films must be oriented normally to the substrate surface. However, tetragonal oxides grown on (001)Si tend to be oriented laterally due to the difference in thermal expansion coefficients. This is not the only challenge that investigators try to overcome. The design and fabrication of a new generation of memory units utilizing LN-based heterostructures are restricted by a range of problems that remain unsolved. One of the major challenges is the synthesis of epitaxial or textured single-phase LN films on semiconductor substrates such that the films have the following combination of properties:

- Low roughness (<1 nm)
- Crack-free, with low mechanical stress at the film/substrate interface
- Low electrical conductivity
- A single-domain twins-free structure
- Low concentrations of defects and grain boundaries to reduce optical losses and leakage.

1.6.1 Synthesis of epitaxial LN films with low mechanical stress

In general, there are two alternative approaches to growing crystalline films: (1) the direct deposition of polycrystalline films under certain conditions; (2) the deposition of amorphous films followed by their crystallization via special post-deposition procedures (annealing, chemical treatment, photon treatment, etc) [127, 128]. The second approach offers a significant advantage: it eliminates deviations from the ideal LN stoichiometry. These deviations typically occur due to the high vapor pressure of lithium and lithium oxide. Unless corrected, they can lead to the formation of a paraelectric phase, $LiNb_3O_8$.

The first problem arising from the optimization of the growth conditions for single-phase textured or epitaxial LN films is their crystallization. Temperature is the major factor affecting the crystallinity of the growing film. The quality of crystallinity depends on several factors, such as the growth technique, the reaction chamber geometry (temperature gradients and thermocouple positions), adatom energy, and the crystallinity, quality, and surface energy of the substrate. Specifically, amorphous films are crystallized at higher temperatures on amorphous substrates compared to crystalline ones. At the same time, the synthesis of epitaxial films requires higher temperatures than the synthesis of polycrystalline layers to maintain high adatom mobility [129].

The authors of [53] studied the deposition kinetics of thin LN films grown by the PLD method on substrates of nanostructured diamond/Si with an intermediate amorphous SiO_2 layer. They revealed that the oxygen pressure greatly influences the orientation, composition, and surface morphology of the synthesized LN films. Under low oxygen pressure, the collision intensity of oxygen atoms with the plasma particles is low, with the result that the sputtered components reach the substrate with high kinetic energy. This leads to the degradation of crystallinity and film orientation. Under relatively high oxygen pressures, the collision probability rises.

Consequently, the low-energy sputtered components reaching the substrate are not capable of forming highly oriented LN films. In addition, intense collisions between O atoms and the lighter component (Li) prevent sufficient transport of the lighter component to the substrate, leading to the formation of the Li-poor phase $LiNb_3O_8$. The author recommends that an oxygen pressure of 40 Pa should be used in the reaction chamber, which promotes the formation of highly oriented LN films with minimal roughness (9 nm). However, this value, combined with a low electro-mechanical coupling coefficient (0.46%), is inadequate for the fabrication of high-quality SAW devices.

The authors of [130] revealed a significant difference between direct crystallization that occurs during the deposition process and solid-phase crystallization in terms of the nucleation and crystallization of the $Li_2O–Nb_2O_5$ system. They revealed that a combination of the Li_3NbO_4 (222), LN (006), and $LiNb_3O_8$ ($\overline{6}02$) orientations yielded the minimum surface free energy. However, in solid-phase crystallization, nucleation was terminated by an excessive presence of oxygen and lithium atoms. The crystallite orientation was controlled by the growth rate of each crystal plane. When the crystallization process exhibited nonequilibrium features, the crystallization of LN and $LiNb_3O_8$ was intensified when the amount of Li_2O was lower than the threshold value.

The presence of mechanical stress in films and at interfaces is key to the functionality of thin film devices, as it degrades their optical and piezoelectric properties. Despite a significant lattice mismatch (8.2%), some authors have reported the successful formation of epitaxial LN films on sapphire. For instance, the authors of [131] synthesized high-quality LN films by the metal–organic chemical vapor deposition (MOCVD) method at a temperature of 650 °C followed by cyclical heating to 860 °C and subsequent cooling (up to six cycles). During the heating process the LN films lost Li_2O (on average, 0.36 mol% per cycle), and this evaporation increased as the film thickness declined. In the as-grown films, polarization reversal was observed only laterally, and twins were not formed in a perpendicular direction. TA activated the $\{01\overline{1}2\}$ twins, eliminating stress in the films. It was revealed that the thermal expansion of the epitaxial films stems from substrate clamping; it does not depend on the residual stress and can be tuned by both the choice of substrate type and the film thickness. Other investigators emphasized that despite the formation of highly oriented LN films on sapphire, significant mechanical stress was observed at the film–substrate interface. This stress was accompanied by a decline in the film lattice constant and deviation of the refractive index from that of the single-crystal LN [19, 132]. In addition, relatively high roughness (40 nm) impeded the formation of heterostructures with low optical losses [19].

To eliminate the negative effect of mechanical stress, LN films were deposited onto Si substrates using buffer layers such as ZnO and SiO_2 [133]. The use of ZnO as an intermediate layer is dictated by several factors. First, ZnO tends to grow along the c-axis, so c-oriented ZnO films can be grown on any substrates, including amorphous SiO_2 layers and substrates that present a significant lattice mismatch. Second, ZnO has a hexagonal structure, which inevitably facilitates the formation of domains with the epitaxial relationships [11$\bar{2}$0] LN ‖[10$\bar{1}$0] ZnO and [11$\bar{2}$0] LN ‖[$\bar{1}$010] ZnO [129]. Third, ZnO can be utilized as a conductive transparent contact. Finally, the use of ZnO allows LN films to be grown on Si at lower temperatures compared to other layers such as MgO and Si_3N_4. Nevertheless, as pointed out in [133], ferroelectric single-axis LN films grown on heated ZnO/Si substrates by the PLD method were not epitaxial. They contained randomly oriented domains, leading to relatively low remnant polarization ($P_r = 0.9\,\mu C\,cm^{-2}$). In addition, single-phase LN films were successfully synthesized only at a temperature of 550 °C, whereas the additional Li-poor $LiNb_3O_8$ phase occurred at higher temperatures. Although the isolating properties of ZnO layers improved when the thickness rose from 15 to 100 nm, the average roughness increased from 5.4 to 12.3 nm, which elevated optical losses.

As an efficient deposition technique for the formation of LN films on a wide range of substrates, RFMS allows the properties of LN-based heterostructures to be tuned by adjusting the sputtering parameters. Since the deposition rate depends on the intensity of collisions between the plasma compound and the atoms of the sputtered material, it is possible to control the degree of mechanical stress in deposited films using the reactive gas pressure. As demonstrated, the mobility of adatoms on the substrate surface is lower at a high sputtering rate when the RFMS process is used [134]. In addition, it has been demonstrated that negative ions (or atoms of electronegative elements) affect film growth depending on their energy in the reaction chamber. Moreover, some researchers have expressed the opinion that negative oxygen ions formed during target sputtering dominate the crystallization of complex oxides such as LN [135]. Thus, the oxygen-balanced reactive gas composition in the reaction chamber, along with the working gas pressure, is the key parameter of the RFMS process for high-quality LN film deposition.

The authors of [70] used the c-oriented growth of a (002)ZnO layer on a Si substrate for the nucleation and growth of (006)LN films using the RFMS process. The Li-rich LN target was sputtered in an Ar(80%) + O_2(20%) gas mixture at a pressure of 1–2 Pa and a temperature of 450 °C. As revealed, the mechanical stress has two components, 'internal' and 'external.' The 'internal' stress stems from the presence of point defects in the film and crystal lattice distortion, which occurs due to the intense bombardment of the substrate surface by plasma particles during RFMS. Specifically, the presence of antisite defects (niobium on lithium sites, Nb_{Li}) can elongate the unit cell in the growth direction (polar axis). As a result, the deposited LN film experiences compression forces under the high gas pressure in the reaction chamber [70]. The external stress primarily originates from the mismatch of lattice parameters and thermal expansion coefficients between the substrate and the film. This study emphasized the dominant role of the reactive gas pressure in the formation of LN films with minimal stress. The authors successfully grew highly oriented (006)LN films with minimal mechanical stress using a unique combination of reactive gas composition (80% Ar + 20% O_2) and pressure (1.33 Pa).

Recently, the dependence of the relative orientation between LN films and a buffer layer (Al:ZnO) was investigated. Specifically, it was revealed that highly oriented (110)LN films grown by the PLD method on Al:ZnO(002) manifested lower mechanical stress compared to (300)LN films deposited on Al:ZnO(103) [136]. In addition, the roughness of LN films decreased from 10 nm to 7 nm and then to 4 nm when fused silica, Al:ZnO/fused silica, and Al:ZnO/Z-cut quartz were used as substrates, respectively. This result demonstrates the crucial role of a substrate in the growth of a bottom layer, which ultimately affects the top-layer morphology and microstructure.

With the aim of creating thin films on conductive substrates, thin LN epitaxial LN films were synthesized on (111)Ag/Si heterostructures by the RFMS method [137]. The authors demonstrated that the deposited films exhibited a two-axis (epitaxial) texture with the orientation relationship (0001), $[11\bar{2}0]$ LiNbO$_3$ ‖(111), $\langle 110 \rangle$ Ag, where the Ag layer additionally served as an electric contact in multi-layered multifunctional memory units.

Epitaxial LN films were grown by the PLD method on GaN substrates with orientation relationships of $[1\bar{1}00]$ LN ‖$[11\bar{2}0]$ GaN and $[10\bar{1}0]$ LN ‖$[1\bar{2}10]$ GaN, despite the occurrence of a thin (3 nm) amorphous intermediate layer at the LN/GaN interface [104]. The authors suggested that two processes take place simultaneously during the formation of an LN/GaN heterojunction: vertical heteroepitaxy (which occurs when LN contacts GaN) and lateral homoepitaxy (which occurs at the surface of amorphous grains). Although the authors performed epitaxial growth, a negative charge with a density of 1.5×10^{13} cm^{-2} present at the LN/GaN interface prevented effective polarization switching in the synthesized heterostructures.

Atomic layer deposition (ALD) is one of the most sophisticated and promising thin film deposition techniques for complex oxides such as LN. The authors of [104] used this method to fabricate thin LN films on various substrates. Films grown under the same conditions had different orientations: they were epitaxial on Al$_2$O$_3$(012), Al$_2$O$_3$(001), and LaAlO$_3$(012) substrates but polycrystalline when Si (100) wafers covered by a natural oxide film were used. They had the following orientation relationships: Al$_2$O$_3$(012)|Al$_2$O$_3$[100] ‖LiNbO$_3$ (012)|LiNbO$_3$[100] and Al$_2$O$_3$(001)|Al$_2$O$_3$[100] ‖LiNbO$_3$ (001)|LiNbO$_3$[100] with a polar axis oriented normally to the film surface. The surface roughness reached the magnitude required for multifunctional memory units. The surface roughnesses were 1.6, 1.3, 1.3, and 1.8 nm for films deposited on Si(100), Al$_2$O$_3$(012), Al$_2$O$_3$(001), and LaAlO$_3$(012), respectively. Although the coercive field was 220 kV cm^{-1} in the fabricated films (which is closest to that for congruent LN), the remnant polarization of 0.4 μC cm^{-2} was still below the corresponding value for bulk LN. As demonstrated, TA of synthesized films in air leads to the evaporation of volatile components (Li) and the formation of the Li-poor phase LiNb$_3$O$_8$, segregated at the grain boundaries. This result agrees with an earlier model proposed by other investigators [84].

1.6.2 Ferroelectric properties and polarization reversal

The majority of ferroelectric-based applications utilize the manipulation of ferro-electric polarization and the associated domain [84]. The synthesis of epitaxial or oriented LN films is vital for effective polarization switching in memory units. Moreover, when the polarization reversal is tuned in single polycrystalline grains, a wide range of options is opened up for the creation of a new generation of nanoscale memory cells [138]. On the other hand, the polarization reversal is closely connected to ferroelectric instability, which negatively affects the functionality of memory units. The strong mechanical stress occurring at the film–substrate interface impairs the structural, optical, and electrical properties of LN-based devices [139]. Ferroelectric properties, as well as polarization reversal, depend on the crystallinity, composition, and orientation of LN films. Specifically, the rotation of the polarization direction from an out-of-plane to an in-plane direction in the film is associated with antisite defects such as niobium in the lithium site (Nb_{Li}) and the presence of $LiNb_3O_8$ with Li ions shifted along the surface plane in the monoclinic lattice framework [140]. Highly oriented thin (001)LN films were deposited onto (111)$SrTiO_3$ by the PLD technique in spite of the fact that (111)$SrTiO_3$ has a large tensile strain with a larger lattice constant than that of bulk LN. The authors of this paper argued that this was the case because both materials have the same in-plane hexagonal symmetry and chemical strain. They observed the continuous rotation of ferroelectric polarization in ordered–disordered LN films [140]. This study revealed that varying the Li vacancy concentration in the thin film could tilt the spontaneous polarization from an out-of-plane to an in-plane direction. The partial inclusion of the non-ferroelectric $LiNb_3O_8$ phase in LN films causes ferroelectric polarization to occur in the in-plane direction. This result is especially beneficial for engineering high levels of ferroelectric polarization.

However, the relatively high concentration of vacancies, as well as the formation of the second paraelectric phase ($LiNbO_3$) in LN films, negatively influences their ferroelectric properties and domain reversal. The composition, orientation, and crystallinity of LN films are influenced by the growing conditions and can be tuned depending on a specific deposition technique. Specifically, films grown using the RFMS and PLD techniques are very sensitive to the reactive gas composition. As reported in [141], LN films sputtered in Ar did not demonstrate any randomly oriented domain reversal, whereas films grown in an As (60%) + O_2 (40%) reactive gas mixture manifested two opposing out-of-plane oriented domain states capable of effective domain poling. The effective piezoelectric coefficient of these films was 6.9 ± 1.2 pm V^{-1}, which is about two thirds of the value for bulk Z-cut LN.

Traditionally, various post-growth techniques, from conventional TA to some specific techniques such as pulsed photon and chemical treatments, are applied to the as-deposited films to improve their structural, electric, and ferroelectric properties.

The authors of [141] revealed that the switching of a box-in-box pattern at ±40 V in annealed LN films grown in an Ar atmosphere led to polarization reversal (figures 1.22(a)–(c)). However, local poling in films fabricated in an Ar + O_2 environment after TA produced polarization reversal similar to that seen in the initial samples (figures 1.22(e) and (f)).

Recently, a new perspective class of erasable nanodevices, such as nonvolatile memories, logic elements, and local strain sensors, has been proposed. The basic paradigm underlying these devices utilizes conducting domain walls (DWs) in an insulating ferroelectric matrix to generate novel electronic ground states that differ from those of the surrounding bulk material [142, 143]. Notably, charged DWs are

Figure 1.22. Local poling in (a–c) LN(Ar) and (d–f) LN(Ar + O$_2$) films after TA: (a and d) topography, as determined by piezoresponse force microscopy (PFM); (b and e) amplitude; and (c and f) phase. LN(Ar) and LN(Ar + O$_2$) correspond to LN films fabricated in Ar and Ar + O$_2$ environments, respectively [141] [2019], reprinted by permission of the publisher (Taylor & Francis Ltd, https://www.tandfonline.com).

more promising because their conducted currents are much higher than those of neutral DWs. This approach was successfully applied in all-ferroelectric nonvolatile LN transistors, which functioned through the redirection of conducting DWs between the drain, gate, and source electrodes [142]. In addition, finer tuning can be achieved by changing the DW configuration. Specifically, the wall conductance increases with its angle of inclination, allowing the DW conduction to be tuned, which is highly desirable in multilevel logic devices [143]. Despite the aforementioned progress, the further evolution of this approach is limited by the fact that the majority of the charged walls are unstable due to their high depolarization energies. Although methods of stabilizing these charged DWs are still lacking, some possible approaches have been proposed to fill this gap. The distinction between the intrinsic conductivities of neutral (N), tail-to-tail (T–T), and head-to-head (H–H) DWs was utilized in an LN-based transistor [144]. The authors stabilized DWs through the control of charge injection, compensating for the domain boundary charge by applying drain–gate, drain–source, and gate–source voltages. They revealed that the electrical conductivity across the H–H DWs is more than 4 orders of magnitude greater than that across the T–T ones. The diagrams of charged DWs and the working principle for the fabricated transistor are shown in figure 1.23.

Shur's leading research group developed another innovative approach to creating and controlling charged domain walls (CDWs) [145]. Two alternative procedures were proposed: (1) forward switching for the liquid electrode at the $Z+$ polar surface and the solid electrode at the $Z-$ polar surface, (2) backward switching for the solid electrode at the $Z+$ polar surface and the liquid electrode at the $Z-$

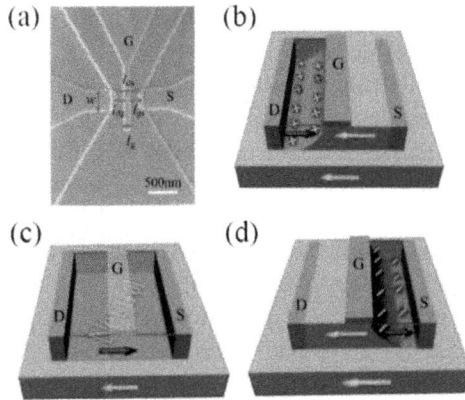

Figure 1.23. (a) Planar scanning electron microscope (SEM) image of a three-terminal LN cell. (b–d) Sketches of charged H–H DWs between drain (D) and gate (G), NDWs between D and source (S), and charged T–T DWs between G and S. Thick arrows show domain orientations and dotted lines delineate walls exposed at the film surfaces. Reprinted from [144], Copyright (2021), with permission from Elsevier.

surface. Within both procedures, the application of a field with a reverse polarity pulse transformed the CDW into a neutral DW. As a result, the tilt of the created CDW was reversibly tuned from 0.2° to 1.2°, leading to a switch between isolated and highly conductive states.

As mentioned earlier, the crystallization of a parent amorphous film of appropriate elemental composition grown on an unheated substrate is one of two alternative approaches in the fabrication of LN films. This technique has the advantage of eliminating the formation of the paraelectric phase $LiNb_3O_8$ due to the deviation of the composition from the stoichiometric LN. The properties of LN films synthesized by this technique greatly depend on the post-growth treatment applied.

As far as FeFET devices are concerned, they are extremely sensitive to any trapped charge in the recording information process [146]. The difference between the dielectric permittivity of LN and a SiO_2 layer, which inevitably forms during the deposition of LN onto silicon substrates, causes a significant portion of the applied gate voltage to drop across this layer [101]. The ferroelectric polarization makes this situation even worse.

1.6.3 Defect concentration and performance of thin LN films

Since the performance of LN-based heterostructures is affected by the presence of defects, a variety of fabrication-related techniques have been proposed to reduce their concentration. Near-stoichiometric Er^{3+} doped LN thin films were synthesized by a post-growth Li-rich vapor transport equilibration (VTE) technique in [147]. This study revealed that antisite Nb_{Li}^{4+} defects constitute a major defect type. The concentration of these defects decreases as the VTE time rises via a mechanism involving vapor transport and solid-state diffusion. The advantage of this technique is that the films are not affected by the annealing temperature. This technique can also be applied to improve the optical properties, which increases the Li_2O content to the near-stoichiometric value of 49.7%.

Defects present at the LN film–substrate interface degrade the optical properties of LN-based optoelectronic devices. As demonstrated, a standard wafer-based PLD technique can be used to deposit 200 nm thick LN thin films with a chemical composition and crystalline quality close to that of a monocrystalline LN substrate [148]. The interface between an LN layer grown by PLD and a Z-cut substrate was found to have a near-zero quantity of visible defects. As pointed out by the authors, the possible source of defects are distortions in the $(11\bar{2}0)$ and (0001) planes of the substrate and the film, respectively, that are not generated by the PLD growth itself. However, the PLD process is sensitive to the deposition regime used and is hardly reproducible.

Some authors have reported that doping with zinc (Zn) ions improved the anti-light-scattering property, light damage resistance, holographic storage image quality, and optical quality of LN films. As demonstrated, Zn^{2+} doping does not alter the crystal structure of LN films [149]. The maximum light transmittance is attained at a doping amount of 3%. The authors of this paper revealed the relationships between the interfacial barrier, the amount of Zn^{2+} dopant used, the migration of oxygen vacancies, and the insulative property. The incorporation of rare-earth ions into LN-based integrated photonics holds great promise for enriching the designer's toolbox in terms of key performance features that are unavailable in existing platforms [150]. For instance, $Er^{3+}/Yb^{3+}:LiNbO_3$ films were spin-coated on sapphire substrates using an alkoxide solution containing polyvinylpyrrolidone [151]. However, despite the simplicity of the sol–gel method, the synthesized films suffered from particle agglomeration, inhomogeneous composition, and crack formation (see figure 1.24). The authors proposed a two-step heat treatment that produced smoother films with better microstructure. Unfortunately, the transparency of the films decreased, and their wave-guiding ability was suppressed.

Oxygen vacancies inevitably generated in most deposition processes make the tuning of LN film properties complicated and uncontrollable. It is generally accepted that oxygen vacancies act as donors in complex metal oxides, and their distribution is highly variable due to the existence of grain boundaries in polycrystalline oxide films. Some investigators proposed the use of low-energy Ar^+ ion irradiation to locally dope LN thin films by controllably introducing oxygen donor vacancies [152]. As demonstrated, due to their low energy, the Ar^+ ions only penetrated the LN to a depth of a few nanometers, generating n-type dopants in the subsurface layer. The formation of this oxygen-deficient layer stems from the dislodging of Li by Ar^+ ions. As proposed, the resistive switching performance of LN-based devices can be tuned by changing the size and the number of irradiated regions below the top electrode. However, since the single-crystal LN films used in the aforementioned work were relatively thick (300 nm), they required a high applied driving voltage, which is a disadvantage for the neuromorphic applications of LN films.

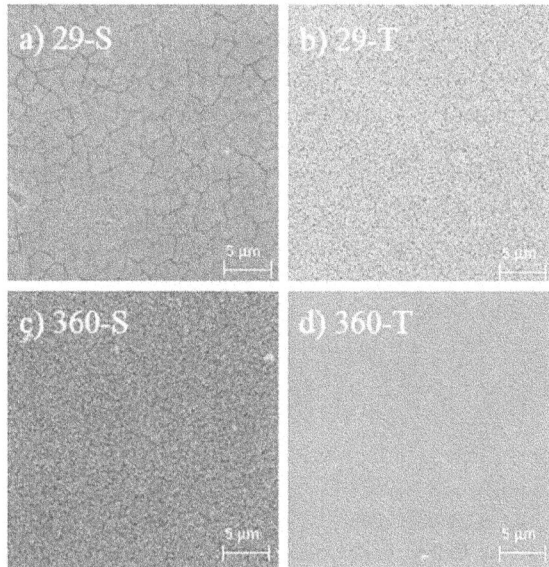

Figure 1.24. SEM images of the as-deposited (a and c) and annealed (b and d) $Er^{3+}Yb^{3+}$:LiNbO$_3$ films; the magnification is 5000×. Reprinted from [151], Copyright (2016), with permission from Elsevier.

1.7 Summary and discussion

The practical interest in the synthesis and properties of LN films, which has boomed in the last few decades, is fully reasonable. The requirements for the structure, morphology, and properties of LN films are formulated only to the 'first approximation' for existing LN films. Limitations originate from the fact that it is impossible to form single-crystal films of complex oxides, even by considering the possible synthesis of epitaxial structures using existing synthesis methods, so it is still impossible to fabricate films that have properties close to those of bulk LN.

In the process of RFMS, the elementary processes of stochastic growth, which are affected by the directional flow of ions and the wide range of their energies, force the formation of highly dispersed mosaic structures.

Thin LN films with arbitrary grain orientation, fabricated by RFMS without the 'ion assist' effect, manifest ferroelectric properties. However, these films have significantly less remnant polarization than bulk lithium niobate. Having said that, films with a single-axis texture $\langle 0001 \rangle$, whose formation is facilitated by the ion assist effect, have remnant polarization which approaches that of bulk LN material.

The oriented crystallization of LN films is unclear, and few substrate–film systems have been investigated. It is worth noting that systematic investigations suggest it is possible to grow oriented LN films for LN-based heterostructures synthesized by various methods.

No systematic data have been reported regarding: (1) the structure of grain boundaries and interfaces and (2) defects. Data on the latter are required to estimate the degree of influence of grain boundaries and interfaces on the film properties.

The nonequilibrium processes used for film fabrication (such as RFMS and PLD) imply a high degree of crystal structure imperfection.

In the last 15 years, the study of LN films has focused on their possible integration with planar technology to create new devices. There is notable progress in the study of the synthesis, structure, and properties of LN-based heterostructures fabricated by effective methods such as RFMS and PLD.

The electrical properties of LN films are significantly influenced by the deposition technique, and they greatly depend on the synthesis conditions. The required characteristics and device reliability are stimulating progress in the fundamental study of the growth and structure of LN films, as well as the development of new synthesis methods and their optimization.

References

[1] Ballman A A 1965 Growth of piezoelectric and ferroelectric materials by the czochralski technique *J. Am. Ceram. Soc.* **48** 112–3

[2] Abrahams S C, Levinstein H J and Reddy J M 1966 Ferroelectric lithium niobate. 5. Polycrystal X-ray diffraction study between 24 °C and 1200 °C *J. Phys. Chem. Solids* **27** 1019–26

[3] Abrahams S C, Reddy J M and Bernstein J L 1966 Ferroelectric lithium niobate. 3. Single crystal X-ray diffraction study at 24 °C *J. Phys. Chem. Solids* **27** 997–1012

[4] Abrahams S C and Marsh P 1986 Defect structure dependence on composition in lithium niobate *Acta Crystallogr.* B **42** 61–8

[5] Rauber A 1978 Chemistry and physics of lithium niobate *Curr. Top. Mater. Sci.* **1** 481–601

[6] Hatano H, Kitamura K and Liu Y 2007 Growth and photorefractive properties of stoichiometric $LiNbO_3$ and $LiTaO_3$ *Photorefractive Materials and Their Applications 2* (New York: Springer) pp 127–64

[7] Nassau K, Levinstein H J and Loiacono G M 1966 Ferroelectric lithium niobate. 2. Preparation of single domain crystals *J. Phys. Chem. Solids* **27** 989–96

[8] Coburn J W and Winters H F 1983 Plasma-assisted etching in microfabrication *Annu. Rev. Mater. Sci.* **13** 91–116

[9] Tamura M and Yoshikado S 2003 Etching characteristics of $LiNbO_3$ crystal by fluorine gas plasma reactive ion etching *Surf. Coat. Technol.* **169** 203–7

[10] Yang W S, Lee H-Y, Kim W K and Yoon D H 2005 Asymmetry ridge structure fabrication and reactive ion etching of $LiNbO_3$ *Opt. Mater.* **27** 1642–6

[11] Ching W Y, Gu Z-Q and Xu Y-N 1994 First-principles calculation of the electronic and optical properties of $LiNbO_3$ *Phys. Rev.* B **50** 1992–5

[12] Schmidt W G, Albrecht M, Wippermann S, Blankenburg S, Rauls E, Fuchs F, Rödl C, Furthmüller J and Hermann A 2008 $LiNbO_3$ ground- and excited-state properties from first-principles calculations *Phys. Rev.* B **77** 035106

[13] Bergman J G, Ashkin A, Ballman A A, Dziedzic J M, Levinstein H J and Smith R G 1968 Curie temperature birefringence, and phase-matching temperature variations in $LiNbO_3$ as a function of melt stoichiometry *Appl. Phys. Lett.* **12** 92

[14] Pogossian S P and Le Gall H 2003 Modeling planar leaky optical waveguides *J. Appl. Phys.* **93** 2337

[15] Fork D K, Armani-Leplingard F and Kingston J J 1996 Application of electroceramic thin films to optical waveguide devices *MRS Bull.* **21** 53–8

[16] Wohltjen H 1984 Mechanism of operation and design considerations for surface acoustic wave device vapour sensors *Sensors Actuators* **5** 307–25

[17] Runde D, Brunken S, Rüter C E and Kip D 2006 Integrated optical electric field sensor based on a Bragg grating in lithium niobate *Appl. Phys.* B **86** 91–5

[18] Lee T C, Lee J T, Robert M A, Wang S and Rabson T A 2003 Surface acoustic wave applications of lithium niobate thin films *Appl. Phys. Lett.* **82** 191–3

[19] Shih W-C, Sun X-Y, Wang T-L and Wu M-S 2009 Growth of c-axis oriented LiNbO$_3$ film on sapphire by pulsed laser deposition for surface acoustic wave applications *Ferroelectrics* **381** 92–9992

[20] Kadota M, Ogami T, Yamamoto K, Tochishita H and Negoro Y 2010 High-frequency lamb wave device composed of MEMS structure using LiNbO$_3$ thin film and air gap *IEEE Trans. Ultrason. Ferroelectr. Freq. Control* **57** 2564–71

[21] Dogheche E, Lansiaux X, Remiens D, Sadaune V, Chauvin S and Gryba T 2003 Growth process and surface acoustic wave characteristics of LiNbO$_3$/diamond/silicon multilayered structures *Jpn. J. Appl. Phys.* **42** 572–4

[22] Tan S, Gilbert T, Hung C-Y, Schlesinger T E and Migliuolo M 1996 Sputter deposited c-oriented LiNbO$_3$ thin films on SiO$_2$ *J. Appl. Phys.* **79** 3548

[23] Gupta V, Bhattacharya P, Yuzyuk Y I, Katiyar R S, Tomar M and Sreenivas K 2004 Growth and characterization of c-axis oriented LiNbO$_3$ film on a transparent conducting Al:ZnO inter-layer on Si *J. Mater. Res.* **19** 2235–9

[24] Lanzhong H 2009 Epitaxial fabrication and memory effect of ferroelectric LiNbO$_3$ film/ AlGaN/GaN heterostructure *Appl. Phys. Lett.* **95** 232907

[25] Scott J F 2000 Ferroelectric memories today *Ferroelectrics* **236** 247–58

[26] Wang S, Wang W, Yakopcic C, Shin E, Subramanyam G and Taha T M 2017 Experimental study of LiNbO$_3$ memristors for use in neuromorphic computing *Microelectron. Eng.* **168** 37–40

[27] Li H, Xia Y, Xu B, Guo H, Yin J and Liu Z 2010 Memristive behaviors of LiNbO$_3$ ferroelectric diodes *Appl. Phys. Lett.* **97** 012902

[28] Greenlee J D, Petersburg C F, Laws Calley W, Jaye C, Fischer D A, Alamgir F M and Alan Doolittle W 2012 In-situ oxygen x-ray absorption spectroscopy investigation of the resistance modulation mechanism in LiNbO2 memristors *Appl. Phys. Lett.* **100** 182106

[29] Pan X *et al* 2016 Rectifying filamentary resistive switching in ion-exfoliated LiNbO$_3$ thin films *Appl. Phys. Lett.* **108** 032904

[30] Strukov D B, Snider G S, Stewart D R and Williams R S 2008 The missing memristor found *Nature* **453** 80–3

[31] Yang J J, Pickett M D, Li X, Ohlberg D A A, Stewart D R and Williams R S 2008 Memristive switching mechanism for metal/oxide/metal nanodevices *Nat. Nanotechnol.* **3** 429–33

[32] Di Ventra M, Pershin Y and Chua L O 2009 Circuit elements with memory: memristors, memcapacitors, and meminductors *Proc. IEEE* **97** 1717–24

[33] Sheng P, Zeng F, Tang G S, Pan F, Yan W S and Hu F C 2012 Structure and ferromagnetism in vanadium-doped LiNbO$_3$ *J. Appl. Phys.* **112** 1–7

[34] Cheng S C, Wang X, Liu F and Zeng F P 2009 Room temperature ferromagnetism in cobalt-doped LiNbO$_3$ single crystalline films *Cryst. Growth Des.* **8** 1235–9

[35] Díaz-Moreno C, Farias R, Hurtado-Macias A, Elizalde-Galindo J and Hernandez-Paz J 2012 Multiferroic response of nanocrystalline lithium niobate *J. Appl. Phys.* **111** 5–8

[36] Ishii M, Ohta D, Uehara M and Kimishima Y 2012 Ferromagnetism of nano-LiNbO$_3$ with vacancies *Trans. Mater. Res. Soc. Jpn.* **37** 443–6

[37] Vasconcelos N S L S *et al* 2003 Epitaxial growth of LiNbO$_3$ thin films in a microwave oven *Thin Solid Films* **436** 213–9

[38] Kondo S, Miyazawa S, Fushimi S and Sugii K 1975 Liquid-phase-epitaxial growth of single-crystal LiNbO$_3$ thin film *Appl. Phys. Lett.* **26** 489

[39] Miyazawa S, Fushimi S and Kondo S 1975 Optical waveguide of LiNbO$_3$ thin film grown by liquid phase epitaxy *Appl. Phys. Lett.* **26** 8

[40] Dubs C, Ruske J-P, Kräußlich J and Tünnermann A 2009 Rib waveguides based on Zn-substituted LiNbO$_3$ films grown by liquid phase epitaxy *Opt. Mater.* **31** 1650–7

[41] Lu Y, Dekker P and Dawes J M 2009 Growth and characterization of lithium niobate planar waveguides by liquid phase epitaxy *J. Cryst. Growth* **311** 1441–5

[42] Margueron S, Bartasyte A, Plausinaitiene V, Abrutis A, Boulet P, Kubilius V and Saltyte Z 2013 Effect of deposition conditions on the stoichiometry and structural properties of LiNbO$_3$ thin films deposited by MOCVD *Proc. SPIE 8626, Oxide-based Materials and Devices IV* **8626** 862612

[43] Dislich H 1971 New routes to multicomponent oxide glasses *Angew. Chem., Int. Ed. Engl.* **10** 363–70

[44] Nashimoto K, Cima M J, McIntyre P C and Rhine W E 1995 Microstructure development of sol–gel derived epitaxial LiNbO$_3$ thin films *J. Mater. Res.* **10** 2564–72

[45] Simões A Z, Ries A, Riccardi C S, Gonzalez A H, Zaghete M A, Stojanovic B D, Cilense M and Varela J A 2004 Potassium niobate thin films prepared through polymeric precursor method *Mater. Lett.* **58** 2537–40

[46] Fakhri M A, Al-Douri Y, Hashim U, Salim E T, Prakash D and Verma K D 2015 Optical investigation of nanophotonic lithium niobate-based optical waveguide *Appl. Phys. B: Lasers Opt.* **121** 107–16

[47] Schwarz H and Tourtellotte H A 1969 Vacuum deposition by high-energy laser with emphasis on barium titanate films *J. Vac. Sci. Technol.* **6** 373–8

[48] Shibata Y, Kaya K, Akashi K, Kanai M, Kawai T and Kawai S 1993 Epitaxial growth of LiNbO$_3$ films on sapphire substrates by excimer laser ablation method and their surface acoustic wave properties *Jpn. J. Appl. Phys.* **32** L745–7

[49] Ogale S B, Nawathey-Dikshit R, Dikshit S J and Kanetkar S M 1992 Pulsed laser deposition of stoichiometric LiNbO$_3$ thin films by using O2 and Ar gas mixtures as ambients *J. Appl. Phys.* **71** 5718

[50] Joshi V, Roy D and Mecartney M L 1993 Low temperature synthesis and properties of lithium niobate thin films *Appl. Phys. Lett.* **63** 1331

[51] Balestrino G *et al* 2001 Epitaxial LiNbO$_3$ thin films grown by pulsed laser deposition for optical waveguides *Appl. Phys. Lett.* **78** 1204–6

[52] Jelínek M, Havránek V, Remsa J, Kocourek T, Vincze A, Bruncko J, Studnička V and Rubešová K 2013 Composition, XRD and morphology study of laser prepared LiNbO$_3$ films *Appl. Phys.* A **110** 883–8

[53] Wang X, Liang Y, Tian S, Man W and Jia J 2013 Oxygen pressure dependent growth of pulsed laser deposited LiNbO$_3$ films on diamond for surface acoustic wave device application *J. Cryst. Growth* **375** 73–7

[54] Wang X, Ye Z, Li G and Zhao B 2007 Influence of substrate temperature on the growth and optical waveguide properties of oriented LiNbO$_3$ thin films *J. Cryst. Growth* **306** 62–7

[55] Foster N F 1969 The deposition and piezoelectric characteristics of sputtered lithium niobate films *J. Appl. Phys.* **40** 420–1

[56] Russo D P G and Kumar C S 1973 Sputtered ferroelectric thin-film electro-optic modulator *Appl. Phys. Lett.* **23** 229–31

[57] Takada S, Ohnishi M, Hayakawa H and Mikoshiba N 1974 Optical waveguides of single-crystal LiNbO$_3$ film deposited by rf sputtering *Appl. Phys. Lett.* **24** 490–2

[58] Hewig G M, Jain K, Sequeda F O, Tom R and Wang P-W 1982 R.F. Sputtering of LiNbO$_3$ thin films *Thin Solid Films* **88** 67–74

[59] Rabson T A, Baumann R C and Rost T A 1990 Thin film lithium niobate on silicon *Ferroelectrics* **112** 265–71

[60] Kanata T, Kobayashi Y and Kubota K 1987 Epitaxial growth of LiNbO$_3$–LiTaO3 thin films on Al2O3 *J. Appl. Phys.* **62** 2989–93

[61] Vook R W 1984 Nucleation and growth of thin films *Opt. Eng.* **23** 343–9

[62] He J and Ye Z 2003 Highly C-axis oriented LiNbO$_3$ thin film on amorphous SiO2 buffer layer and its growth mechanism *Chin. Sci. Bull.* **48** 2290

[63] Shtansky D V, Kulinich S A, Terashima K and Yoshida T 2001 Crystallography and structural evolution of LiNbO$_3$ and LiNb$_{1-x}$Ta$_x$O$_3$ films on sapphire prepared by high-rate thermal plasma spray chemical vapor deposition *J. Mater. Res.* **16** 2271–9

[64] Veignant F 1998 Epitaxial growth of LiNbO$_3$ on αAl2O3(0001) *Thin Solid Films* **336** 163–7

[65] Akazawa H 2009 Target-quality dependent crystallinity of sputter-deposited LiNbO$_3$ films: observation of impurity segregation *Thin Solid Films* **517** 5786–92

[66] Griffel G, Ruschin S and Croitoru N 1989 Linear electro-optic effect in sputtered polycrystalline LiNbO$_3$ films *Appl. Phys. Lett.* **54** 1385

[67] Park S-K, Baek M-S, Bae S-C, Kim K-W, Kwun S-Y, Kim Y-J and Kim J-H 1999 (012) Preferred orientation of LiNbO$_3$ thin films by RF-magnetron sputtering *Jpn. J. Appl. Phys.* **38** 4167–71

[68] Curtis B J and Brunner H R 1975 The growth of thin films of lithium niobate by chemical vapour de position *Mater. Res. Bull.* **10** 515–20

[69] Lee T-H, Hwang F-T, Lee C-T and Lee H-Y 2007 Investigation of LiNbO$_3$ thin films grown on Si substrate using magnetron sputter *Mater. Sci. Eng.* B **136** 92–5

[70] Shandilya S, Tomar M and Gupta V 2012 Deposition of stress free c-axis oriented LiNbO$_3$ thin film grown on (002) ZnO coated Si substrate *J. Appl. Phys.* **111** 10–6

[71] Park S K, Baek M S, Bae S C, Kwun S Y, Kim K T and Kim K W 1999 Properties of LiNbO$_3$ thin film prepared from ceramic Li–Nb–K–O target *Solid State Commun.* **111** 347–52

[72] Rost T A, Lin H, Rabson T A, Baumann R C and Callahan D L 1992 Deposition and analysis of lithium niobate and other lithium niobium oxides by rf magnetron sputtering *J. Appl. Phys.* **72** 4336–43

[73] Tsirlin M 2004 Influence of gas phase composition on the defects formation in lithium niobate *J. Mater. Sci.* **39** 3187–9

[74] Gordillo-Vázquez F J and Afonso C N 2002 Influence of Ar and O2 atmospheres on the Li atom concentration in the plasma produced by laser ablation of LiNbO$_3$ *J. Appl. Phys.* **92** 7651

[75] Wang X, He J, Huang J, Zhao B and Ye Z 2003 Effects of oxygen pressure on the c-axis oriented growth of LiNbO$_3$ thin film on SiO2/Si substrate by pulsed laser deposition *J. Mater. Sci. Lett.* **22** 225–7

[76] Bornand V and Papet P 2005 LiNbO$_3$-based ferroelectric heterostructures *J. Phys. IV* **126** 89–92

[77] Kiselev D A, Zhukov R N, Bykov A S, Voronova M I, Shcherbachev K D, Malinkovich M D and Parkhomenko Y N 2014 Effect of annealing on the structure and phase composition of thin electro-optical lithium niobate films *Inorg. Mater.* **50** 419–22

[78] Simões A Z, Zaghete M A, Stojanovic B D, Gonzalez A H, Riccardi C S, Cantoni M and Varela J A 2004 Influence of oxygen atmosphere on crystallization and properties of LiNbO$_3$ thin films *J. Eur. Ceram. Soc.* **24** 1607–13

[79] Simões A Z, Zaghete M A, Stojanovic B D, Riccardi C S, Ries A, Gonzalez A H and Varela J A 2003 LiNbO$_3$ thin films prepared through polymeric precursor method *Mater. Lett.* **57** 2333–9

[80] Bornand V, Huet I and Papet P 2002 LiNbO$_3$ thin films deposited on Si substrates: a morphological development study *Mater. Chem. Phys.* **77** 571–7

[81] Edon V, Rèmiens D and Saada S 2009 Structural, electrical and piezoelectric properties of LiNbO$_3$ thin films for surface acoustic wave resonators applications *Appl. Surf. Sci.* **256** 1455–60

[82] Simões A Z, Gonzalez A H, Zaghete M A, Stojanovic B D, Cavalheiro A A, Moeckli P, Setter N and Varela J A 2002 Influence of oxygen flow on crystallization and morphology of LiNbO$_3$ thin films *Ferroelectrics* **271** 33–8

[83] Akazawa H and Shimada M 2007 Mechanism for LiNb3O8 phase formation during thermal annealing of crystalline and amorphous LiNbO$_3$ thin films *J. Mater. Res.* **22** 1726–36

[84] Akazawa H and Shimada M 2006 Precipitation kinetics of LiNbO$_3$ and LiNb3O8 crystalline phases in thermally annealed amorphous LiNbO$_3$ thin films *Phys. Status Solidi* **203** 2823–7

[85] Volk T and Wöhlecke M 2008 *Lithium Niobate : Defects, Photorefraction and Ferroelectric Switching* (Berlin: Springer)

[86] Wong K KInstitution of Electrical Engineers and INSPEC (Information Service) 2002 *Properties of Lithium Niobate* (London: INSPEC/Institution of Electrical Engineers)

[87] Wilkinson A P, Cheetham A K and Jarman R H 1993 The defect structure of congruently melting lithium niobate *J. Appl. Phys.* **74** 3080

[88] Donnerberg H, Tomlinson S M, Catlow C R A and Schirmer O F 1989 Computer-simulation studies of intrinsic defects in LiNbO$_3$ crystals *Phys. Rev.* B **40** 11909–16

[89] Schirmer O F, Imlau M, Merschjann C, Schoke B and D 2009 Electron small polarons and bipolarons in LiNbO$_3$ *J. Phys.: Condens. Matter* **21** 123201

[90] Herth P, Granzow T, Schaniel D, Woike T, Imlau M and Krätzig E 2005 Evidence for light-induced hole polarons in LiNbO$_3$ *Phys. Rev. Lett.* **95** 067404

[91] Reyher H-J, Schulz R and Thiemann O 1994 Investigation of the optical-absorption bands of Nb^{4+} and Ti^{3+} in lithium niobate using magnetic circular dichroism and optically detected magnetic-resonance techniques *Phys. Rev.* B **50** 3609–19

[92] Sugak D, Zhydachevskii Y, Sugak Y, Buryy O, Ubizskii S, Solskii I, Schrader M and Becker K-D 2007 *In situ* investigation of optical absorption changes in LiNbO$_3$ during reducing/oxidizing high-temperature treatments *J. Phys.: Condens. Matter* **19** 086211

[93] Schirmer O F 2006 O−bound small polarons in oxide materials *J. Phys.: Condens. Matter* **18** R667–704

[94] Lim D, Jang B, Moon S, Won C and Yi J 2001 Characteristics of LiNbO$_3$ memory capacitors fabricated using a low thermal budget process *Solid-State Electron.* **45** 1159–63

[95] Sumets M 2017 Charge transport in LiNbO$_3$-based heterostructures *J. Nonlinear Opt. Phys. Mater.* **26** 1750011

[96] Simmons J G 1967 Poole–Frenkel effect and Schottky effect in metal–insulator–metal systems *Phys. Rev.* **155** 657–60

[97] Hao L-Z, Liu Y-J, Zhu J, Lei H-W, Liu Y-Y, Tang Z-Y, Zhang Y, Zhang W-L and Li Y-R 2011 Rectifying the current–voltage characteristics of a LiNbO$_3$ Film/GaN heterojunction *Chin. Phys. Lett.* **28** 107703

[98] Guo S M, Zhao Y G, Xiong C M and Lang P L 2006 Rectifying *I–V* characteristic of LiNbO$_3$/Nb-doped SrTiO$_3$ heterojunction *Appl. Phys. Lett.* **89** 223506

[99] Ievlev V, Sumets M, Kostyuchenko A, Ovchinnikov O, Vakhtel V and Kannykin S 2013 Band diagram of the Si-LiNbO$_3$ heterostructures grown by radio-frequency magnetron sputtering *Thin Solid Films* **542** 289–94

[100] Choi S-W, Choi Y-S, Lim D-G, Moon S-I, Kim S-H, Jang B-S and Yi J 2000 Effect of RTA treatment on LiNbO$_3$ MFS memory capacitors *Korean J. Ceram* **6** 138–42

[101] Iyevlev V, Kostyuchenko A, Sumets M and Vakhtel V 2011 Electrical and structural properties of LiNbO$_3$ films, grown by RF magnetron sputtering *J. Mater. Sci., Mater. Electron.* **22** 1258–63

[102] Maissel L I and Glang R 1970 *Handbook of Thin Film Technology* (New York: McGraw-Hill)

[103] Ievlev V, Sumets M and Kostyuchenko A 2013 Conduction mechanisms in Si-LiNbO$_3$ heterostructures grown by ion-beam sputtering method *J. Mater. Sci.* **48** 1562–70

[104] Hao L *et al* 2012 Integration and electrical properties of epitaxial LiNbO$_3$ ferroelectric film on n-type GaN semiconductor *Thin Solid Films* **520** 3035–8

[105] Akazawa H 2007 Observation of both potential barrier-type and filament-type resistance switching with sputtered LiNbO$_3$ thin films *Jpn. J. Appl. Phys.* **46** L848–50

[106] Hao L Z, Zhu J, Luo W B, Zeng H Z, Li Y R and Zhang Y 2010 Electron trap memory characteristics of LiNbO$_3$ film/AlGaN/GaN heterostructure *Appl. Phys. Lett.* **96** 032103

[107] Padovani F A and Stratton R 1966 Field and thermionic-field emission in Schottky barriers *Solid-State Electron.* **9** 695–707

[108] Sumets M, Ievlev V, Kostyuchenko A, Vakhtel V, Kannykin S and Kobzev A 2014 Electrical properties of Si-LiNbO$_3$ heterostructures grown by radio-frequency magnetron sputtering in an Ar + O2 environment *Thin Solid Films* **552** 32–8

[109] Simmons J G 1971 Conduction in thin dielectric films *J. Phys. D: Appl. Phys.* **4** 202

[110] Hill R M 1971 Poole–Frenkel conduction in amorphous solids *Philos. Mag.* **23** 59–86

[111] Joshi V, Roy D and Mecartney M L 1995 Nonlinear conduction in textured and non textured lithium niobate thin films *Integr. Ferroelectr.* **6** 321–7

[112] Iyevlev V, Kostyuchenko A and Sumets M 2011 Fabrication, substructure and properties of LiNbO 3 films *Proc SPIE* **7747** 77471J

[113] Iyevlev V, Sumets M and Kostyuchenko A 2012 Current–voltage characteristics and impedance spectroscopy of LiNbO$_3$ films grown by RF magnetron sputtering *J. Mater. Sci., Mater. Electron.* **23** 913–20

[114] Mott N F 1969 Conduction in non-crystalline materials *Philos. Mag.* **19** 835–52

[115] Shandilya S, Tomar M, Sreenivas K and Gupta V 2009 Purely hopping conduction in c-axis oriented LiNbO$_3$ thin films *J. Appl. Phys.* **105** 094105

[116] Easwaran N, Balasubramanian C, Narayandass S A K and Mangalaraj D 1992 Dielectric and AC conduction properties of thermally evaporated lithium niobate thin films *Phys. Status Solidi* **129** 443–51

[117] Ievlev V, Sumets M, Kostyuchenko A and Bezryadin N 2013 Dielectric losses and ac conductivity of Si-LiNbO$_3$ heterostructures grown by the RF magnetron sputtering method *J. Mater. Sci., Mater. Electron.* **24** 1651–7

[118] Mott N F and Davis E A 1972 *Electronic Processes in Non-Crystalline Materials* (Oxford: Clarendon)

[119] Sumets M, Ievlev V, Kostyuchenko A and Kuz'mina V 2014 Influence sputtering conditions on electrical characteristics of Si-LiNbO$_3$ heterostructures formed by radio-frequency magnetron sputtering *Mol. Cryst. Liq. Cryst.* **603** 202–15

[120] Macdonald J R 1992 Impedance spectroscopy *Ann. Biomed. Eng.* **20** 289–305

[121] Sumets M, Kostyuchenko A, Ievlev V and Dybov V 2016 Electrical properties of phase formation in LiNbO$_3$ films grown by radio-frequency magnetron sputtering method *J. Mater. Sci., Mater. Electron.* **27** 7979–86

[122] Lanfredi S and Rodrigues A C M 1999 Impedance spectroscopy study of the electrical conductivity and dielectric constant of polycrystalline LiNbO$_3$ *J. Appl. Phys.* **86** 2215–9

[123] Graça M P F, Prezas P R, Costa M M and Valente M A 2012 Structural and dielectric characterization of LiNbO$_3$ nano-size powders obtained by Pechini method *J. Sol–Gel Sci. Technol.* **64** 78–85

[124] Perentzis G, Horopanitis E, Pavlidou E and Papadimitriou L 2004 Thermally activated ionic conduction in LiNbO$_3$ electrolyte thin films *Mater. Sci. Eng.* B **108** 174–8

[125] Chen R H, Chen L-F and Chia C-T 2007 Impedance spectroscopic studies on congruent LiNbO$_3$ single crystal *J. Phys.: Condens. Matter* **19** 086225

[126] Sumets M P, Dybov V A and Ievlev V M 2017 LiNbO$_3$ films: potential application, synthesis techniques, structure, properties *Inorg. Mater.* **53** 1361–77

[127] Sumets M, Ievlev V, Dybov V, Kostyuchenko A, Serikov D, Kannykin S, Kotov G and Belonogov E 2019 Electrical properties of amorphous films and crystallization of Li–Nb–O system on silicon *J. Mater. Sci., Mater. Electron.* **30** 15662–9

[128] Nakano H and Suyama Y 2010 In-situ TEM observation of crystallization process for LiNbO$_3$ and NaNbO$_3$ *Adv. Sci. Technol.* **63** 47–51

[129] Bartasyte A, Margueron S, Baron T, Oliveri S and Boulet P 2017 Toward high-quality epitaxial LiNbO$_3$ and LiTaO3 thin films for acoustic and optical applications *Adv. Mater. Interfaces* **4** 1600998

[130] Akazawa H 2014 Nucleation and crystallization of Li2O–Nb2O5 ternary compound thin films co-sputtered from LiNbO$_3$ and Li2O targets *Thin Solid Films* **556** 74–80

[131] Bartasyte A, Plausinaitiene V, Abrutis A, Stanionyte S, Margueron S, Kubilius V, Boulet P, Huband S and Thomas P A 2015 Thickness dependent stresses and thermal expansion of epitaxial LiNbO$_3$ thin films on C-sapphire *Mater. Chem. Phys.* **149** 622–31

[132] Shandilya S, Sharma A, Tomar M and Gupta V 2012 Optical properties of the c-axis oriented LiNbO$_3$ thin film *Thin Solid Films* **520** 2142–6

[133] Hao L, Li Y, Zhu J, Wu Z, Deng J, Liu X and Zhang W 2013 Fabrication and electrical properties of LiNbO$_3$/ZnO/n-Si heterojunction *AIP Adv.* **3** 042106

[134] Haier P, Hermann B A, Esser N, Pietsch U, Lüders K and Richter W 1998 Influence of the deposition rate on the structure of thin metal layers *Thin Solid Films* **318** 223–6

[135] Ellmer K 2000 Magnetron sputtering of transparent conductive zinc oxide: relation between the sputtering parameters and the electronic properties *J. Phys. D: Appl. Phys.* **33** R17–32

[136] Tumuluri A, Bharati M S S, Rao S V and James Raju K C 2017 Structural, optical and femtosecond third-order nonlinear optical properties of $LiNbO_3$ thin films *Mater. Res. Bull.* **94** 342–51

[137] Sumets M, Kostyuchenko A, Ievlev V, Kannykin S and Dybov V 2015 Sputtering condition effect on structure and properties of $LiNbO_3$ films *J. Mater. Sci., Mater. Electron.* **26** 4250–6

[138] Shur V Y, Akhmatkhanov A R and Baturin I S 2015 Micro- and nano-domain engineering in lithium niobate *Appl. Phys. Rev.* **2** 040604

[139] Bartasyte A, Plausinaitiene V, Abrutis A, Murauskas T, Boulet P, Margueron S, Gleize J, Robert S, Kubilius V and Saltyte Z 2012 Residual stresses and clamped thermal expansion in $LiNbO_3$ and LiTaO3 thin films *Appl. Phys. Lett.* **101** 122902

[140] Yoo T S *et al* 2018 Ferroelectric polarization rotation in order–disorder-type $LiNbO_3$ thin films *ACS Appl. Mater. Interfaces* **10** 41471–8

[141] Turygin A P, Abramov A S, Alikin D O, Sumets M P, Dybov V A, Kostyuchenko A V, Belonogov E K, Ievlev V M and Ya. Shur V 2020 The domain structure and local switching of $LiNbO_3$ thin films deposited on Si(001) by radio-frequency magnetron sputtering *Ferroelectrics* **560** 86–94

[142] Chai X, Jiang J, Zhang Q, Hou X, Meng F, Wang J, Gu L, Zhang D W and Jiang A Q 2020 Nonvolatile ferroelectric field-effect transistors *Nat. Commun.* **11** 2811

[143] Lu H *et al* 2019 Electrical tunability of domain wall conductivity in $LiNbO_3$ thin films *Adv. Mater.* **31** 1902890

[144] Chai X, Lian J, Wang C, Hu X, Sun J, Jiang J and Jiang A 2021 Conductions through head-to-head and tail-to-tail domain walls in $LiNbO_3$ nanodevices *J. Alloys Compd.* **873** 159837

[145] Esin A A, Akhmatkhanov A R and Shur V Y 2019 Tilt control of the charged domain walls in lithium niobate *Appl. Phys. Lett.* **114** 092901

[146] Yurchuk E, Muller J, Muller S, Paul J, Pesic M, Van Bentum R, Schroeder U and Mikolajick T 2016 Charge-trapping phenomena in HfO2-based FeFET-type nonvolatile memories *IEEE Trans. Electron Devices* **63** 3501–7

[147] Xue S D and Zhang D L 2023 Near-stoichiometric Er3 + doped $LiNbO_3$ thin films prepared by vapor transport equilibration treatment *Opt. Mater.* **136** 113476

[148] Sauze L C, Vaxelaire N, Templier R, Rouchon D, Pierre F, Guedj C, Remiens D, Rodriguez G, Bousquet M and Dupont F 2023 Homo-epitaxial growth of $LiNbO_3$ thin films by pulsed laser deposition *J. Cryst. Growth* **601** 126950

[149] Liu J, Li X, Li X, Bai J and Xu Y 2023 *J–V* characteristics of sol–gel-prepared Zn2 + doped $LiNbO_3$ *Mater. Sci. Semicond. Process.* **156** 107304

[150] Chen K, Zhu Y, Liu Z and Xue D 2021 State of the art in crystallization of $LiNbO_3$ and their applications *Molecules* **26** 7044

[151] Rubešová K, Mikolášová D, Hlásek T, Jakeš V, Nekvindová P, Bouša D and Oswald J 2016 Waveguiding Er3 +/Yb3 + :$LiNbO_3$ films prepared by a sol–gel method using polyvinylpyrrolidone *J. Lumin.* **176** 260–5

[152] Pan X *et al* 2019 Ar + ions irradiation induced memristive behavior and neuromorphic computing in monolithic $LiNbO_3$ thin films *Appl. Surf. Sci.* **484** 751–8

IOP Publishing

Lithium Niobate-Based Heterostructures (Second Edition)
Synthesis, properties, and electron phenomena
Maxim Sumets

Chapter 2

Synthesis, structure, and surface morphology of LiNbO$_3$ films

This chapter presents original results regarding the influence of the synthesis conditions used in radio-frequency magnetron sputtering (RFMS) and ion-beam sputtering (IBS) methods on the structure, composition, and surface morphology of as-grown LiNbO$_3$ (LN) films. The RFMS method offers several advantages over the IBS method, including greater flexibility in sputtering regimes, higher growth rates, and the capability to fabricate highly oriented films on various substrates. We propose optimal RFMS parameters for the fabrication of single-phase c-oriented LN films and Si–LiNbO$_3$ heterostructures. Special attention is given to the evolution of the structure, composition, and morphology of amorphous LN during crystallization under thermal annealing (TA). Detailed descriptions of the early stages of LN film growth are provided, which are crucial for the controllable fabrication of LN-based heterostructures.

2.1 Technological regimes for the synthesis of thin LiNbO$_3$ films by radio-frequency magnetron sputtering and ion-beam sputtering methods

In this chapter, I describe the basic technological stages utilized by our research group for the deposition of LN films. RFMS and IBS were used in the sputtering system, as schematically shown in figure 2.1.

Single-crystal lithium niobate wafers with a diameter of 65 mm were used as targets[1]. To study the effect of synthesis conditions on the structure and properties of the deposited films, various sputtering parameters were used, as detailed in table 2.1.

---

[1] Single crystals were grown at the I.V. Tananaev Institute of Chemistry and Technology of Rare Elements and Mineral Raw Materials at the Russian Academy of Sciences Kola Science Center.

doi:10.1088/978-0-7503-6305-1ch2

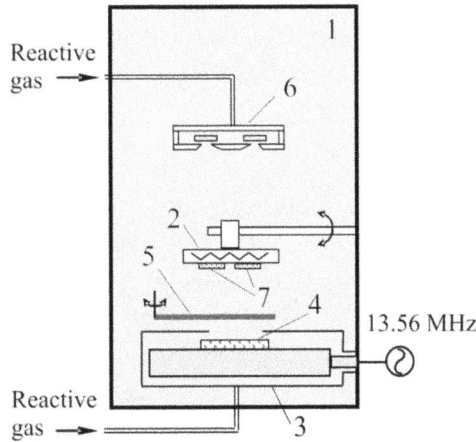

Figure 2.1. Schematic diagram of a multifunctional sputtering system: 1—reaction chamber, 2—substrate holder, 3—magnetron, 4—water-cooled target, 5—shutter, 6—ion source, 7—substrate.

Table 2.1. Technological and sputtering parameters used in the initial stage of our study.

Type of substrate	Magnetron power/ (supply power)	Working pressure (Pa)	Reactive gas composition	Substrate temperature (°C)	Substrate position
• (001)Si (n-type conductivity with $\rho = 4.5\ \Omega\cdot cm$ and p-type conductivity with $\rho = 20\ \Omega\cdot cm$); • (111)Si; • (001)Si–SiO$_2$; • Fluorphlogopite; • Fluorphlogopite–epitaxial Ag film	• 100 W/ (2 kW)	• $1.5\cdot10^{-1}$ ('low' pressure) • $5\cdot10^{-1}$ ('high' pressure)	• Ar • Ar + O$_2$ (at Ar/O$_2$ ratios of 60:40 and 80:20)	• Unheated • 550	• Coaxial with the target ('Over the target erosion zone') at different distances • Shifted along the horizontal plane ('Offset from the target erosion zone')

Figure 2.2 illustrates the relative positions of the target and substrate in the 'offset from the target erosion zone' regime.

The optimal magnetron power was chosen according to the recommendations provided in chapter 1. The sample preparation for the Si wafers and Si–SiO$_2$ heterostructures is described in the following. It is known that the use of mechanical polishing methods for sample surfaces leads to significant disruption of the crystal structure at depths of up to 50 μm. To remove this disrupted layer, wet etching was

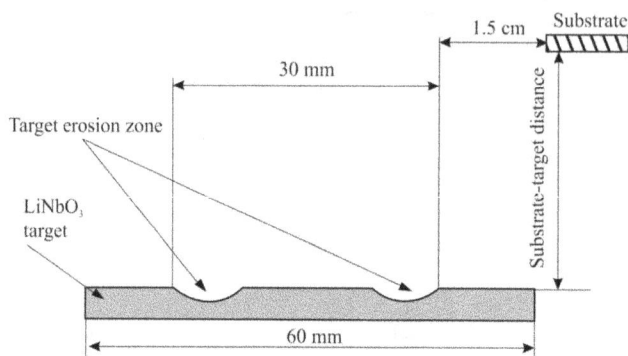

Figure 2.2. Schematic diagram of the relative target and substrate positions in the 'offset from the target erosion zone' regime.

used. The most common etching solution, $H_2SO_4:H_2O_2:H_2O = 5:1:1$, would have allowed us to obtain a substrate with a high-quality mirror surface. Nevertheless, we used a plasma etching approach to achieve the same effect in a single cycle. The substrates were etched in the same sputtering system in an Ar plasma for 2 min. This method effectively removes the disrupted silicon layer and surface contaminants if $Si–SiO_2$ substrates are used. Furthermore, the thin SiO_2 film present on the Si substrate influences subsequent technological processes and affects the electrical parameters of synthesized LN-based heterostructures. This issue was also resolved by plasma etching.

Thickness control for the films thus deposited was achieved using an atomic force microscopy (AFM) profile for a control sample fabricated under the same conditions. TA of the studied heterostructures was conducted in a coaxial furnace in air at a temperature of 650 °C for 1 h.

Our study of the structure, elemental compositions, and surface morphology of these thin films was conducted using transmission electron microscopy (TEM, EMV-100BR), AFM (Solver P47), Rutherford backscattering spectrometry (RBS, α-particles with an energy of 2.3 MeV), Raman spectroscopy (Raman Microscope RamMics M532 EnSpectr, in the range of 350–3650 cm^{-1} using a laser operating at a wavelength of 532 nm), and x-ray diffraction (XRD, ARL X'TRA Thermo Techno with a Cu Kα source operated at 40 kV and 35 mA). The accuracy of the Raman spectral measurements was ±2 cm^{-1}. The cross-sections of the specimens' heterostructure were studied using a Philips EM-430 ST microscope.

For TEM, it was necessary to separate the film from the Si substrate. Wet etching in a mixture of hydrofluoric and nitric acids (1:5) was used for this purpose. For Ag substrates, a 50% solution of nitric acid was used.

2.2 Direct growth of polycrystalline LiNbO$_3$ films

In the initial phase of our study, we assessed the feasibility of depositing single-phase LN films onto the substrates detailed in table 2.1. We achieved this using through thermal evaporation and condensation in a vacuum (8×10^{-4} Pa) to depositing epitaxial Ag films onto fluorphlogopite plates. The plates were heated to heated to 450 °C, and we created films with thicknesses of up to 1 μm.

2.2.1 Films deposited by RFMS

Films with thicknesses of up to 4 μm were deposited using RFMS in an Ar environment at a working pressure of 0.1 Pa and a magnetron power of 100 W. At substrate–target distances of 4 cm and 5 cm, the growth rates were approximately 10 nm min^{-1} and 17 nm min^{-1}, respectively.

One of the most significant properties of RFMS is the effect of spatial plasma inhomogeneity during the synthesis process. This phenomenon arises when the composition and texture of the films depend on the position of the substrate relative to the target erosion zone.

To assess the extent of this effect, we studied thin (0.1 μm) films deposited at a substrate temperature of $T = 550$ °C on (111)Si and (001)Si wafers, as well as on the surface of SiO$_2$. The substrate positions relative to the target erosion zone were as follows: positions 1 and 2—directly over the target erosion zone at distances of 4 and 10 cm, respectively; positions 3 and 4—offset from the target erosion zone within the same plane.

Based on TEM patterns, amorphous films formed on unheated substrates ($T = 20$ °C–300 °C) at a working power of 100 W. When exposed to an electron beam (in the electron microscope), these films crystallized, forming nanocrystalline LN films composed of grains up to 50 nm in size, predominantly oriented in the (110) direction parallel to the substrate surface (see figure 2.3).

The compositions of thick films (up to 4 μm) deposited at a temperature of 550 °C on various substrates positioned over the target erosion zone are shown in figure 2.4 and table 2.2.

As indicated in table 2.2, the elemental composition in the bulk of the films corresponds to stoichiometric LN. In addition, for the films synthesized on the substrates in position A, carbon is present in the near-surface area, which is typical for films deposited in sputtering systems with steam-oil pumping.

Figure 2.5 displays electron diffraction patterns and micrographs of films with a thickness of 0.1 μm deposited at a temperature of $T = 550$ °C: (a)–(d) on Si wafers (in position 1); (e), (f) on Si–SiO$_2$ heterostructures (in position 4). Analysis of the electron diffraction patterns (a) and (c) in figure 2.5 indicates that single-phase LN films are formed under these sputtering conditions, and there is no evidence of

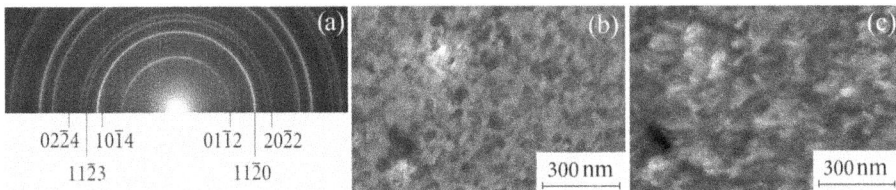

Figure 2.3. TEM patterns of LN films (a) TEM diffraction pattern, (b) the bright-field TEM image and (c) the dark-field TEM image.

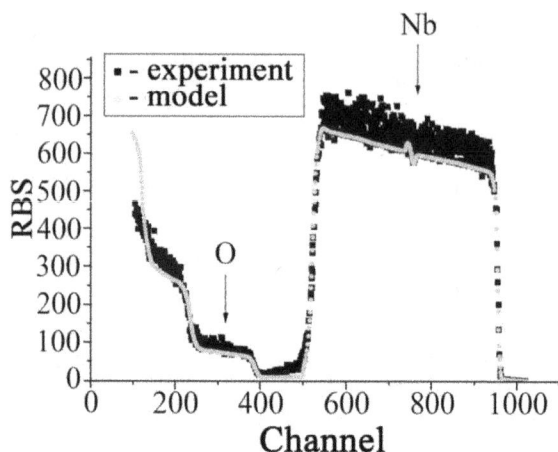

Figure 2.4. RBS spectra for a film with a thickness of 2 μm, deposited on a Si–SiO$_2$ substrate. Reproduced from [1], with permission from Springer Nature.

Table 2.2. Elemental compositions of films deposited by RFMS on (001)Si wafers (A), (111)Si–SiO$_2$ heterostructures (B), and fluorphlogopite-epitaxial (111)Ag film heterostructures (C).

Substrate	Depth (nm)	Relative concentration of elements (at. %)						
		Li	Nb	O	Si	C	Ag	K
A	566	20.0	18.0	52.00	0.0	10.0	0.0	0.0
	1216	20.0	20.0	60.0	0.0	0.0	0.0	0.0
	1888	20.0	20.0	60.0	0.0	0.0	0.0	0.0
	1954	0.0	0.0	0.0	100.0	0.0	0.0	0.0
	42 139	0.0	0.0	0.0	100.0	0.0	0.0	0.0
B	216	20.0	20.0	60.0	0.0	0.0	0.0	0.0
	650	20.0	20.0	60.0	0.0	0.0	0.0	0.0
	1257	0.0	0.0	66.6	33.3	0.0	0.0	0.0
	19 439	0.0	0.0	0.0	100.0	0.0	0.0	0.0
C	1712	20.0	20.0	60.0	0.0	0.0	0.0	0.0
	1747	0.0	50.0	0.0	0.0	0.0	50.0	0.0
	2153	0.0	0.0	0.0	0.0	0.0	100.0	0.0
	13 646	0.0	0.0	59.0	34.7	0.0	0.0	6.2

epitaxial growth. The relatively intense reflection indicates that single-axis $\langle 0001 \rangle$-textured LN films are formed on the (001)Si and (111)Si substrates.

As shown in figure 2.5, the substructure of the LN films deposited on Si substrates of both (111) and (001) orientations is identical. This could be due to the presence of a thin natural SiO$_2$ layer on the Si surface, which passivates the orienting effect of the substrate. It is noteworthy that the grain and subgrain sizes range from 10 to 50 nm, while the largest grains are found in films synthesized on (111)Si–SiO$_2$

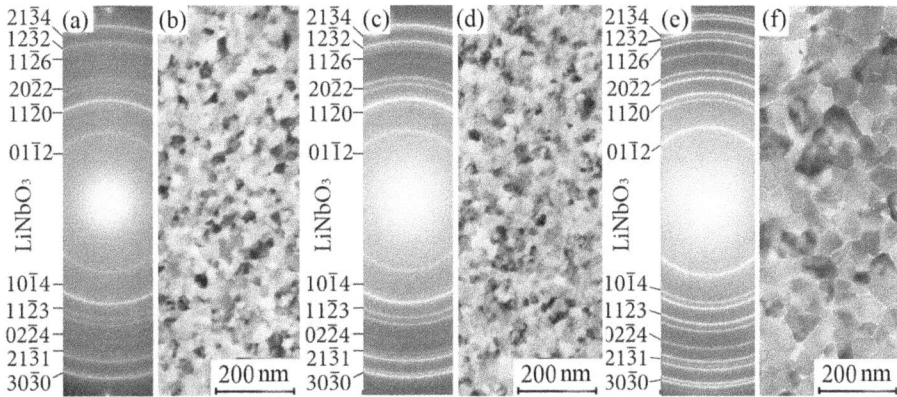

Figure 2.5. Electron diffraction patterns and micrographs of LN film substructure: (a) and (b) films deposited on (111)Si substrates (in position 1); (c) and (d) films deposited on (001)Si substrates (in position 1); (e) and (f) films deposited on (111)Si–SiO$_2$ heterostructures (in position 4).

heterostructures offset from the target erosion zone. These single-phase LN films with arbitrary grain orientation suggest that thermal activation at the given temperature ($T = 550$ °C) is insufficient for reactions at the SiO$_2$/LN interface.

Figure 2.6 shows the XRD patterns of the films with thicknesses of around 1 μm on (001)Si and about 2 μm on the (001)Si–SiO$_2$ heterostructure.

All peaks in these patterns correspond to LN, and the pronounced $\langle 0001 \rangle$ texture evolves as the thickness increases. The presence of only peaks corresponding to LN with the preferred orientations (001) and (104) (as indicated by TEM data) is consistent with the results of other studies [2, 3]. XRD patterns of the films deposited on the (111)Ag substrate show peaks corresponding to LN with the (018) preferred orientation (see figure 2.7).

Figure 2.8 demonstrates bright-field cross-sections of a (111)Si–LN heterostructure (a) (with an LN film thickness of 2.0 μm) along with the high-resolution structure (b). The LN film is clearly seen along with the (111)Si substrate, for which the interplanar distance is $d_{111} = 3.21$ Å. The atomic planes visible in the film correspond to the following planes, according to the Crystallographic Database (see appendix): (01$\bar{1}$2) LN ($d = 3.74$ Å) and (0224) LN ($d = 1.87$ Å). Furthermore, as shown in figure 2.8, a thin (5 nm) amorphous SiO$_2$ layer exists at the interface, passivating the orienting effect of the (111)Si and (001)Si substrates on the growth of LN films.

Figure 2.9 presents bright-field cross-sectional patterns of a (111)Ag–LN heterostructure (a and b) and the diffraction patterns of the Ag substrate (c) and the LN film (d) for parallel zones $\langle 110 \rangle$ and $<11\bar{2}0>$, respectively. In this sputtering regime, a two-axis (epitaxial) texture is formed with the orientation ratio (0001), [11 $\bar{2}$ 0] LiNbO$_3$ || (111), $\langle 110 \rangle$ Ag, producing a mosaic substructure in the fabricated LN films. Thus, it is advisable to use epitaxial Ag films as substrates for the deposition of LN films, particularly for the synthesis of films separated from a substrate.

In the following stage, we studied the surface morphology of the as-grown films. Figure 2.10 shows AFM images obtained in topography mode and the height

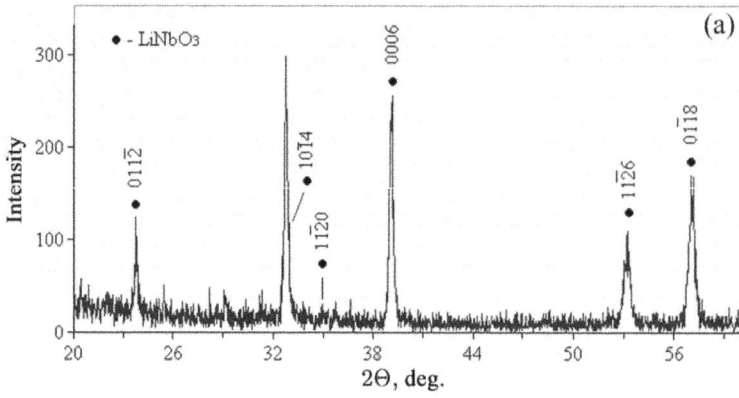

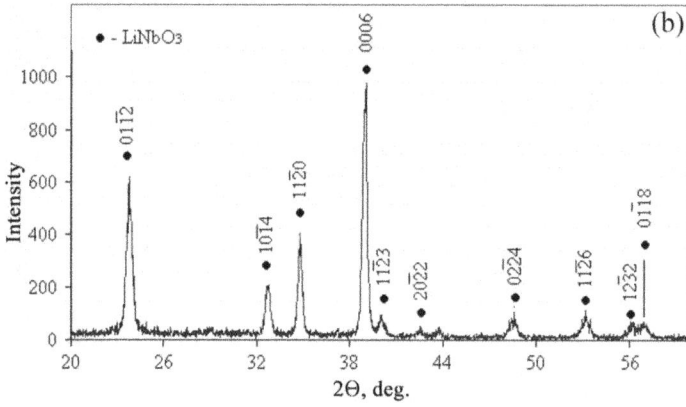

Figure 2.6. XRD patterns of films with thicknesses of about 1.0 μm on (001)Si (a) and 2.0 μm on Si–SiO$_2$ (b). Reproduced from [1], with permission from Springer Nature.

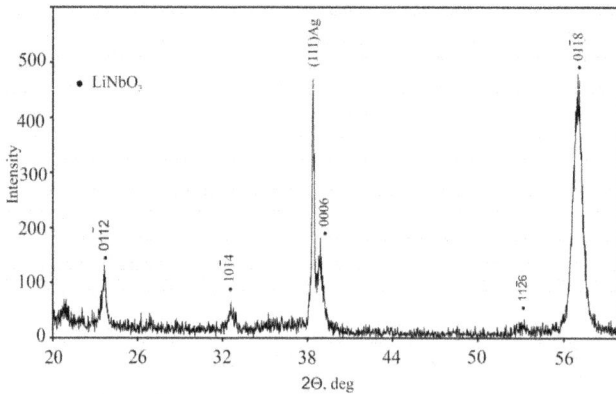

Figure 2.7. XRD pattern of LN films deposited on Ag(111).

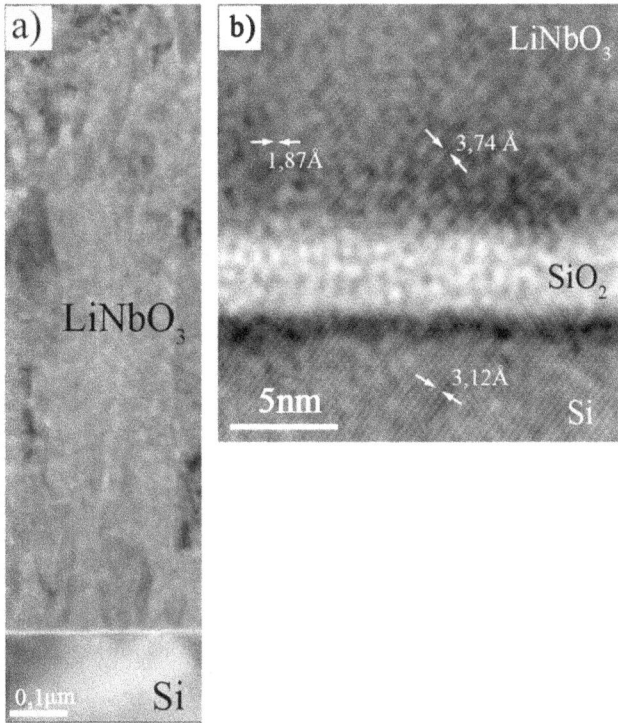

Figure 2.8. Bright-field photomicrographs of the cross-sectional pattern of a (111)Si–LN heterostructure (a) (the thickness of the LN film is 2.0 μm) along with its high-resolution structure (b). Reproduced from [1], with permission from Springer Nature.

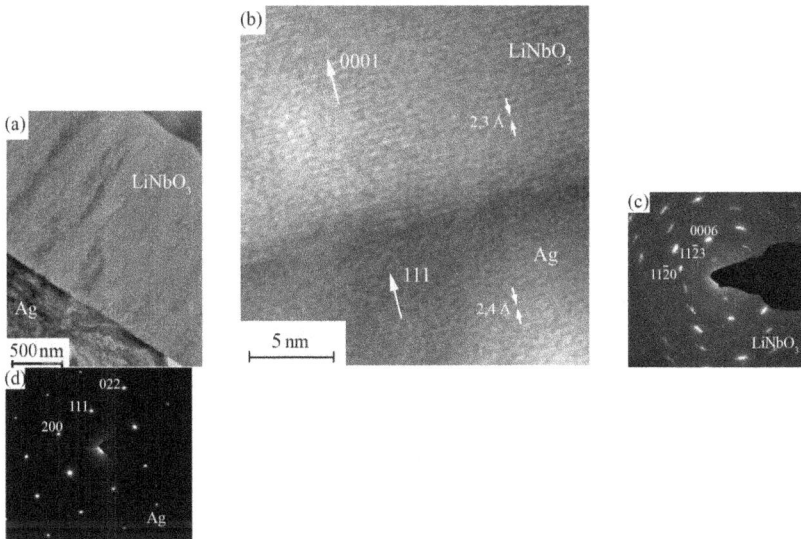

Figure 2.9. Bright-field photomicrographs of the cross-sectional patterns of a (111)Ag–LN heterostructure ((a) and (b)). Electron diffraction patterns ((c) and (d)) of an LN film (c), showing parallel $\langle 110 \rangle$ and $<11\bar{2}0>$ zones, and an (111)Ag film (d). Reproduced from [4], with permission from Springer Nature.

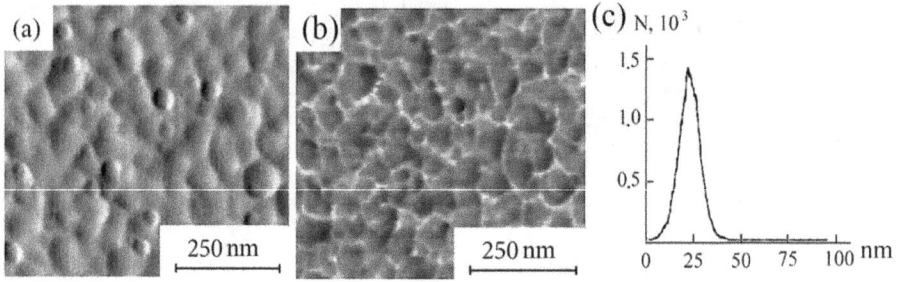

Figure 2.10. AFM images obtained using the phase contrast mode (a); registration of the change of resonant frequency oscillations of a cantilever (b); and the height distribution curve (c), showing the surface roughness of an LN films deposited on (001)Si substrates by the RFMS method. Part (a) reproduced from [5], with permission from Springer Nature.

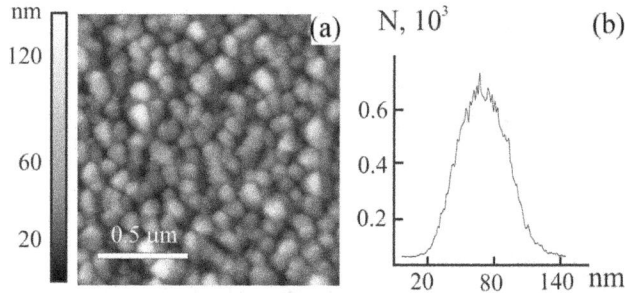

Figure 2.11. AFM images showing the surface morphology (a) and the height distribution curve (b) of an LN film with a thickness of 1μm deposited onto a Si–SiO$_2$ substrate offset from the target erosion zone.

distribution curve for LN films with a thickness of 1 μm deposited on (001)Si substrates at a temperature of $T = 550$ °C. Analysis of these patterns revealed that the lateral surface relief corresponds to the grain substructure of the studied films. As shown by comparison with figure 2.8(b), this structure does not change when the thickness increases from 0.1 to 1.0 μm. This indicates that the films synthesized in this sputtering regime do not have a single orientation throughout their thickness. The average size of the inhomogeneities is 25 nm.

For the films deposited onto Si–SiO$_2$ substrates, analysis of AFM patterns (see figure 2.11) indicates that the lateral size of inhomogeneities is about 70 nm, which coincides with the TEM study (see figure 2.5). The average roughness of the films is 16 nm.

The Raman spectra of films with a thickness of 1.5 μm, deposited on (001)Si (a) and Si–SiO$_2$ (b) substrates at a temperature of $T = 550$ °C, are shown in figure 2.12. All spectra correspond to the Raman spectra of LN [6]. Spectral lines with frequencies of 153, 233, 264, 314, 361, 429, 589, 630, 778, and 867 cm^{-1} correspond to polycrystalline lithium niobate fabricated by high-temperature synthesis. However, almost all observed lines have lower frequencies (by a few cm^{-1}) compared to the spectra attributed to poly- and single-crystal lithium niobate, which can be attributed to deviations from stoichiometry and the film structure.

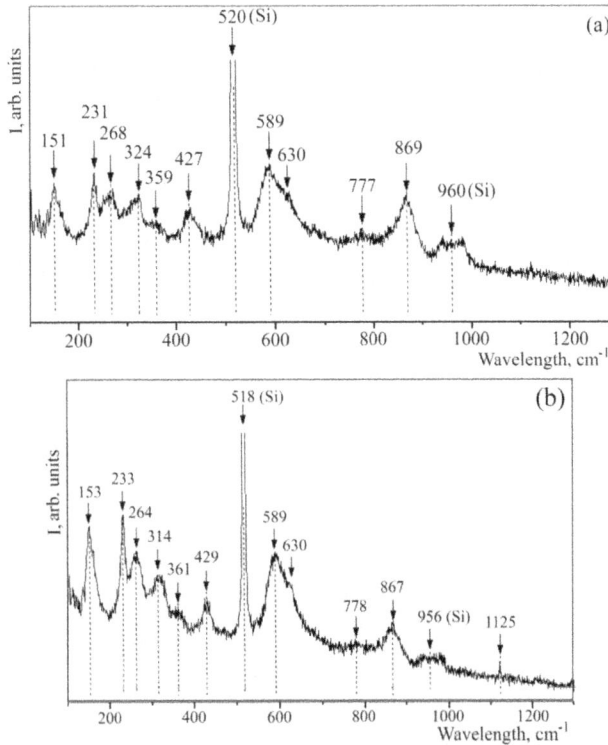

Figure 2.12. Raman spectra of films with a thickness of $1.5\,\mu m$ grown on (001)Si (a) and Si–SiO$_2$ (b) at a temperature of $T = 550\ °C$.

The lines with frequencies of 518, 956, and $1125\ cm^{-1}$ correspond to the Raman spectra of silicon, whereas the line with a frequency of $956\ cm^{-1}$ can be attributed to the valence bridge oscillations of oxygen atoms in the 'partially destroyed' NbO$_6$ octahedra formed during RFMS. Evidently, a line with a frequency of $867\ cm^{-1}$ is a double line, which may be caused by differences in the positions of Nb atoms in different octahedra.

2.2.2 Films deposited by ion-beam sputtering

Thin films with a thickness of up to $1\,\mu m$ were deposited on (001) Si wafers using the IBS technique in an Ar atmosphere under a gas pressure of 1×10^{-1} Pa and a working power of 2 kW. Under these conditions, the film growth rate was $3\ nm\ min^{-1}$ at a distance of 4.0–5.0 cm between the substrate and the target.

The elemental compositions of the synthesized films, obtained through RBS spectra, are provided in table 2.3.

The results indicate that the O/Nb ratio in the studied films is lower than that of stoichiometric lithium niobate, suggesting a relatively high concentration of oxygen vacancies is present in films fabricated by the IBS method. As demonstrated in our previous work, these vacancies affect the electrical conductivity of LN-based

Table 2.3. Elemental compositions of films deposited by the IBS method on (001)Si wafers.

Layer depth (nm)	Relative concentration of elements (%)					
	Li	Nb	O	Si	C	W
51	21.5	17.3	43	0	18	0.3
910	26.2	21.1	52.4	0	0	0.3
1232	25.3	21.7	52.7	0	0	0.3
1555	26.5	20.3	52.9	0	0	0.3
1662	22.6	20.2	53	3.9	0	0.3
1748	19.9	15	49.9	15	0	0.3
1853	12.2	5.9	30.6	50.9	0	0.3
1976	6	2.2	11.1	80.4	0	0.3
16 843	0	0	0	100	0	0

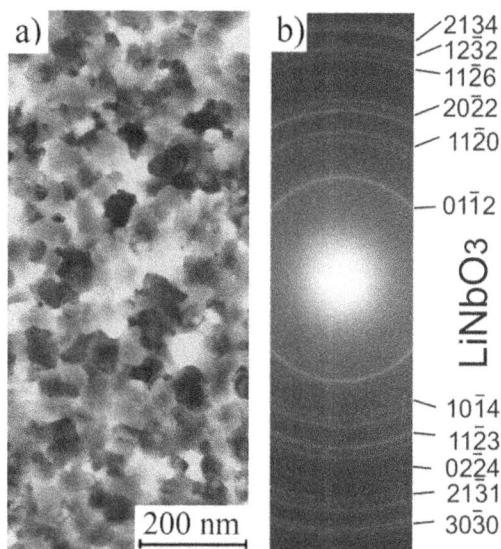

Figure 2.13. Micrographs (a) and electron diffraction patterns (b) of LN films with a thickness of about 0.1 μm grown on a heated (550 °C) (001)Si substrate using the IBS method. Reproduced from [7], with permission from Springer Nature.

heterostructures [7]. In addition, table 2.3 reveals that an extended intermediate layer with variable composition is formed at the substrate–film interface. The presence of W atoms in the film is attributed to the use of a tungsten spiral to generate thermionic emission to reduce the positive charge on the target during the IBS process.

A TEM study showed that, as in the case of films deposited by the RFMS method, films synthesized by IBS are amorphous. The stable formation of nano-crystalline films is feasible only at temperatures no lower than 550 °C. Figure 2.13

shows TEM patterns of films with a thickness of 0.1 μm deposited onto (001) Si substrates at a temperature of 550 °C.

The electron diffraction pattern in figure 2.13(b) indicates that single-phase LN films with arbitrary grain orientation are formed under these sputtering conditions. The micrograph in figure 2.13(a) shows that the average grain size does not exceed 50 nm. XRD patterns, shown in figure 2.14, also demonstrate the single-phase nature of the studied LN films. All peaks in this diffraction pattern correspond to polycrystalline lithium niobate with an average grain size of 50 nm, as calculated based on the Debye–Scherrer method.

Figure 2.15 shows an AFM image of the surface and a bar graph illustrating the height distribution of as-grown LN film on (001) Si. The average surface roughness within a scanned area of $2.0 \times 2.0\ \mu m^2$ is 40 nm, and the average lateral size of the crystalline blocks is around 200 nm. However, considering that the maximum height difference is considerably lower than the integral thickness of the films (see figure 2.16), we can conclude that LN films fabricated using IBS are continuous and have high porosity.

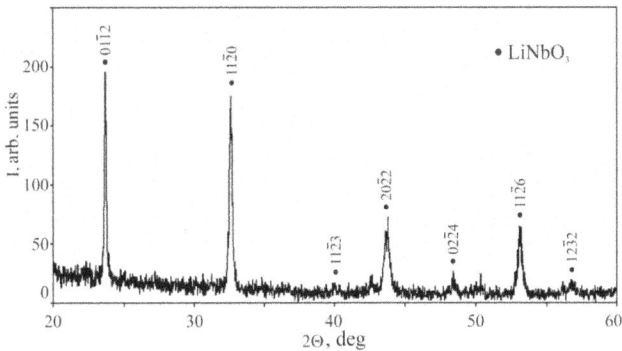

Figure 2.14. XRD pattern of a film with a thickness of 1 μm deposited using the IBS method on a (001)Si substrate at $T = 550$ °C.

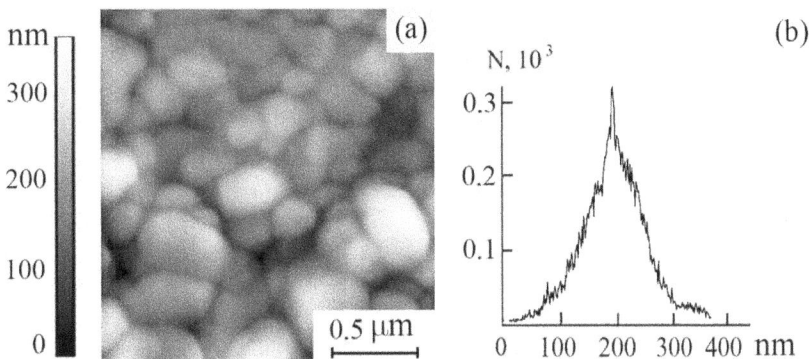

Figure 2.15. AFM image obtained in the topography mode (a) and height distribution curve (b) showing the surface roughness of LN film with a thickness of 1.0 μm formed on a (001)Si substrate. Reproduced from [7], with permission from Springer Nature.

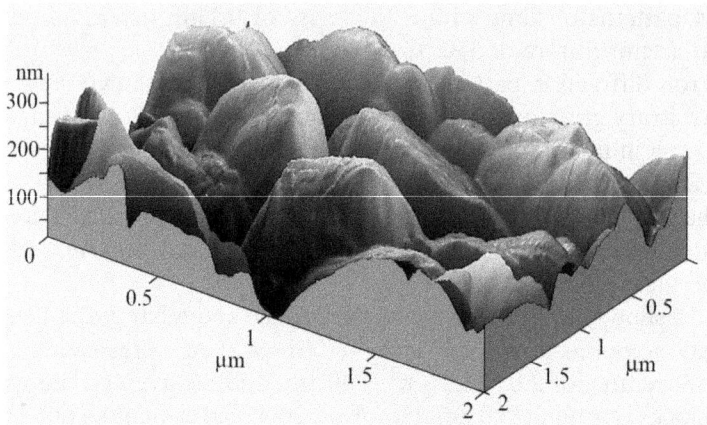

Figure 2.16. Three-dimensional AFM image of the film surface fabricated using the IBS method.

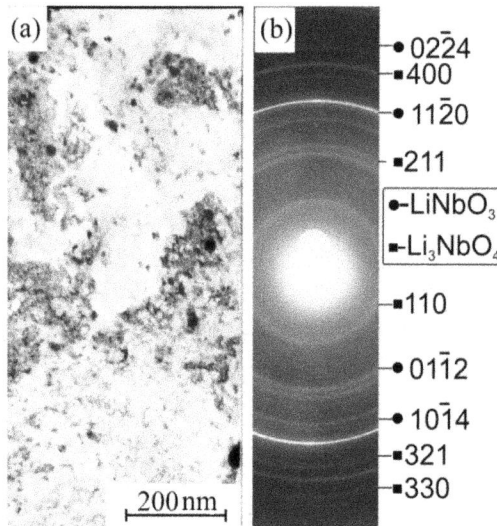

Figure 2.17. Micrograph (a) and electron diffraction pattern (b) of LN films with a thickness of about 0.1 μm grown on a heated (550 °C) fluorphlogopite-epitaxial (111)Ag film substrate using the IBS method.

Figure 2.17(a) shows a TEM image of a film with a thickness of 0.1 μm deposited on a heated ($T = 550$ °C) fluorphlogopite-epitaxial (111)Ag film heterostructure. The diffraction pattern (figure 2.7(b)) indicates that the Li_3NbO_4 phase, along with the LN phase, is formed in films deposited on the (111)Ag epitaxial film using IBS. The relatively high intensity of the $11\bar{2}0$ reflex suggests the formation of the $\langle 0001 \rangle$ texture.

Based on the results of this preliminary study, we can conclude that both RFMS and IBS are effective deposition methods for LN films that preserve the elemental composition. Thin polycrystalline LN films with arbitrarily oriented grains sized at

50 nm are formed on the heated ($T = 550\,°C$) (001) Si substrates by the IBS method. At a distance of 4–5 cm between the target and the substrate, the growth rate is $3\,\mathrm{nm\,min}^{-1}$.

For the RFMS method, at the same substrate–target distance and a working power of 100 W, the growth rate is $10\,\mathrm{nm\,min}^{-1}$. On the (001) Si and (111) Si substrates, single-phase $\langle 0001 \rangle$-textured LN films with a grain size of 10 nm are formed over the target erosion zone. The oriented effect of the substrate is not observed due to the formation of a thin (5 nm) amorphous SiO_2 layer at the substrate–film interface.

Single-phase LN films with randomly oriented grains (50 nm in size) are formed on substrates over the target erosion zone. At the same time, epitaxial LN films are fabricated on (111) Ag substrates by the RFMS method. A two-axis (epitaxial) texture is formed with the orientation relationship of (0001), [$11\bar{2}0$] $LiNbO_3$ ∥ (111), $\langle 110 \rangle$ Ag, resulting in a mosaic substructure in the fabricated LN films. However, films deposited by the IBS technique do not exhibit these epitaxial properties and contain the 'parasitic' Li_3NbO_4 phase.

These differences in properties highlight the necessity of studying the effect of the technological regime on the structure, composition, and surface morphology of the synthesized LN films.

2.3 Influence of the synthesis regime and subsequent annealing on the composition and structural properties of $LiNbO_3$ films

As demonstrated in chapter 1, sputtering conditions such as the reactive gas pressure and composition, substrate temperature, substrate position relative to the target, etc. greatly influence the composition and structure of deposited films. Complexity arises because different combinations of technological parameters can affect the properties of synthesized films in various ways, making these parameters unique. In this section, we assess the influence of the most significant technological factors on the structure, composition, and surface morphology of synthesized films to propose the optimal sputtering regimes.

We synthesized films using the RFMS and IBS methods under various regimes on the following heated (550 °C) substrates: (001)Si, (111)Si, and fluorphlogopite-epitaxial (111)Ag film heterostructures. Given the strong dependence of plasma properties on the relative positions of the substrate and the target, we employed the following two basic substrate positions: 'above the target erosion zone' and 'offset from the target erosion zone,' with a variation of vertical distance. This change in plasma conditions illustrates the 'ion assist effect' on film properties. The deposition regimes used in this study are provided in table 2.4.

2.3.1 Plasma effect

First, we study the influence of an Ar plasma with a constant composition. We then discuss the influence of oxygen on the plasma and consequently on the properties of the films. Figure 2.18 shows the typical XRD patterns of films fabricated by the RFMS method under regimes 1, 2, and 3 (see table 2.4) [4].

Table 2.4. Deposition regimes used in our study of the effects of the plasma properties on the structure and composition of LN films.

Regime #	Deposition method	Magnetron (supply) power	Reactive gas pressure (Pa)	Substrate–target distance (cm)	Substrate position
1	RFMS	100 W	1.5×10^{-1}	5	Above the target erosion zone
2				5	Offset from the target erosion zone
3				10	Offset from the target erosion zone
4			5×10^{-1}	5	Offset from the target erosion zone
5	IBS	2 kW	1×10^{-1}	5	

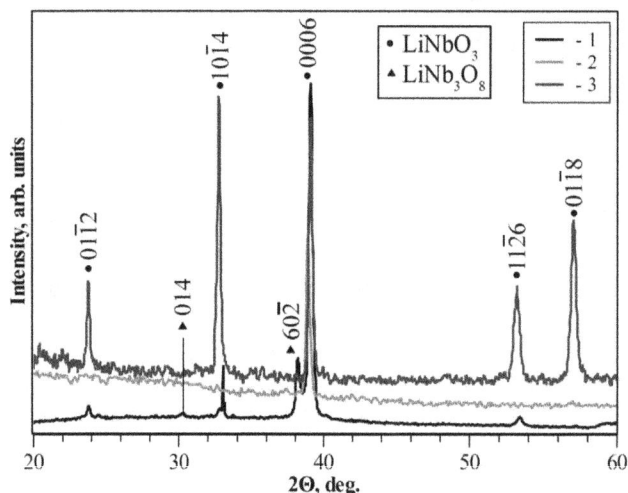

Figure 2.18. XRD patterns of films deposited by RFMS on (001)Si substrates under three regimes (see table 2.4). 1—regime 1, 2—regime 2, and 3—regime 3.

As can be seen in figure 2.18, single-phase c-oriented LN films are formed under regimes 1 and 2. However, when the substrates are above the target erosion zone (regime 1), the effect of intense plasma influence is not as pronounced as it is for films deposited under a moderate plasma effect. These results align with the findings of other investigators [8]. Conversely, films synthesized on substrates positioned offset from the target erosion zone, which experience a minimal plasma effect, do not

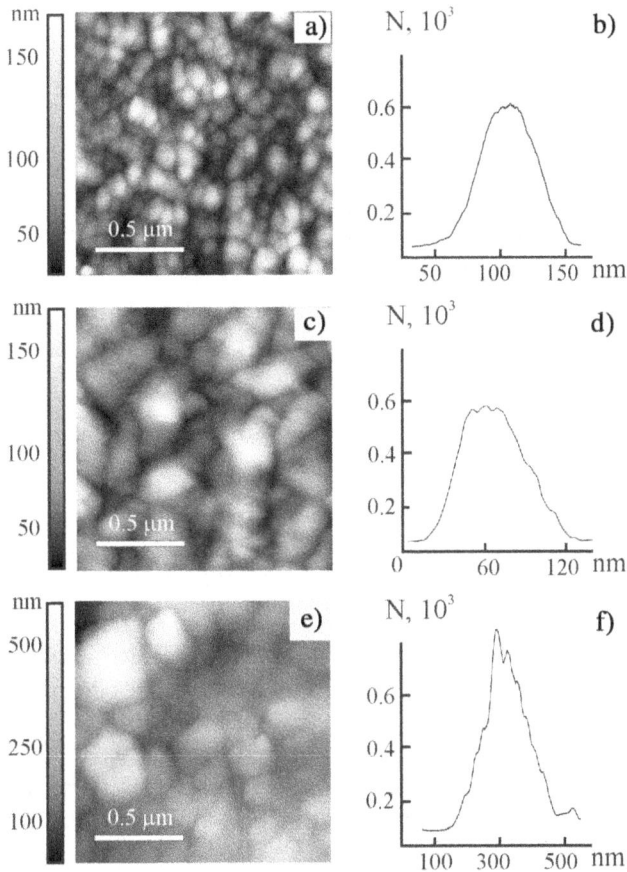

Figure 2.19. AFM images obtained using the topography mode (a) and height distribution curves showing surface roughnesses (b) of LN films formed on (001)Si substrates under conditions 2 ((a) and (b)), 3 ((c) and (d)) and 5 ((e) and (f)) (see table 2.4). Reproduced from [4], with permission from Springer Nature.

demonstrate directional growth (see figure 2.18, regime 3). Similarly, films fabricated by the IBS method (regime 5) also show arbitrary grain orientation (see figure 2.14), indicating a weak plasma effect. Thus, removing the substrates from the target erosion zone or using the IBS method reduces the plasma effect on the films' properties, decreasing their growth rate.

Figure 2.19 shows AFM images of the morphologies of films deposited by RFMS under regimes 2, 3, and 5 [4]. Analyzing the surface morphology reveals that the average surface roughness increases from 10 nm (regime 1) to 30 nm (regime 5) as the plasma effect decreases, consistent with our recent paper [4].

TEM patterns of films with a thickness of 0.1 μm, fabricated by the RFMS method under regimes 1 (figure 2.20 parts (a) and (b)), 2 (parts (c) and (d)), and 3 (parts (e) and (f)), show that all reflexes in the electron diffraction patterns can be attributed to LN.

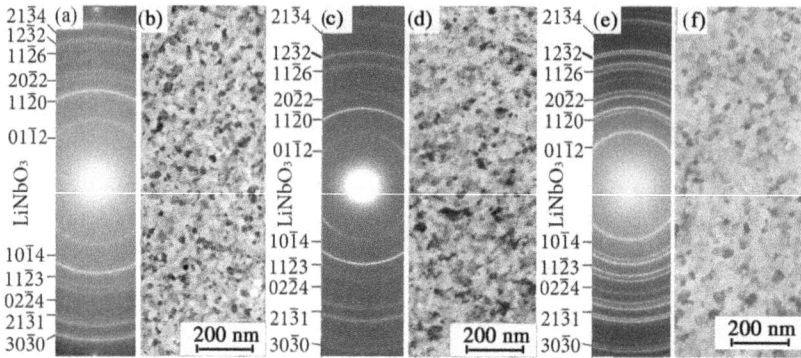

Figure 2.20. Electron diffraction patterns (a), (c), and (e) and micrographs (b), (d), and (f) of LN films with a thickness of 0.1 µm, grown on (001)Si substrates under regime 1 ((a) and (b)), regime 2 ((c) and (d)), and regime 3 ((e) and (f)).

The relatively high intensity of reflex (01 $\bar{1}$ 2) in figure 2.20(a) and (c) indicates that a single-axis ⟨0001⟩ texture is formed under regimes 1 and 2. In contrast, electron diffraction patterns of films deposited under a minimal plasma effect (regime 3) exhibit arbitrary grain orientation. Micrographs of the film substructures (see figures 2.20(b), (d), and (f)) show that the crystallite size increases from 20 nm (regime 1) to 50 nm (regime 3). Thus, an increase in the plasma effect leads to a decrease in the average crystallite size and the formation of single-axis ⟨0001⟩ textured LN films, as shown in our previous works [4, 8, 9].

RBS data reported in [4] indicate that stoichiometric LN films are formed under regime 3 (see table 2.4), accompanied by the formation of a thin intermediate layer at the Si–LN interface with a thickness of 20 nm, which is likely caused by elemental diffusion at that temperature. Therefore, the plasma effect does not influence the elemental composition of LN films deposited by the RFMS method under the studied sputtering regimes. From this perspective, using the RFMS method is more favorable compared to the IBS method, which does not support the formation of stoichiometric LN films (see table 2.3).

2.3.2 Effect of reactive gas pressure

To study this effect, we investigated films fabricated by the RFMS technique on substrates situated according to regime 2 in table 2.4. Figure 2.21 shows the XRD pattern of a film with a thickness of 0.5 µm fabricated by RFMS at higher argon pressure in a reaction chamber (regime 4 in table 2.4).

As shown in figure 2.21, a fivefold increase in reactive gas pressure compared to regime 2 results in the formation of films containing two phases: LN and $LiNb_3O_8$. The lithium-deficient phase $LiNb_3O_8$ (lithium triniobate) is a non-ferroelectric phase, often referred to as 'parasitic.' As emphasized in chapter 1, this phase forms when the lithium concentration in the plasma or film is reduced. At high working pressure in a reaction chamber, Li atoms are more volatile compared to niobium atoms [10, 11]. Furthermore, at higher working pressure, the mean free path of Li

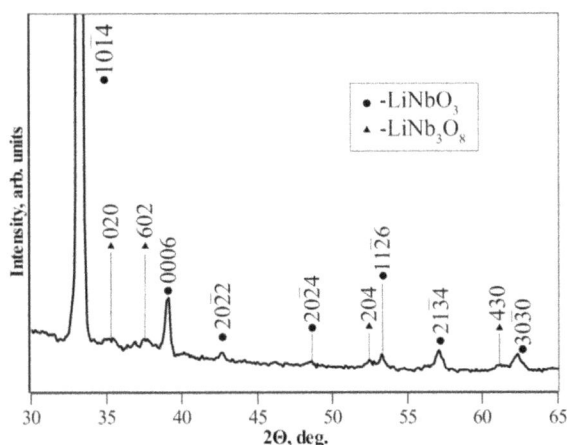

Figure 2.21. XRD pattern of a film deposited on a (001)Si substrate by the RFMS method under regime 4 (see table 2.4).

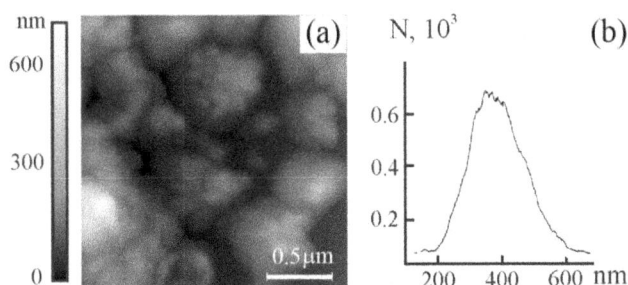

Figure 2.22. AFM image obtained using the topography mode (a) and height distribution curve showing the surface roughness (b) of LN films deposited on (001)Si substrates by the RFMS method under condition 4 (see table 2.4).

atoms in plasma declines significantly, preventing them from reaching the substrate and leading to the formation of the lithium-deficient $LiNb_3O_8$ phase. Consequently, this results in the generation of Li vacancies.

Figure 2.22 shows the morphology of a film fabricated by RFMS in an Ar atmosphere at high reactive gas pressure.

Analysis of AFM images indicates that the surface morphology of the films changes significantly with increasing working pressure in the reaction chamber. The average surface roughness of films synthesized in this regime is around 60 nm, which is almost three times higher than those deposited at lower working pressure (regime 2 in table 2.4). Furthermore, as shown by the study of elemental composition (RBS data) in figure 2.23, an intermediate layer with a thickness of about 0.5 μm forms at the Si–LN interface. Apparently, high working pressure in the reaction chamber not only decreases the concentration of Li atoms in the plasma but also causes intensive

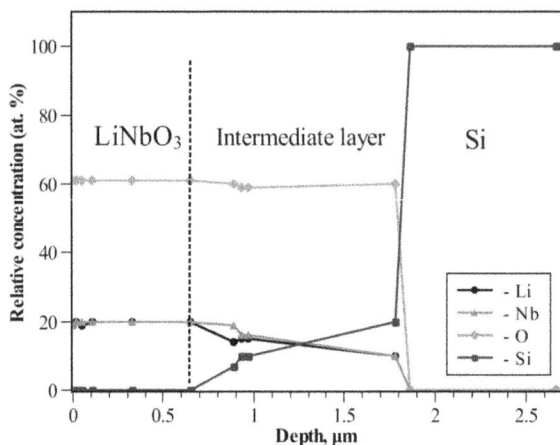

Figure 2.23. RBS spectrum of a film synthesized on a (001)Si substrate under regime 4 (see table 2.4).

bombardment of the film surface, forming a defect layer. This layer facilitates interface reactions and the interdiffusion of atoms, resulting in the formation of the intermediate layer.

To study how reactive plasma influences film structure during the initial growth stages, we synthesized thin films with a thickness of 100 nm on (001)Si substrates by the RFMS method according to regimes 1, 2, and 3 in table 2.4, varying the substrate location in the horizontal plane, as shown in figure 2.24. Figure 2.24 presents the results of the TEM study for as-grown films.

The reactive gas pressure is higher directly over the target erosion zone (regime 1 in table 2.4, position 1 in figure 2.24(a)) than when a substrate is offset from it (regimes 2 and 3 in table 2.4). This variation is due to spatial plasma inhomogeneity. Consequently, the TEM patterns shown in figures 2.24(b)–(g) correspond to films deposited as the reactive gas pressure increases. The TEM patterns show that when the reactive gas pressure is low (figures 2.24(b) and (c)), polycrystalline films with a grain size of 70 nm are formed. The clear-cut character of the grain boundaries indicates that an equilibrium structure is formed. All reflexes in the microdiffraction pattern (figure 2.24(c)) correspond to the rhombohedral lattice of LN. The relatively high intensity of the reflex $11\bar{2}0$ indicates the formation of textured $\langle 0001 \rangle$ LN films. When the reactive gas pressure increases, the grain size decreases to 30 nm (figures 2.24(b), (d), and (f)), accompanied by the disappearance of film texture. Moreover, in the second and third cases (figures 2.24(e) and (g)), a halo is observed, which is attributed to inelastic electron scattering. This indicates the formation of an amorphous LN layer due to the effect of bombardment of the film surface by plasma particles, which is consistent with results reported in [8] and [4]. In addition, it is important to note that films fabricated at the highest reactive gas pressure contain two phases (in addition to the amorphous phase): LN and the oxygen-deficient $LiNbO_2$ phase.

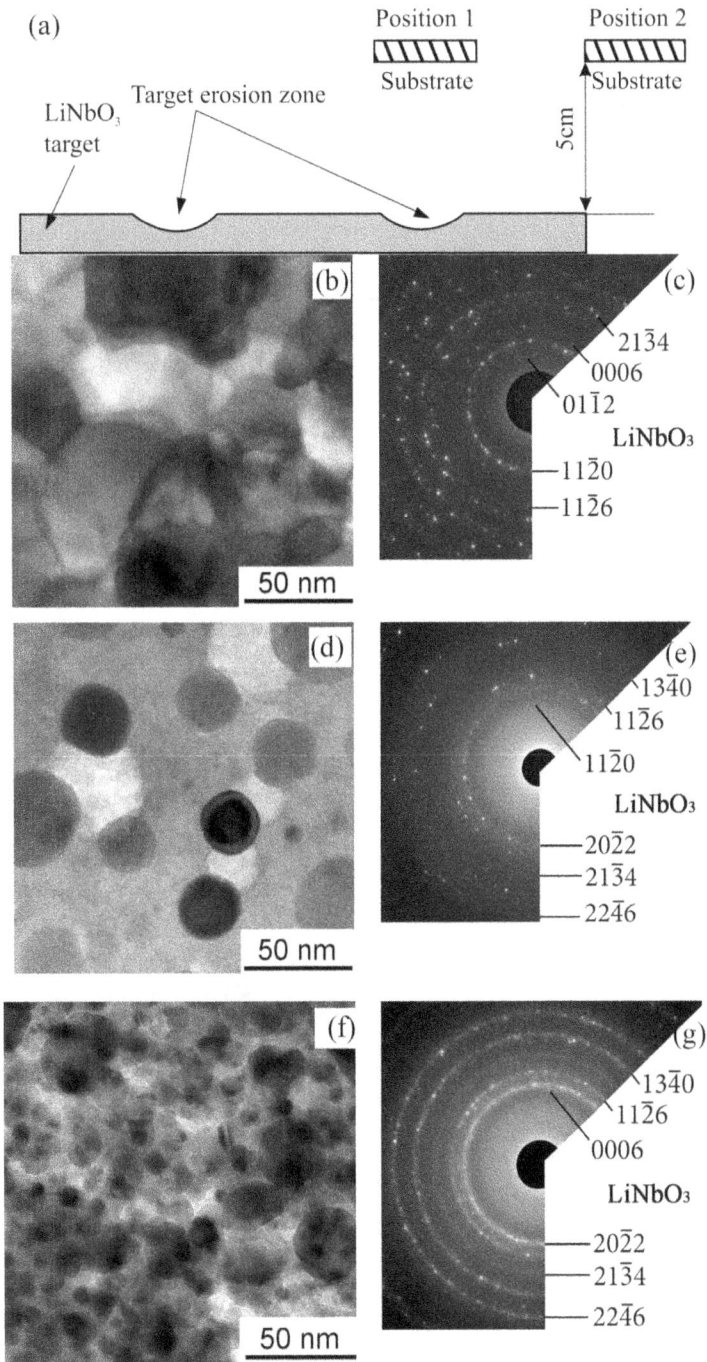

Figure 2.24. Positions of the samples during RFMS sputtering (a), TEM patterns (b, d, and f) and microdiffraction patterns (c, e, and g) of films formed under different reactive gas pressures (see comments in the text). Reproduced from [4], with permission from Springer Nature.

2.3.3 Effect of substrate type

As demonstrated earlier, Si substrates do not exhibit an orienting effect on the deposited LN films due to the formation of a thin (5 nm) amorphous SiO_2 layer at the Si–LN interface. A similar effect was reported in [12], where a thin amorphous intermediate layer was observed during the deposition of LN films onto Si substrates. However, it remains unclear why textured LN films form on an amorphous SiO_2 layer. Some investigators believe that the key factor influencing this phenomenon is that the $\langle 0001 \rangle$ axis is the polarization axis for lithium niobate. It was stressed in [13] that the conditions for the directional growth of $\langle 0001 \rangle$-textured LN films occur when deposition is conducted under external electric fields. Evidently, in our case, the intrinsic electric field attributed to the plasma in the RFMS method, directed perpendicular to the substrate surface, ensures the directional growth of LN films in the field direction. Since, within the framework of the Volmer–Weber growth mechanism, atomic migrations along a surface are greatly affected by charged particles near this surface, this accounts for the decrease in crystallite size when the working pressure increases. This phenomenon is due to the strong dependence of particle energy on reactive gas pressure and substrate–target distance.

If a sufficient number of oxygen atoms is supplied by the plasma to the substrate surface, accompanied by the migration of atoms along the surface activated by the plasma, the result is the sequential 'stacking' of lithium, niobium, and oxygen atoms in two-dimensional O–Li–O–Nb–O layers (layer by layer) along the polar axis. At higher working pressures, the surface is intensively bombarded by ions, inducing the formation of a disordered surface layer with a high concentration of vacancies. Numerous vacancies in the amorphous layer serve as centers, capturing Li atoms and intensifying reactions at the interface, which is enhanced by the capture of electrons by traps in this layer. Simultaneously, the bombardment leads to the breaking of Si–O–Si bonds in the SiO_2 layer, enhancing the diffusion of Li atoms into the substrate and enabling the effusion of Si atoms toward the film surface. All these factors create conditions for the formation of an intermediate layer of complex composition and the $LiNbO_2$ phase at the substrate–film interface, which is observed in our case at high reactive gas pressures.

Regarding epitaxial (111)Ag substrates, a very strong orienting effect occurs because, under this regime, a two-axis (epitaxial) texture is formed with the orientation ratio of (0001), $[11\bar{2}0]$ $LiNbO_3$ || (111), $\langle 110 \rangle$ Ag, giving rise to the mosaic substructure of fabricated LN films. The mechanism of this phenomenon is unclear and requires further study.

2.3.4 Effect of oxygen presence in the gas reaction environment

Based on the review presented in chapter 1, we understand that the presence of oxygen in the gas reaction chamber positively affects the structure and composition of deposited films. Nevertheless, the impact of introducing oxygen atoms into the reaction chamber can vary due to the mutual influence of various technological parameters of the RFMS method in each specific regime.

To investigate how the presence of oxygen in a reactive plasma influences the properties of synthesized LN-based heterostructures, we deposited films using the RFMS method in an Ar + O_2 plasma. The oxygen content chosen was Ar: O_2 = 60:40, based on the recommendations discussed in chapter 1. For comparison, we also synthesized films under the same sputtering conditions but in a pure Ar environment.

Films with a thickness of 0.5 µm were fabricated by the RFMS technique in an Ar (60%) + O_2 (40%) atmosphere at a magnetron power of 100 W. The reactive gas pressure of $P = 1.5 \times 10^{-1}$ Pa was selected because, at this pressure, highly oriented single-phase LN films are formed under these conditions, as indicated by the experimental results discussed earlier. During the RFMS process, substrates were positioned offset from the target erosion zone but were still subjected to the ion assist effect. Film growth under the optimal working pressure of 1.5×10^{-1} Pa with ion assist was accompanied by a higher degree of flux laminarity, resulting in the formation of a single-axis texture. The growth rate of the films under these conditions, at a distance of 5 cm between the heated (001)Si substrate ($T = 550$ ° C) and the target, was 10 nm min^{-1}.

Figure 2.25 shows comparative results of the study of the elemental composition of films deposited in an Ar + O_2 gas mixture and in a pure Ar atmosphere.

As shown in figure 2.25, the elemental composition of the films fabricated in both environments corresponds to stoichiometric LN. However, an intermediate SiO_2 layer with a thickness of about 0.6 µm forms when the films are deposited in an Ar + O_2 reactive gas mixture. This layer inevitably grows due to the presence of oxygen in the reactive plasma, resulting in the oxidation of the Si substrate surface and the formation of the SiO_2 layer.

The XRD pattern shown in figure 2.26 indicates that highly oriented, single-axis $\langle 0001 \rangle$-textured LN films are formed during the RFMS process in an Ar + O_2 reactive gas atmosphere. This finding aligns well with the work of other authors [14, 15] and our own results [16].

Calculations using the Selyakov–Scherrer method reveal that the average grain size is around 40 nm, which is similar to that of films deposited in a pure Ar atmosphere. Figures 2.27 and 2.28 present AFM images of LN films fabricated in an Ar atmosphere and an Ar + O_2 gas mixture, respectively.

The surface relief of films grown in an Ar atmosphere consists of elongated inhomogeneities with dimensions of up to 50 nm in width and up to 150 nm in length. The average surface roughness is 10 nm for the scanned area of $2 \times 2 \, \mu m^2$ (see figure 2.27). In contrast, the surface relief of films grown in an Ar + O_2 atmosphere shows inhomogeneities with a lateral dimension of 60 nm (see figure 2.28). The average surface roughness is 3 nm, which is considerably lower than that of films fabricated in an Ar environment. The lateral dimensions of the studied films correlate with the average size of the coherent scattering region calculated from XRD patterns. Low surface roughness is crucial for electrical contacts, making RFMS in an Ar + O_2 gas mixture more efficient from this perspective.

In our opinion, the formation of a single-axis texture is primarily caused by the plasma effect (the ion assist effect), while the presence of oxygen in the reaction

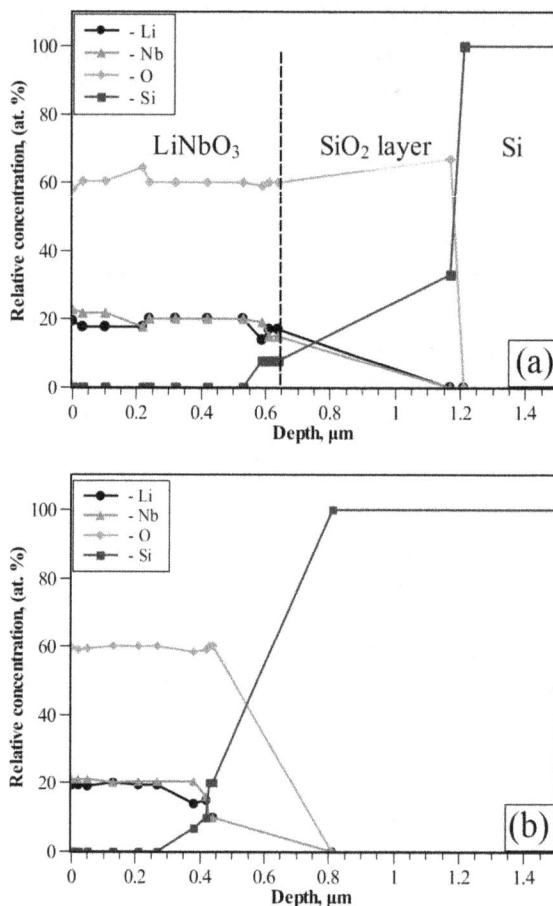

Figure 2.25. RBS spectrums for (001)Si–LN heterostructures grown in an Ar + O_2 gas mixture (a) and in an Ar gas environment (b).

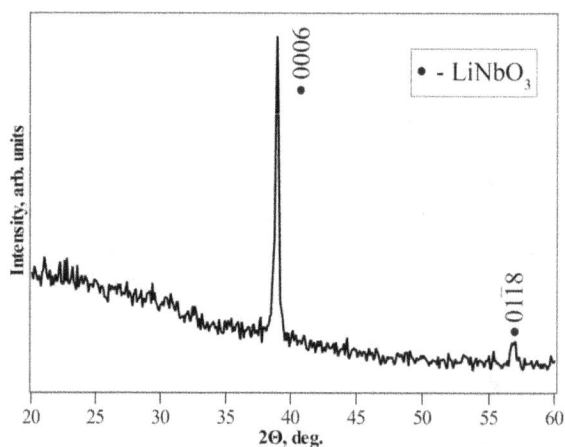

Figure 2.26. XRD pattern of a film with a thickness of about 1.0 μm fabricated in an Ar + O_2 environment.

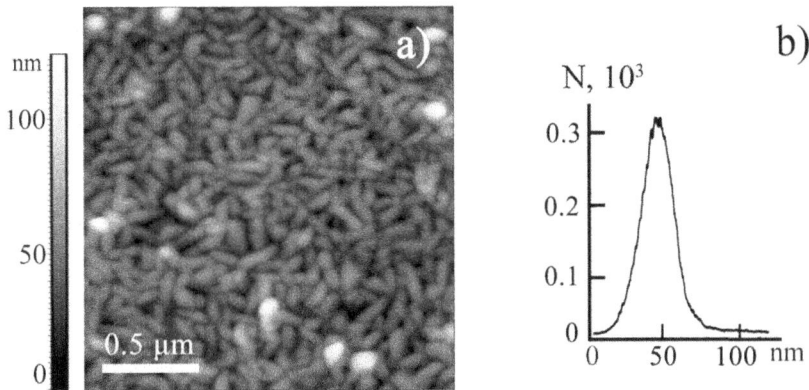

Figure 2.27. AFM image obtained using the topography mode (a) and height distribution curve showing the surface roughness (b) of LN films fabricated on a (001)Si substrate by the RFMS method in an Ar environment. Reprinted from [16], Copyright (2014), with permission from Elsevier.

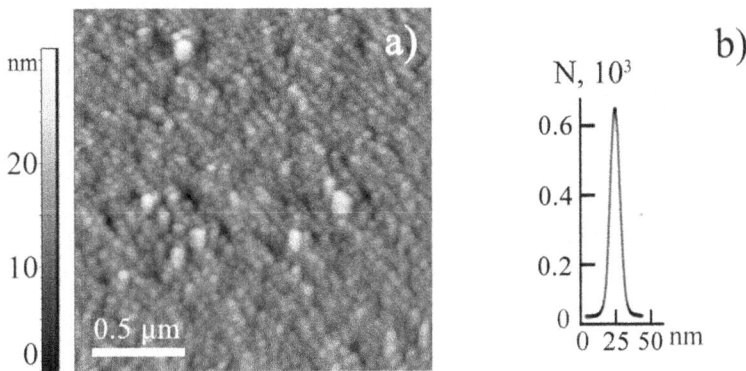

Figure 2.28. AFM image obtained using the topography mode (a) and height distribution curve showing the surface roughness (b) of LN films deposited by RFMS method on a (001)Si substrate in an Ar + O_2 environment. Reprinted from [16], Copyright (2014), with permission from Elsevier.

chamber plays a secondary role. Supporting this view, some authors emphasize the critical role of oxygen in the formation of oriented LN films. For instance, in [17], it is noted that LN films fabricated in an Ar atmosphere on atomically clean Si surfaces without a natural oxide have arbitrary grain orientation. In contrast, when oxygen is present in the reaction chamber, it covers the growing surface with a two-dimensional atomic layer. Atomic migration, activated by plasma, facilitates the formation of such a layer and, consequently, a single-axis texture.

The extended SiO_2 layer prevents the formation of this texture. As suggested in [17], Li atoms can react with the SiO_2 film, forming a double-atomic layer with a top layer of LiNbSiO compound. This top layer inhibits the formation of the LN phase. This observation is consistent with our results, where only LN films with random grain orientation are formed on Si–SiO_2 heterostructures. The formation of oxygen atomic monolayers on the surface creates favorable conditions for the migration of adatoms, minimizing potential barriers between islands of a new phase at the beginning of film growth. This process facilitates the formation of a smoother surface compared to films grown in a pure Ar environment. Supporting this view, results from [14] demonstrate a strong dependence of surface roughness on oxygen pressure in the reaction chamber, with the minimal surface roughness of 9 nm corresponding to an oxygen pressure of 40 Pa.

Thus, based on our study, the optimal sputtering regimes for the fabrication of single-phase LN films are as shown in table 2.5.

2.3.5 Effect of thermal annealing

TA in an air or oxygen atmosphere is one of the most effective post-growth treatments for as-grown films. Two processes play a significant role in TA: recrystallization and the diffusion of oxygen, leading to further reactions within the bulk and at the interfaces.

To study the effect of TA, we fabricated films on (001)Si substrates using the ion assist effect, offset from the target erosion zone. The deposition regimes are provided in table 2.6. An annealing temperature of 600 °C was chosen because, at this temperature, evaporation of components begins from the surface of LN films, allowing their properties to be effectively influenced through TA [9].

Figure 2.29 presents the results of our composition study, which used the RBS method for samples LN1-T, LN2-T, and LN3-T.

As shown by comparing the RBS spectra of as-grown films (figures 2.23 and 2.25) with those of films after TA (figure 2.29), TA in an air environment does not significantly influence the elemental composition of the studied films.

Figure 2.30 shows the post-TA XRD patterns of the films deposited under the regimes given in table 2.5.

As can be seen in figure 2.30, compared to as-grown films (see figures 2.18, 2.21, and 2.26), TA results in the recrystallization of the films and an average grain size increase of up to 80–100 nm, doubling the original size. In addition, it is important to note that TA leads to the formation of the Li-poor $LiNb_3O_8$ phase in films synthesized at a reactive gas pressure of 0.15 Pa (see samples LN2-T and LN3-T). Furthermore, in as-grown films with $\langle 0001 \rangle$texture (samples LN1 and LN2), TA results in the disappearance of this texture, forming polycrystalline films with a random grain orientation. This experimental observation aligns well with data from numerous investigators [18–22], who have also observed the formation of this 'parasitic' phase after TA in both vacuum and oxygen environments. As noted in chapter 1, this phenomenon is caused by the loss of lithium and oxygen from the film due to the desorption of Li_2O during TA.

The results of the morphological study of the films after TA are provided in figures 2.31–2.33 [23].

Analysis of the AFM images shown in figures 2.31–2.33 indicates a twofold increase in the average surface roughness of the studied films after TA. The average surface roughnesses are 70, 14, and 15 nm for samples LN1-T, LN2-T, and LN3-T, respectively. The most pronounced increase was observed for films fabricated in an

Table 2.5. Optimal sputtering regimes for thin LN films fabricated using the IBS and RFMS methods[†].

Deposition method	Substrate	Reactive gas environment	Magnetron (supply) power	Substrate temperature	Reactive gas pressure (Pa)	Substrate–target distance and position (cm)	Properties of LN films
RFMS	Si	Ar	100 W	550 °C	0.15	4–5 (offset of the target erosion zone)	Single-phase ⟨0001⟩-textured LN films. Average grain size is 40 nm. Average surface roughness is 10 nm.
	Si	Ar				8–10 (offset of the target erosion zone)	Single-phase LN films with random grain orientation. Average grain size is 50 nm. Average surface roughness is 16 nm.
	Si	Ar + O$_2$ (60/40)				4–5 (offset of the target erosion zone)	Single-phase ⟨0001⟩-textured LN films with minimal surface roughness (3 nm). Average grain size is 40 nm.
	(111)Ag					4–5 (offset of the target erosion zone)	Single-phase LN films with epitaxial texture with the orientation ratio (0001), [11$\bar{2}$0] LiNbO$_3$ ‖ (111), ⟨110⟩ Ag.
IBS	Si	Ar	2 kW			4–5 (offset of the target erosion zone)	Single-phase LN films with random grain orientation. Average grain size is 50 nm. Average surface roughness is 40 nm.

[†]Shaded cells correspond to the films, fabricated without the plasma effect.

Table 2.6. Synthesis regimes used for samples that underwent TA treatment.

Sample #	Reactive gas composition	Reactive gas pressure (Pa)	TA at temperature of 600 °C	Substrate–target distance and position (cm)
LN1	Ar	0.5	–	5
LN1-T			+	
LN2		0.15	–	
LN2-T			+	
LN3	Ar + O_2	0.15	–	
LN3-T			+	

Ar + O_2 gas mixture, rising from 3 nm for as-grown films (see figure 2.28 for sample LN3) to 14 nm for the films after TA (see figure 2.33). It is also worth noting the increase in relief inhomogeneities after TA, which are 300, 200, and 150 nm for samples LN1-T, LN2-T, and LN3-T, respectively [23]. These sizes are consistent with estimates made using XRD patterns.

For LN films fabricated by the IBS method, the TEM patterns shown in figure 2.34 indicate that TA of as-grown films results in the formation of the lithium triniobate phase ($LiNb_3O_8$) and an increase in the average grain size from 50 to 100 nm [24].

As revealed for films synthesized by the RFMS technique, LN films fabricated by the IBS method exhibit arbitrary grain orientation after TA along with an increase in surface roughness from 50 to 100 nm [25].

To summarize, some general trends can be attributed to all LN-based hetero-structures after TA, regardless of the deposition method. One finding is that the average grain size and surface roughness increase after TA, along with the disappearance of the texture in as-grown LN films. In addition, the 'parasitic' $LiNb_3O_8$ phase is formed in the films after TA as a consequence of the desorption of volatile compounds such as Li_2O. Some authors argue that the $LiNb_3O_8$ phase forms due to the separation of the $Li_{1-x}Nb_{1+y}O_3$ phase into two rival phases, LN and $LiNb_3O_8$ [26]. In contrast, others believe that the $LiNb_3O_8$ phase forms on the surface due to the desorption of LiO_2, as a high concentration of oxygen vacancies near the surface promotes the separation of the LN and $LiNb_3O_8$ phases [22].

2.3.6 Early stages of lithium niobate film growth

The integration of LN thin films into optoelectronic devices and memory units has become a focal point of research, as it opens avenues for miniaturization, enhanced performance, and novel functionalities. However, the reproducible and controlled growth of high-quality LN thin films remains a formidable challenge, primarily due to the intricate processes involved in their nucleation and early stages of growth. Understanding the early stages of LN film growth is paramount for tailoring film properties to meet specific device requirements. Nucleation, the initial formation of crystalline clusters on a substrate, sets the stage for subsequent growth processes and

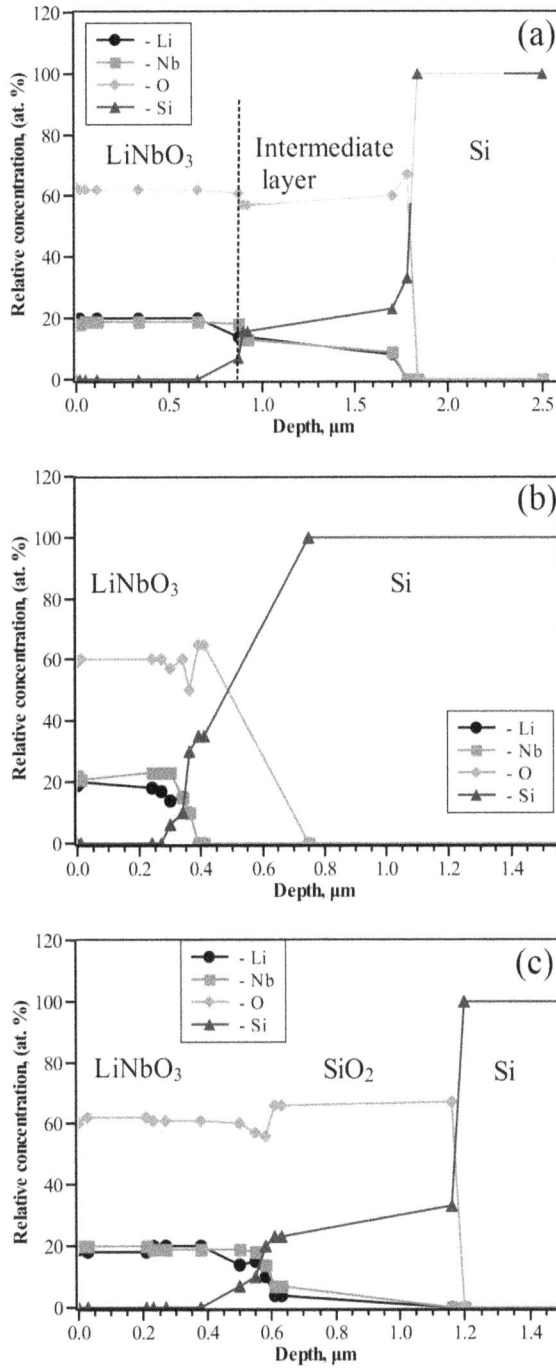

Figure 2.29. RBS spectrums of films fabricated by the RFMS method and subjected to TA. (a)—sample LN1-T, (b)—sample LN2-T, (c)—sample LN3-T.

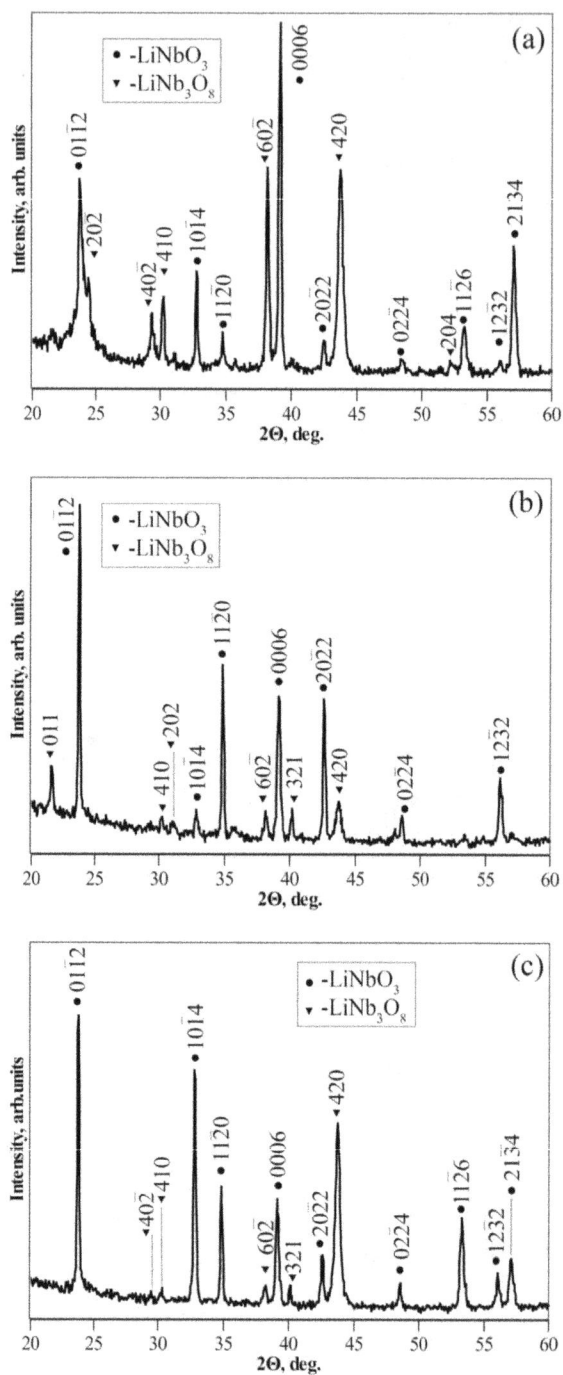

Figure 2.30. XRD patterns of the samples LN1-T (a), LN2-T (b), and LN3-T (c) (see table 2.6).

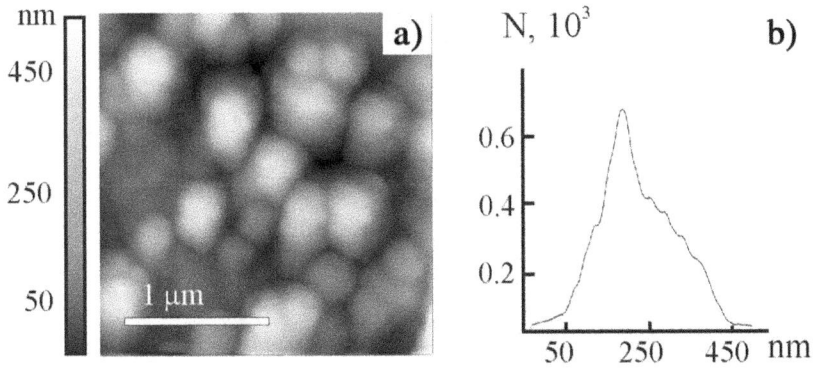

Figure 2.31. AFM image obtained using the topography mode (a) and height distribution curve showing the surface roughness (b) of LN films related to the sample LN1-T. Reproduced from [23], with permission from Springer Nature.

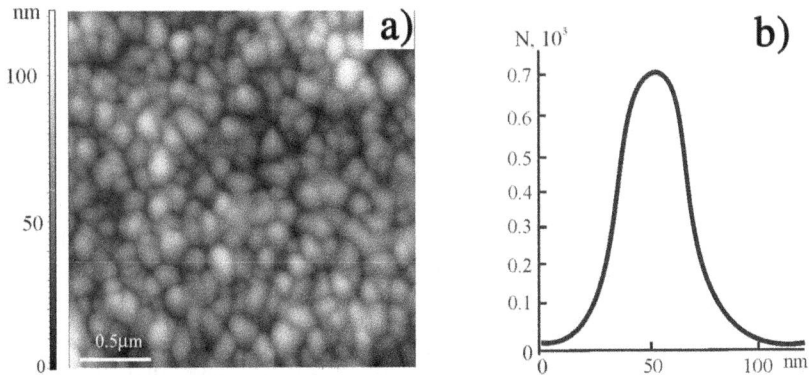

Figure 2.32. AFM image obtained using the topography mode (a) and height distribution curve showing the surface roughness (b) of LN films related to the sample LN2-T. Reproduced from [23], with permission from Springer Nature.

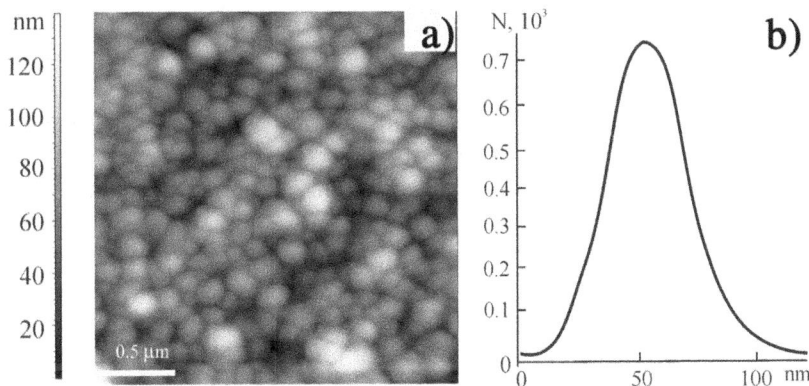

Figure 2.33. AFM image obtained using the topography mode (a) and height distribution curve showing the surface roughness (b) of LN films related to the sample LN3-T. Reproduced from [23], with permission from Springer Nature.

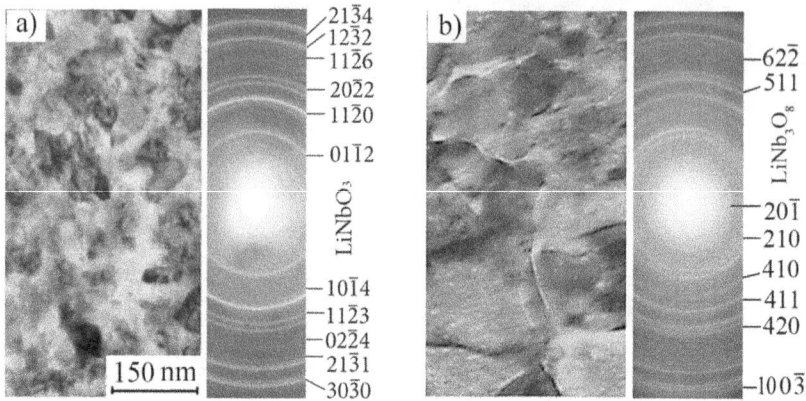

Figure 2.34. Micrographs and electron diffraction patterns of LN films deposited by the IBS method on (001) Si substrates. (a)—as-grown films, (b)—films after TA. Reproduced with permission from [24]. © (2011) COPYRIGHT Society of Photo-Optical Instrumentation Engineers (SPIE).

significantly influences the ultimate film quality. Despite the extensive research dedicated to LN thin film deposition, several challenges persist, impeding the realization of reproducible and high-quality films. The primary hurdles lie in the nucleation and early growth stages, where intricate interfacial processes govern the evolution of the film.

Controlling nucleation density, promoting uniform crystalline orientation, and mitigating defect formation during the early stages are critical objectives that demand a comprehensive understanding of the underlying mechanisms. Nucleation is a pivotal step in thin film growth, acting as the inception point for the development of a continuous crystalline film. In the case of LN, nucleation is influenced by a myriad of factors, including substrate choice, deposition technique, and process parameters. The kinetics of nucleation dictate the density and distribution of nucleation sites, ultimately shaping the overall film structure. Unraveling the nucleation kinetics of LN thin films is essential for designing strategies to control nucleation and enhance film quality. Achieving epitaxial growth poses challenges, particularly in the early stages where the formation of a well-defined nucleation layer is critical. Defects in thin films can arise during nucleation and early growth, leading to adverse effects on device performance. Common defects include dislocations, grain boundaries, and point defects, each of which influences the structural, electrical, and optical properties of the film.

It is well-known that the sputtering conditions (reactive gas composition and pressure, magnetron power, spatial plasma inhomogeneity, etc.) are powerful tools for changing the structure and texture of films with complex compositions [27]. The synthesis of c-oriented LN films on Si substrates was shown to be possible in [16]. As reported, plasma compounds are some of the most influential factors driving texture formation in a growing film. Atoms of a sputtered target are not simply fixed at certain positions on the substrate surface. By contrast, they can either diffuse or move along the surface. Moving along the surface, they can combine with other atoms, resulting in nucleation. Due to variation in adhesion energy, epitaxial growth follows three models: the Volmer–Weber (island) model, the Frank and van der Merwe (layer) model, and the Stranski–Krastanov (layer followed by island) model [28]. Some authors have reported the nucleation of islands with a size of 105 nm during the early growth stages of LN films deposited by the chemical vapor deposition (CVD) process [29, 30]. Further deposition leads to coalescence and the formation of thin LN films. This mechanism corresponds to the Volmer–Weber model. Another research team also suggested that the growth mechanism of LN films grown by the pulsed laser deposition (PLD) method is likely analogous to the Volmer–Weber model [31]. In regard to the RFMS process, some authors have utilized the advantages of this technique to achieve a high nucleation density for the formation and control of highly oriented LN films by combining this method with spray pyrolysis techniques [20]. On the other hand, the initial growth stages of LN films fabricated by the RFMS technique on oriented and non-oriented substrates have not yet been studied comparatively. To fill this gap, the initial growth stages of LN films on oriented (Si) and non-oriented (amorphous a-C) substrates were investigated [32].

The studied LN films with a thickness of up to 0.1 μm were fabricated by the RFMS method with sputtering times from 0.5 to 7 min. The oriented and non-oriented substrates were made from (001)Si wafers and $Si-SiO_2$–amorphous carbon heterostructures, respectively. The substrates were heated to $T_{sub} = 550$ °C. The amorphous carbon acted as a sublayer in the sample preparation technique for TEM investigation. The RFMS process was carried out in an Ar atmosphere at a pressure and RF power of $P = 0.8$ Pa and 15 W cm^{-2}, respectively. The substrates were located above and offset from the target erosion zone. The following two sample preparation procedures were used to detach an island-like film from a substrate. To study the initial growth process on the oriented (001)Si substrates, a thin carbon film was deposited on an as-grown LN film. This carbon film (along with the studied LN film) then underwent a lift-off process in an acid solution consisting of $H_2O + HNO_3 + HF$. To investigate the initial growth process on non-oriented substrates, we used $Si-SiO_2$–(a-C) heterostructures as a substrate. After the RFMS process, a second amorphous a-C carbon layer was deposited onto the as-grown LN films, followed by a lift-off process in an $H_2O + HF$ solution. As a result, we obtained an (a-C)–LN–(a-C) sandwich structure.

This section has been reproduced with permission from [32].

Initial growth stages on oriented substrates
In figure 2.35, SEM and TEM patterns, along with a microdiffraction pattern, illustrate a film's growth via the RFMS process over periods from 0.5 to 7 min on a heated, oriented (001)Si substrate positioned above the target erosion zone. Analysis of figure 2.35 reveals the formation of 10–20 nm-sized LN islands after 30 s of the RFMS process. As the RFMS duration extends to 4 min, the average size of the islands increases to 40–50 nm. Extending the RFMS process to 7 min results in the creation of a continuous LN film with an average grain size of 50–60 nm and an axial ⟨0001⟩texture. Figure 2.36 depicts the variation of the average grain size with

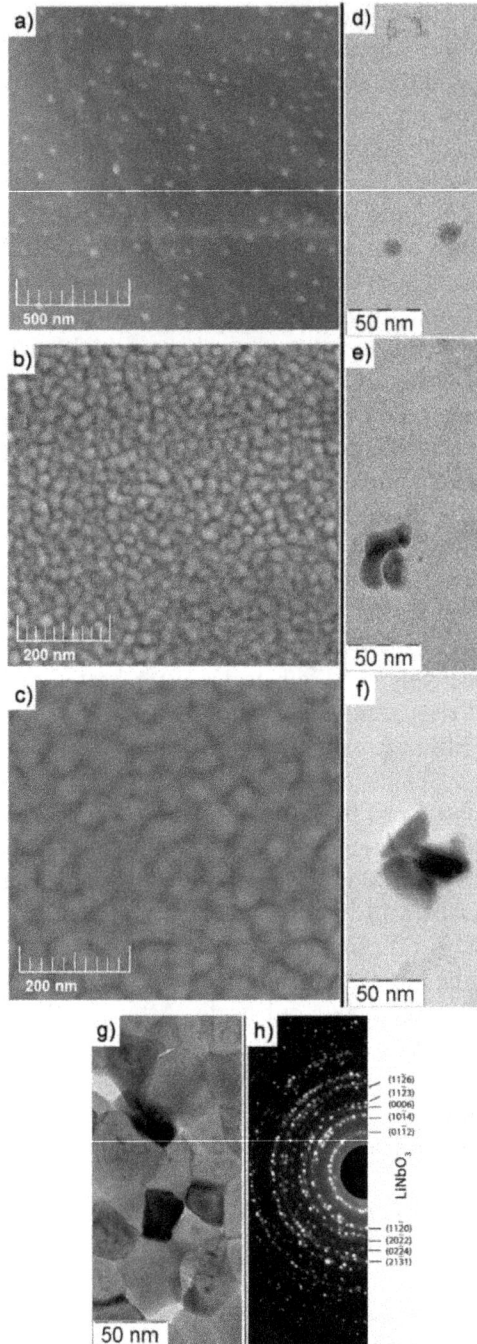

Figure 2.35. SEM (a)–(c) and TEM (d)–(g) patterns and a microdiffraction pattern (h) of LN films fabricated by the RFMS method. The sputtering times were as follows: a, d—30 s; b, e—2 min; c, f—4 min; g—7 min. Reprinted from [32], Copyright (2020), with permission from Elsevier.

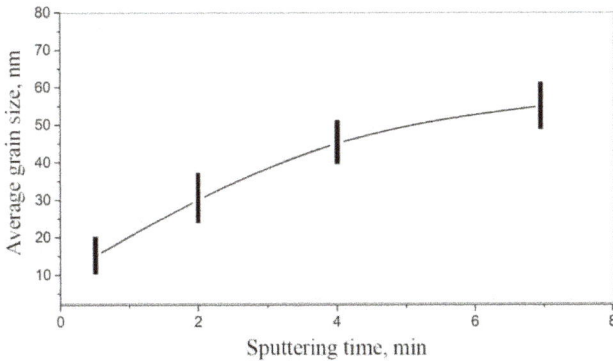

Figure 2.36. The average grain size as a function of sputtering time for LN films grown on (001)Si. Reprinted from [32], Copyright (2020), with permission from Elsevier.

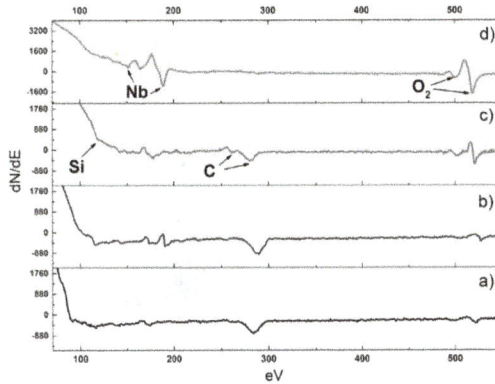

Figure 2.37. Auger spectra of LN films deposited by RFMS onto (001)Si in 30 s (a), 2 min (b), 4 min (c), and 7 min (d). Reprinted from [32], Copyright (2020), with permission from Elsevier.

sputtering time. The observed low density of crystallites in TEM images of samples grown for 4 min is attributed to the erosion of certain crystallites on the carbon replica during the lift-off process described above.

Figure 2.37 demonstrates the Auger spectra of the films deposited onto (001)Si for various sputtering times. The presence of all peaks, attributed to Nb and O, indicates the formation of LN films after 30 s. The absence of a Li peak is caused by its location in the area of secondary electron emission. Thus, the registration of Li is impossible with this technique. Considering the emission depth of secondary electrons, the presence of the Si peak indicates that the LN films are not continuous. The obtained results suggest that the structural and morphological evolution of LN films under the RFMS process is influenced by island nucleation followed by coalescence. This result indicates that the Volmer–Weber model of film growth is more plausible in our case. In contrast to our results, the authors of [30] reported that isolated pyramids limited by the $\{01\bar{1}2\}$ family of crystal planes were formed at the earliest stages of LN deposition on α-Al$_2$O$_3$. The coalescence of these pyramids then gave rise to flat aggregates, promoting the formation of continuous films in the PLD process.

Initial growth stages on non-oriented substrates
Figure 2.38 displays TEM patterns, microdiffraction patterns, and histograms depicting the island size distribution for films grown on non-oriented amorphous substrates. The sputtering conditions employed are akin to those utilized for

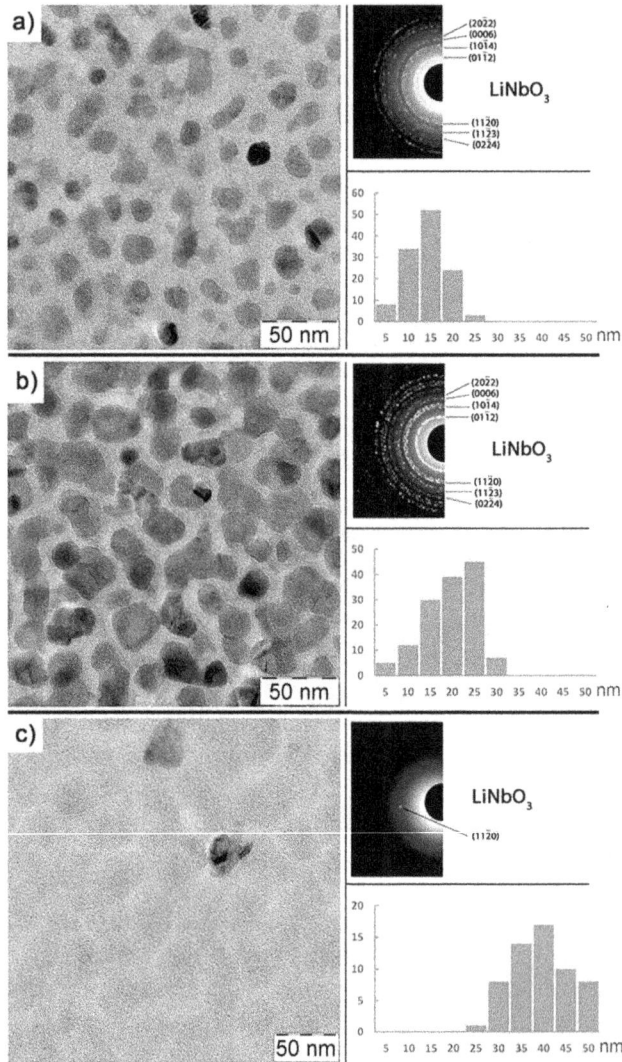

Figure 2.38. (a)–(c) TEM images, microdiffraction patterns, and island size distribution histograms of LN films fabricated by the RFMS method on (a-C)-SiO$_2$ heterostructures (non-oriented substrates). The sputtering times were: a– 30 s; b—2 min; c—4 min. Reprinted from [32], Copyright (2020), with permission from Elsevier.

oriented (001)Si substrates. Examining figure 2.38 reveals that all the grown islands consist of single-phase LN with a preferential $\langle 0001 \rangle$-axial texture, as evidenced by the high intensity of the $(11\bar{2}0)$ reflex (as inferred from both figures 2.38 and 2.39). Unlike LN films grown using various methods, no formation of other phases such as LiO_2, Nb_2O_5, or $LiNb_3O_8$ is observed during the studied growth stages [33]. The structural and morphological evolution of LN films under the RFMS process is characterized by island nucleation followed by coalescence, as elucidated earlier for the oriented substrates.

As seen in figure 2.38, the growth process exhibits distinct stages. At a sputtering time of 30 s, there is the nucleation and growth of a relatively low number of spatially separated islands. When the sputtering time extends to 2 min, there is a noticeable coalescence of single grains. Progressing further to a sputtering time of 4 min, we observe a stage of collective recrystallization leading to a twofold increase in grain size. Notably, material erosion occurred after lifting off a coating deposited during the 4 min interval. However, an analysis of the island morphology of the studied film could be conducted using imprints on the carbon replica. Consequently, the findings suggest that, following the nucleation stage, the growth of LN films primarily occurs through the evolution of existing islands rather than the formation of new ones. This observation points to the relatively high mobility of adatoms.

Our prior work [4, 27] examined the impact of plasma on the formation of thin LN films. In this study, we comparatively examine the structure, morphology, and composition of LN films produced via RFMS during the initial growth stages. Figure 2.40 presents TEM patterns, island size distribution histograms, and electron diffraction patterns of LN films deposited on amorphous (a-C)-SiO_2 substrates.

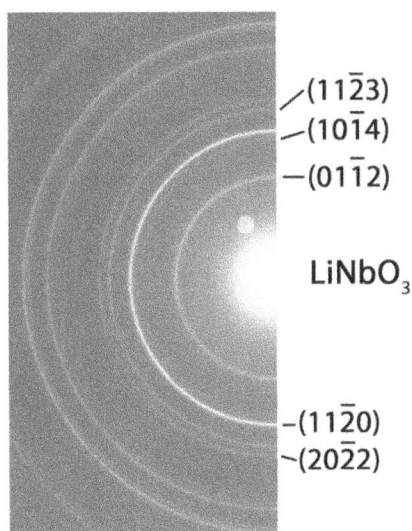

Figure 2.39. General diffraction pattern of a film fabricated by RFMS in an Ar atmosphere on an amorphous (C-SiO_2) substrate exposed to plasma for 4 min. Reprinted from [32], Copyright (2020), with permission from Elsevier.

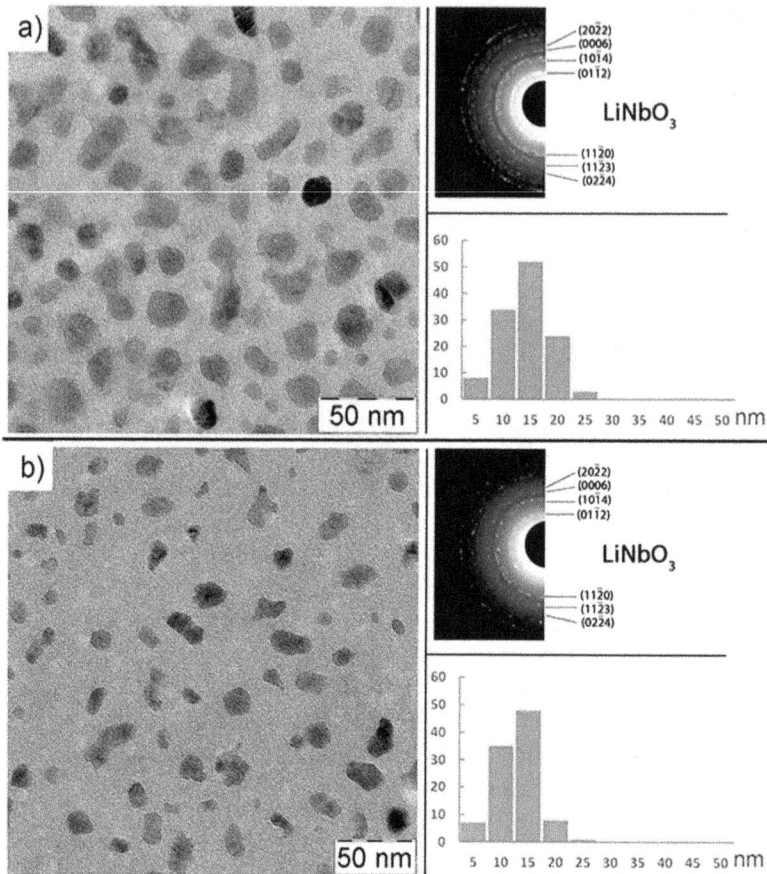

Figure 2.40. (a), (b) TEM images, microdiffraction patterns, and the island size distribution histograms of LN films fabricated by the RFMS method in 2 min on (a-C)–SiO$_2$ heterostructures under the plasma effect (a) and without this effect (b). Reprinted from [32], Copyright (2020), with permission from Elsevier.

RFMS was conducted both under the influence of plasma (with substrates positioned above the target erosion zone) and without this effect (with substrates positioned away from the target erosion zone). As depicted in figure 2.40, fewer nucleations with smaller sizes formed without the plasma effect compared to those formed under the influence of intense RF plasma. Two factors contribute to this effect: (1) the higher mobility of sputtered material adatoms, leading to their 'diffusion' toward peripheral substrate areas; (2) the accumulation of a significant amount of sputtered material over the plasma discharge area.

2.3.7 Crystallization of amorphous Li–Nb–O films

Achieving epitaxial film growth necessitates maintaining a relatively elevated substrate temperature. Experimental findings suggest that, when implementing optimal RFMS conditions for LN, the substrate temperature should be set at a minimum of 550 °C [34]. Due to the elevated vapor pressure of lithium and lithium oxide, the elemental composition of the developing film significantly diverges from the stoichiometry of LN. This deviation is concomitant with the emergence of the paraelectric phase $LiNb_3O_8$.

An alternative approach involves the crystallization of a primary amorphous film with an appropriate elemental composition deposited on a substrate without heating [35]. It is noteworthy that this method of growing LN films has received limited research attention. Various aspects, such as the structure of amorphous Li–Nb–O films resembling the composition of LN, the kinetics governing their crystallization, the texture and substructure, as well as the morphology of surfaces formed during both LN film crystallization and recrystallization, remain largely unexplored. The crystallization of amorphous LN films stands at the forefront of materials science, ushering in a new era of technological advancements with profound implications for various applications. This intricate process involves the transformation of disordered, noncrystalline structures into well-defined crystalline formations, unlocking a myriad of possibilities in the realm of electronic devices, photonics, and beyond. The crystallization process can be induced through conventional TA or pulsed photon treatment (PPT) utilizing xenon lamps with a continuous spectrum in the 0.2–1.2 μm range. The efficacy of laser processing (LP), demonstrating a significant acceleration of processes by 200–300 times, has been established in various applications, including metal film recrystallization [36], oxide and silicide film growth [37, 38], and the crystallization of amorphous metallic alloys. As researchers delve into the complexities of this crystallization phenomenon, they are not only unraveling the fundamental principles that govern the transition from amorphous to crystalline states but also paving the way for innovations that could revolutionize fields such as telecommunications, sensors, and integrated optics. This section embarks on a comprehensive exploration of the crystallization dynamics of amorphous LN films, examining the underlying mechanisms, some experimental techniques, and potential applications that underscore the significance of this research frontier.

Thin films of amorphous Li–Nb–O, with thicknesses of either 70 or 1500 nm, were fabricated through two distinct methods: RFMS and IBS [39]. The deposition process involved depositing these films onto the surface of a (111) monocrystalline n-type silicon wafer or a fracture surface of a NaCl single crystal. The latter choice allowed free-standing thin films to be obtained by dissolving the substrate.

In the RFMS process, the primary conditions were set as follows: a sputtering environment of $Ar:O_2$ in a 4:1 ratio was maintained at a pressure of approximately 0.3 Pa; the magnetron power density on the target was maintained at around $\sim15\ W\ cm^{-2}$. On the other hand, the IBS process involved utilizing a turbomolecular pump system, with a working gas mixture of $Ar:O_2$ in a 19:1 ratio, a working pressure of 0.05 Pa, an accelerating voltage of 2.1 kV, an ion-beam current of 0.1 A, and charge compensation on the target.

Subsequently, the films underwent heat treatment in a vacuum for 1 h at temperatures ranging from 300 to 550 °C. Films grown on NaCl were separated from the substrate in water and positioned on an electron microscope support grid. The structural analysis of films grown on NaCl involved utilizing reflection high-energy electron diffraction (RHEED) both before and after TA. EG-100M and

PREM-200 instruments were deployed for this purpose. The phase composition and texture of the films on silicon were examined using a combination of RHEED (EG-100M) and XRD (ARL X'TRA Thermo Techno diffractometer).

Films deposited by RFMS

Figure 2.41 depicts segments of electron diffraction patterns for 70-nanometer-thick Li–Nb–O films both before and after TA at various temperatures. The first electron diffraction pattern characterizes the initial film structure as 'amorphous,' according to the contemporary definition of amorphous substances. This definition hinges on

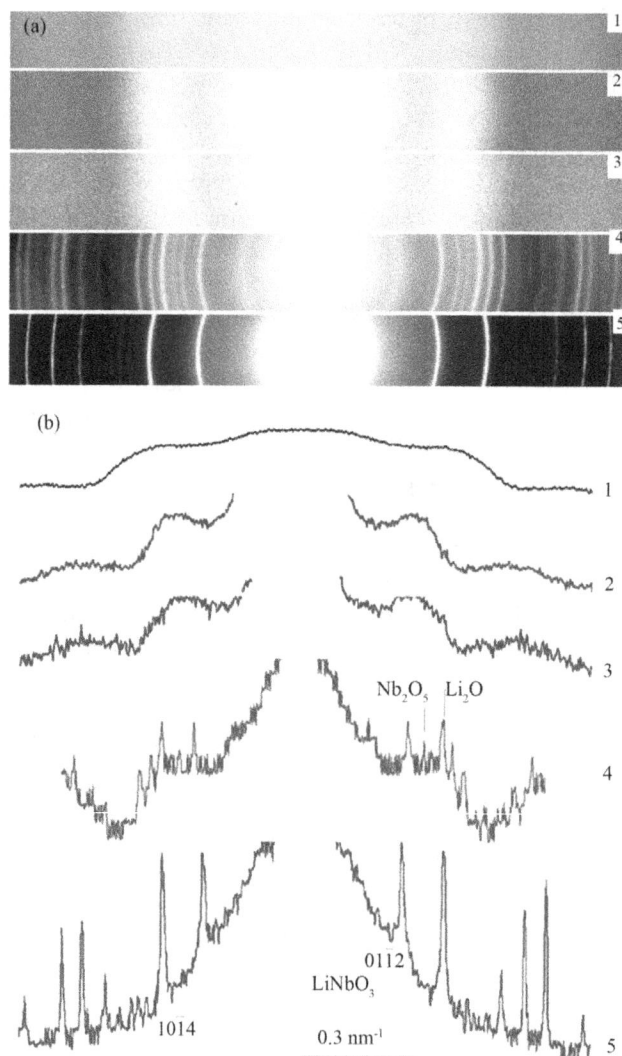

Figure 2.41. (a) Fragments of electron diffraction patterns and (b) intensity profiles of thin films before (1) and after TA (2–5) in vacuum for 1 h at (2) 300 °C, (3) 350 °C, (4) 400 °C, and (5) 550 °C. Reproduced from [39], with permission from Springer Nature.

the presence of short-range structural order and the absence of long-range order [40]. There is a noteworthy two-level intensity distribution profile within the halo, suggesting intricacies in the structure of the 'amorphous' film. Abrosimova [41] interpreted similar XRD patterns of amorphous Ni–Mo–B metallic alloys as indicative of a biphasic amorphous structure, though without specifying the phases. In contrast, we propose an alternative interpretation of electron diffraction patterns 1–3 (figure 2.41(a)). This interpretation hinges on a common feature observed across all the 'amorphous' substances we have studied: the position of the halo in the XRD (electron diffraction) patterns of 'amorphous samples' corresponds to that of the strongest reflections predicted by the structure factor for crystalline samples with the same elemental composition.

The high intensity of halos 1–3 (figure 2.41(b)) spans the angular range where the most robust diffraction lines, corresponding to the structures of Nb_2O_5, NbO_2, Li_2O, and LN, should be present [42]. Following growth and TA treatment at 300 and 350 °C, the broad electron diffraction rings persist (figure 2.41(a)), but their width slightly decreases, indicating an enlargement in the size of the scattering domains of the crystalline phase. The observed diffraction pattern results from the interference between waves that are elastically scattered on the basic atoms of structural units. Broad diffraction rings, resembling halos, arise due to the size effect in diffraction. This effect is attributed to two key factors: (1) the small size and shape of a coherently scattering elementary volume of the material; (2) the size and shape of reciprocal lattice sites. These factors align with the spatial intensity distribution of a wave. This effect occurs when diffraction take place due to infinitesimally small nuclei of crystalline phases with arbitrary mutual orientations [42].

Specifically, in the case of amorphous iron-based solid solutions, the maximum of the halo coincides with the position of the (110) line of a crystalline solid solution with the corresponding composition. This alignment is expected, given the definition of the structure factor as the sum of the amplitudes of elastically scattered waves on the basic atoms of the corresponding crystal structure, particularly in the context of diffraction by a body-centered cubic (BCC) structure.

When elastically scattered waves in the examined film are embedded within crystals of sufficient size, the wave coherence becomes constrained by the exceedingly minute dimensions of crystalline nuclei with various orientations. This constraint leads us to the deduction that amorphous Li–Nb–O films comprise nanostructured precursors (cluster nuclei) that ultimately give rise to crystalline phases of niobium oxides, lithium oxides, and LN as the synthesized products. Sections 4 and 5 of the electron diffraction patterns in figure 2.41(a) and the corresponding intensity profiles in figure 2.41(b) depict the progression of crystallization during high-temperature treatment at 400 °C and 550 °C.

At 400 °C, observation reveals not only the presence of LN but also Li_2O and $LiNb_2O_5$, which indicates the film's initial elemental composition and structure. Consequently, the fundamentally limited translational symmetry emerges as the most discernible characteristic of these ostensibly 'amorphous' structures.

Figure 2.42 presents a segment of the XRD pattern for a Li–Nb–O film 1.5 μm thick. The intensity profile within the halo region aligns well with two normal

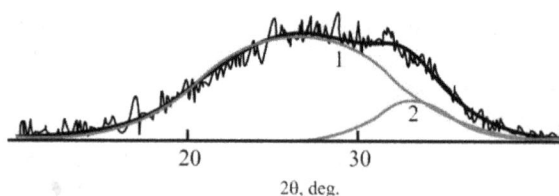

Figure 2.42. Portion of the XRD pattern of an 'amorphous' LN film grown by RFMS on the surface of a silicon wafer. Two Gaussian curves approximating the experimental XRD spectra are shown in red. Reproduced from [39], with permission from Springer Nature.

distributions (curves 1 and 2), whose maxima correspond to angles mirroring the positions of interference peaks with the highest intensity in the XRD patterns of the crystal structures of niobium and lithium oxides [42]. This alignment is consistent with RHEED results obtained for thin films.

Films deposited by IBS

Figure 2.43 illustrates XRD patterns that characterize the structural features of a film in its as-grown state (scan 1), after TA for 1 h (scan 2), and following 2 h of TA (scan 3) at a temperature of 425 °C in an air environment. Scan 1 exhibits three distinct levels of halos covering the spectrum of strong lines corresponding to crystal structures such as Nb_2O_5 (levels A and B), NbO_2, LN (levels B and C), and Li_2O (level C).

Throughout the TA, the primary reflexes from crystalline LN become prominent, and their intensity increases during subsequent annealing. Notably, the reduction in the background intensity level of the XRD pattern implies a decline in the quantity of the initial nanophases during the synthesis of LN. This is supported by the observed variation in the intensity of the halo at $2\theta \sim 14°$, attributed to nano-crystalline Nb_2O_5. When TA is used, the synthesis of LN is complete in 2.5 h, as depicted in figure 2.43(a).

For comparative analysis, figure 2.43(b) presents the XRD pattern of a Li–Nb–O/ Si heterostructure treated with PPT for 1 s, corresponding to an energy fluence of approximately $90\,J\,cm^{-2}$. Evidently, the time required for the complete synthesis of a single-phase LN film through PPT is three orders of magnitude shorter than the time required for TA. This expedited synthesis is attributed to the rapid delivery of energy to the surface layer of the Li–Nb–O/Si heterostructure.

Energy absorption is facilitated by the initial film structure, ensuring the filling of the bandgap, which has a width of approximately 4 eV in single-crystal LN. Brief PPT triggers the subsequent synthesis of the final phase and initiates a secondary recrystallization process. This process can be enhanced by exciting the electronic subsystem and generating a high concentration of nonequilibrium vacancies. A comparison between scan 3 in figure 2.43(a) and the XRD pattern of the film after PPT (figure 2.43(b)) reveals that TA produces a $20\bar{2}2$ texture in the film. In contrast, there is no discernible texture after PPT, which is attributable to the rapid rate of LN synthesis. The estimated crystallite size in the fully crystalline film after TA ranges

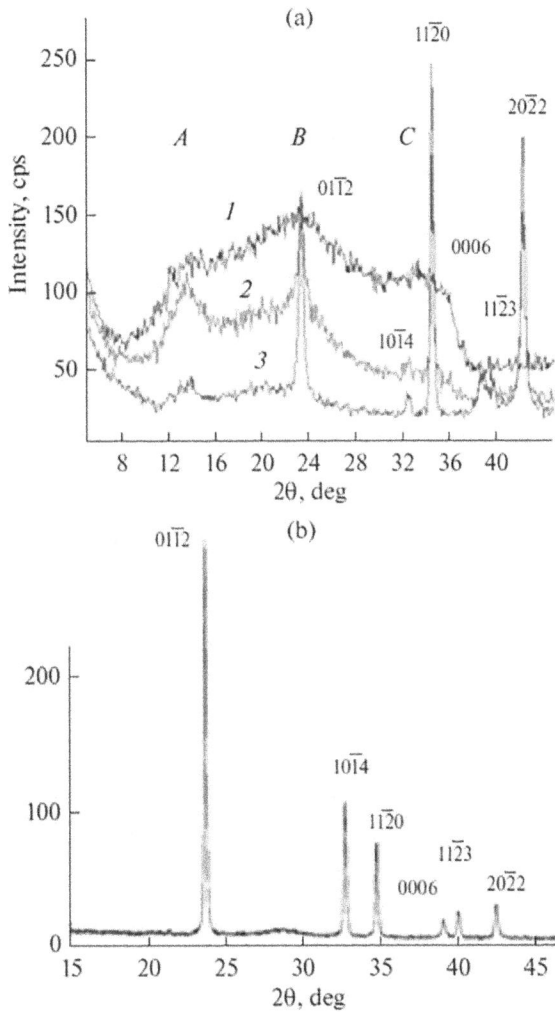

Figure 2.43. Portions of the XRD patterns of films grown by IBS on the surface of a silicon wafer: (a) as-grown film (line 1); after TA in air at 425 °C for 1 h (line 2) and 2 h (line 3); (b) after PPT in an argon atmosphere for 1 s. Reproduced from [39], with permission from Springer Nature.

from 20 to 35 nm. Conversely, in the crystallized through PPT, the crystallite size falls within the range of 50–90 nm. This discrepancy signifies that the crystallites in the latter film are considerably larger than those formed during the TA process, indicating the occurrence of a secondary recrystallization process.

In contrast to other authors who characterize films deposited onto non-heated Si substrates as amorphous [21, 43], our investigation has revealed that Li–Nb–O films fabricated on unheated substrates through RFMS and IBS techniques can be regarded as *highly nanostructured*. This structure is composed of suitable nuclei of crystalline lithium oxides, niobium oxides, and LN and is formed due to the reaction between these constituents.

2.4 Summary and discussion

1. The possibility of fabricating single-phase LN films by RFMS and IBS methods in an Ar and an Ar + O_2 reactive environment has been demonstrated. The RFMS method has more advantages than the IBS method due to the higher flexibility of its sputtering regimes, higher growth rate, and the possibility of fabricating highly oriented films on various substrates.

2. The RFMS method possesses a range of critical parameters that greatly influence structure, composition, and surface morphology. The following patterns have been revealed:

    - *Plasma effect.* When an appropriate relative position of a substrate and a suitable target–substrate distance are chosen, plasma particles cause directional growth of LN film on all types of substrates, influencing the deposition process. The $\langle 0001 \rangle$ texture is formed on (001)Si, (111)Si wafers, and on Si–SiO$_2$ heterostructures. From this point of view, the optimal substrate position is offset from the target erosion zone at a distance of 4–5 cm under a working pressure and power of 0.15 Pa and 100 W, respectively.

    - *Effect of reactive gas pressure.* When the reactive gas pressure increases, only films with a random grain orientation are formed, and grain size decreases with pressure. This is caused by the heavy bombardment of the surface by plasma particles, making it amorphous in the initial growth stages and forming an intermediate layer at the substrate–film interface. The strong bombardment of the film's surface and the decline in the mean free path of the plasma particles lead to a decrease in the concentration of Li atoms near the growing surface. The $LiNb_3O_8$ phase is formed along with LN in the films.

    - *Types of substrates.* Films deposited onto (001)Si and (111)Si wafers do not manifest the orienting substrate effect due to the formation of a thin (5 nm) buffer SiO$_2$ layer, which passivates the substrates. In contrast, epitaxial LN films are fabricated on (111)Ag substrates by the RFMS method. The two-axis (epitaxial) texture is formed with an orientation ratio of (0001), $[11\bar{2}0]$ LiNbO$_3$ || (111), $\langle 110 \rangle$ Ag, yielding a mosaic substructure in the fabricated LN films. Thus, it is advisable to use epitaxial Ag films as substrates for the deposition of LN films (and specifically for the synthesis of films separated from a substrate).

3. The presence of oxygen and Ar in the reaction chamber results in the formation of highly oriented $\langle 0001 \rangle$ LN films with minimal surface roughness. An extended SiO$_2$ layer is formed at the substrate–film interface due to the oxidation of the surface in the presence of oxygen.

4. TA of as-grown textured LN films in air leads to their recrystallization, increasing the average surface roughness and grain size twofold (up to 15–70 nm and 80–100 nm, respectively). TA results in the formation of the $LiNb_3O_8$ phase in films fabricated under a working pressure of $1.5 \cdot 10^{-1}$ Pa.

The texture present in as-grown films is leveled, and the grain orientation becomes arbitrary.

5. The structural and morphological evolution seen in the initial growth stages originates from island nucleation followed by coalescence and does not depend on the substrate type. After the nucleation stage, the existing islands, rather than the formation of new ones, feed the further growth of LN films. This growth scenario occurs due to the high mobility of adatoms in the initial stages of LN film deposition and corresponds to the Volmer–Weber model.

6. The structure of Li–Nb–O films grown on unheated substrates by RFMS and IBS can be considered extremely nanostructured, comprising appropriate nuclei of crystalline lithium oxides, niobium oxides, and LN as products of the reactions between them. We have demonstrated the possibility of achieving the synthesis of polycrystalline LN films through TA or brief PPT of amorphous Li–Nb–O films. In the case of brief PPT, the LN synthesis process is three orders of magnitude faster than in the case of TA.

References

[1] Iyevlev V, Kostyuchenko A, Sumets M and Vakhtel V 2011 Electrical and structural properties of LiNbO$_3$ films, grown by RF magnetron sputtering *J. Mater. Sci., Mater. Electron.* **22** 1258–63

[2] Simões A Z, Zaghete M A, Stojanovic B D, Riccardi C S, Ries A, Gonzalez A H and Varela J A 2003 LiNbO$_3$ thin films prepared through polymeric precursor method *Mater. Lett.* **57** 2333–9

[3] Rabson T A, Baumann R C and Rost T A 1990 Thin film lithium niobate on silicon *Ferroelectrics* **112** 265–71

[4] Sumets M, Kostyuchenko A, Ievlev V, Kannykin S and Dybov V 2015 Sputtering condition effect on structure and properties of LiNbO$_3$ films *J. Mater. Sci., Mater. Electron.* **26** 4250–6

[5] Sumets M P, Dybov V A and Ievlev V M 2017 LiNbO$_3$ films: potential application, synthesis techniques, structure, properties *Inorg. Mater.* **53** 1361–77

[6] Yang X, Lan G, Li B and Wang H 1987 Raman spectra and directional dispersion in LiNbO$_3$ and LiTaO$_3$ *Phys. Status Solidi* **142** 287–300

[7] Ievlev V, Sumets M and Kostyuchenko A 2013 Conduction mechanisms in Si-LiNbO$_3$ heterostructures grown by ion-beam sputtering method *J. Mater. Sci.* **48** 1562–70

[8] Rost T A, Lin H, Rabson T A, Baumann R C and Callahan D L 1992 Deposition and analysis of lithium niobate and other lithium niobium oxides by rf magnetron sputtering *J. Appl. Phys.* **72** 4336–43

[9] Park S K, Baek M S, Bae S C, Kwun S Y, Kim K T and Kim K W 1999 Properties of LiNbO$_3$ thin film prepared from ceramic Li–Nb–K–O target *Solid State Commun.* **111** 347–52

[10] Kong Y, Xu J, Chen X, Zhang C, Zhang W and Zhang G 2000 Ilmenite-like stacking defect in nonstoichiometric lithium niobate crystals investigated by Raman scattering spectra *J. Appl. Phys.* **87** 4410

[11] Blümel J, Born E and Metzger T 1994 Solid state NMR study supporting the lithium vacancy defect model in congruent lithium niobate *J. Phys. Chem. Solids* **55** 589–93

[12] Akazawa H and Shimada M 2004 Correlation between interfacial structure and c-axis-orientation of LiNbO$_3$ films grown on Si and SiO$_2$ by electron cyclotron resonance plasma sputtering *J. Cryst. Growth* **270** 560–7

[13] Hu W S, Liu Z G, Lu Y-Q, Zhu S N and Feng D 1996 Pulsed-laser deposition and optical properties of completely (001) textured optical waveguiding LiNbO$_3$ films upon SiO$_2$/Si substrates *Opt. Lett.* **21** 946

[14] Wang X, Liang Y, Tian S, Man W and Jia J 2013 Oxygen pressure dependent growth of pulsed laser deposited LiNbO$_3$ films on diamond for surface acoustic wave device application *J. Cryst. Growth* **375** 73–7

[15] Tan S, Gilbert T, Hung C-Y, Schlesinger T E and Migliuolo M 1996 Sputter deposited c-oriented LiNbO$_3$ thin films on SiO$_2$ *J. Appl. Phys.* **79** 3548

[16] Sumets M, Ievlev V, Kostyuchenko A, Vakhtel V, Kannykin S and Kobzev A 2014 Electrical properties of Si-LiNbO3 heterostructures grown by radio-frequency magnetron sputtering in an Ar + O$_2$ environment *Thin Solid Films* **552** 32–8

[17] Akazawa H and Shimada M 2006 Factors driving c-axis orientation and disorientation of LiNbO$_3$ thin films deposited on TiN and indium tin oxide by electron cyclotron resonance plasma sputtering *J. Appl. Phys.* **99** 124103

[18] Simões A Z, Zaghete M A, Stojanovic B D, Gonzalez A H, Riccardi C S, Cantoni M and Varela J A 2004 Influence of oxygen atmosphere on crystallization and properties of LiNbO$_3$ thin films *J. Eur. Ceram. Soc.* **24** 1607–13

[19] Kiselev D A, Zhukov R N, Bykov A S, Voronova M I, Shcherbachev K D, Malinkovich M D and Parkhomenko Y N 2014 Effect of annealing on the structure and phase composition of thin electro-optical lithium niobate films *Inorg. Mater.* **50** 419–22

[20] Bornand V, Huet I and Papet P 2002 LiNbO$_3$ thin films deposited on Si substrates: a morphological development study *Mater. Chem. Phys.* **77** 571–7

[21] Akazawa H and Shimada M 2006 Precipitation kinetics of LiNbO$_3$ and LiNb$_3$O$_8$ crystalline phases in thermally annealed amorphous LiNbO$_3$ thin films *Phys. Status Solidi* **203** 2823–7

[22] Akazawa H and Shimada M 2007 Mechanism for LiNb$_3$O$_8$ phase formation during thermal annealing of crystalline and amorphous LiNbO$_3$ thin films *J. Mater. Res.* **22** 1726–36

[23] Sumets M, Kostyuchenko A, Ievlev V, Kannykin S and Dybov V 2015 Influence of thermal annealing on structural properties and oxide charge of LiNbO$_3$ films *J. Mater. Sci., Mater. Electron.* **26** 7853–9

[24] Iyevlev V, Kostyuchenko A and Sumets M 2011 Fabrication, substructure and properties of LiNbO$_3$ films *Proc. SPIE* **7747** 77471J–8

[25] Ievlev V M, Sumets M P and Kostyuchenko A V 2012 Effect of thermal annealing on electrical properties of Si-LiNbO$_3$ *Mater. Sci. Forum* **700** 53–7

[26] Esdaile R J 1985 Comment on 'Characterization of TiO$_2$, LiNb$_3$O$_8$, and (Ti$_{0.65}$Nb$_{0.35}$)O$_2$ compound growth observed during Ti:LiNbO$_3$ optical waveguide fabrication' *J. Appl. Phys.* **58** 1070–1

[27] Kuzmina V O, Sinelnikov A A, Soldatenko S A and Sumets M 2018 Activation energy of subgrain growth process and morphology evolution in β-SiC/Si(111) heterostructures synthesized by pulse photon treatment method in a methane atmosphere *J. Mater. Sci., Mater. Electron.* **29** 20097–103

[28] Vook R W 1984 Nucleation and growth of thin films *Opt. Eng.* **23** 343–9

[29] Shtansky D V, Kulinich S A, Terashima K and Yoshida T 2001 Crystallography and structural evolution of LiNbO$_3$ and LiNb$_{1-x}$Ta$_x$O$_3$ films on sapphire prepared by high-rate thermal plasma spray chemical vapor deposition *J. Mater. Res.* **16** 2271–9

[30] Veignant F, Gandais M, Aubert P and Garry G 1998 Epitaxial growth of LiNbO$_3$ on αAl$_2$O$_3$(0001) *Thin Solid Films* **336** 163–7

[31] He J and Ye Z 2003 Highly C-axis oriented LiNbO$_3$ thin film on amorphous SiO$_2$ buffer layer and its growth mechanism *Chin. Sci. Bull.* **48** 2290–4 https://www.sciengine.com/Sci% 20Bull%20Chin/doi/10.1360/03ww0053

[32] Dybov V, Serikov D, Ryzhkova G and Sumets M 2020 Early stages of lithium niobate films growth fabricated by radio-frequency magnetron sputtering on crystalline (001)Si and amorphous carbon substrates *Surf. Interfaces* **19** 100530

[33] Akazawa H and Shimada M 2004 Electron cyclotron resonance plasma sputtering growth of textured films of c-axis-oriented LiNbO$_3$ on Si(100) and Si(111) surfaces *J. Vac. Sci. Technol. A: Vac. Surf. Films* **22** 1793–8

[34] Sumets M, Ievlev V, Dybov V, Kostyuchenko A, Serikov D, Kannykin S and Belonogov E 2019 Synthesis and properties of multifunctional Si–LiNbO$_3$ heterostructures for non-volatile memory units *J. Mater. Sci., Mater. Electron.* **30** 16562–70

[35] Atak G and Coşkun Ö D 2018 LiNbO$_3$ thin films for all-solid-state electrochromic devices *Opt. Mater.* **82** 160–7

[36] Ievlev V M, Turaeva T L, Latyshev A N, Sinel'Nikov A A and Selivanov V N 2007 Effect of photon irradiation on the process of recrystallization of thin metallic films *Phys. Met. Metallogr.* **103** 58–63

[37] Ievlev V M, Kannykin S V, Kushchev S B, Sinel'Nikov A A and Soldatenko S A 2012 Synthesis of rutile films activated by photon treatment *Inorg. Mater. Appl. Res.* **3** 189–92

[38] Ievlev V M, Kushchev S B and Sanin V N 2002 Solid-phase synthesis of silicides during pulsed photonic processing of Si-Me heterostructures (Me: Pt, Pd, Ni, Mo, Ti) *FHOM* 27–31

[39] Ievlev V M, Belonogov E K, Dybov V A, Kannykin S V, Serikov D V, Sitnikov A V and Sumets M P 2019 Synthesis of lithium niobate during crystallization of amorphous Li–Nb–O film *Inorg. Mater.* **55** 1237–41

[40] Glezer A M and Shurygina N A 2017 *Amorphous-Nanocrystalline Alloys* (Boca Raton, FL: CRC Press)

[41] Abrosimova G E 2011 Evolution of the structure of amorphous alloys *Phys. Usp.* **54** 1227–42

[42] Vaĭnshteĭn B K 1964 *Structure Analysis by Electron Diffraction* (Oxford: Pergamon)

[43] Sauze L C, Vaxelaire N, Rouchon D, Templier R, Remiens D, Rodriguez G and Dupont F 2021 Effect of the annealing treatment on the physical and structural properties of LiNbO$_3$ thin films deposited by radio-frequency sputtering at room temperature *Thin Solid Films* **726** 138660

IOP Publishing

Lithium Niobate-Based Heterostructures (Second Edition)
Synthesis, properties, and electron phenomena
Maxim Sumets

Chapter 3

Electron phenomena in LiNbO$_3$-based heterostructures

In this chapter, we examine the electrical properties of Si–LiNbO$_3$ heterostructures fabricated by the RFMS and IBS methods without the ion assist effect. Through an analysis of current–voltage and capacitance–voltage characteristics, we propose an energy band diagram for the fabricated Si–LiNbO$_3$ heterostructures, detailing all main parameters. Regardless of the conductivity type of Si wafers, a positive oxide charge is formed, which limits the functionality of LiNbO$_3$-based heterostructures. The concentration of localized charge centers varies with the deposition method. The remnant polarization in the synthesized films is significantly lower than that in bulk lithium niobate, negatively impacting the application of the fabricated heterostructures in memory units. This reduction in polarization is likely due to the arbitrary orientation of grains.

We investigate the charge transport mechanisms affecting the DC conductivity of the studied heterostructures in detail, considering both temperature and applied voltage. Using impedance spectroscopy (IS), we demonstrate that dielectric relaxation in LiNbO$_3$ films is influenced by the Maxwell–Wagner relaxation phenomenon.

3.1 Basic electrical properties of LiNbO$_3$ thin films in Si–LiNbO$_3$ heterostructures

As demonstrated in chapter 1, the electrical properties of thin LiNbO$_3$ films (such as the dielectric constant, conductivity, bandgap, and remnant polarization) significantly influence the functionality of integrated electronic and optoelectronic devices. In this chapter, we examine the main electrical properties of LiNbO$_3$ films and Si–LiNbO$_3$ heterostructures.

We fabricated (001)Si–LiNbO$_3$ and (001)Si–SiO$_2$–LiNbO$_3$ heterostructures using RFMS and IBS methods without the ion assist effect, following the optimal regimes proposed in chapter 2. It is important to note that all results reported in this chapter pertain to heterostructures fabricated without the plasma effect. Silicon wafers ((001) Si of n- and p-type conductivity with $\rho = 20\,\Omega\cdot$ cm and $\rho = 4.5\,\Omega\cdot$ cm, respectively) and Si–SiO$_2$ heterostructures were used as substrates. The Si–SiO$_2$ heterostructures were fabricated by annealing Si wafers in a coaxial furnace at 700 °C.

Using the RFMS method, deposition was performed on substrates heated to 550 °C and located offset from the target erosion zone (8–10 cm apart), which eliminates the plasma effect. Analysis of the composition and structure of thin LiNbO$_3$ films, as discussed in chapter 2, suggests that these regimes form single-phase polycrystalline LiNbO$_3$ films with arbitrary grain orientation.

The electrical properties of the fabricated heterostructures were studied using techniques based on current–voltage (I–V) and high-frequency (1 MHz) capacitance–voltage (C–V) characteristics, as well as tangent loss frequency dependence and IS in the frequency range of 30–10^4 Hz and temperatures ranging from 77 K to 400 K. I–V characteristics were measured at a voltage change rate of dV/d$t = 0.1$ V s^{-1}. Ferroelectric properties were studied by recording hysteresis loops using the Sawyer–Tower method. The top contacts used for electrical measurements, with an area of $S = 1 \times 10^{-6}$ m^2, were formed by thermal evaporation and condensation of Al in a vacuum (1 × 10^{-4} Pa). The bottom electrodes were created using an In/Ga eutectic alloy on the Si substrates, which formed ohmic contacts [1].

3.1.1 Capacitance–voltage and current–voltage characteristics of LiNbO$_3$-based heterostructures

The typical high-frequency ($f = 1$ MHz) C–V characteristics of (001)Si–SiO$_2$–LiNbO$_3$–Al heterostructures, fabricated by the RFMS method on Si substrates of p-type conductivity, correspond to the C–V characteristics of metal–insulator–semiconductor (MIS) systems, as shown in figure 3.1. Analysis of C–V characteristics was conducted according to the standard methods of C–V spectroscopy [2]. The studied (001)Si–SiO$_2$–LiNbO$_3$–Al heterostructures can be characterized as an MIS structure with a double-layered dielectric, SiO$_2$–LiNbO$_3$. In this case, the capacitance of this structure in the accumulation regime, C_i, is equivalent to the capacitance of two capacitors, C_{LN} and C_{ox}, connected in series that represent the capacitances of the LiNbO$_3$ and SiO$_2$ layers, respectively.

Using the formulas for a parallel-plate capacitor and for the capacitance of two capacitors in series, we have the following system of equations:

$$C_{LN} = \frac{\varepsilon_{LN}\varepsilon_0 S}{d}$$

$$C_{ox} = \frac{\varepsilon_{ox}\varepsilon_0 S}{d_{ox}} \tag{3.1}$$

$$\frac{1}{C_i} = \frac{1}{C_{LN}} + \frac{1}{C_{ox}}.$$

Here, C_i is the capacitance of an MIS capacitor in the accumulation regime; S is the contact area; $\varepsilon_0 = 8.85 \times 10^{-12}$ F m^{-1} is the electric constant; ε_{LN} and ε_{ox} are the

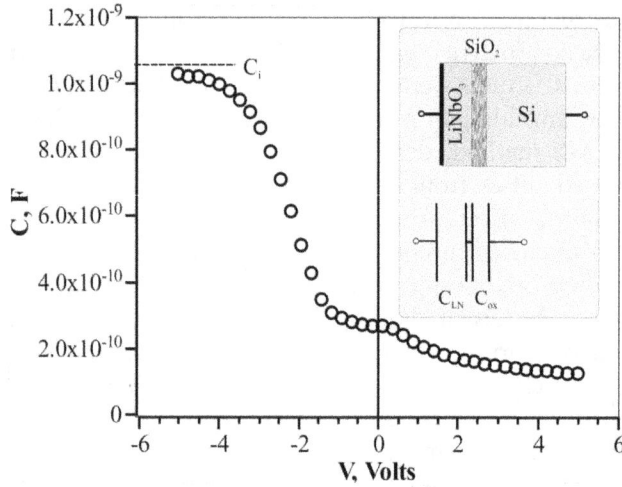

Figure 3.1. Typical high-frequency (1 MHz) $C\text{–}V$ characteristics of a (001)Si–SiO$_2$–LiNbO$_3$–Al heterostructure. The insert schematically shows the studied heterostructure and its equivalent circuit in the accumulation regime for a measured capacitance of C_i.

dielectric constants of LiNbO$_3$ and SiO$_2$, respectively; and d and d_{ox} are the thicknesses of the LiNbO$_3$ film and the silicon dioxide layer, respectively. Solving these equations for ε_{LN}, we obtain:

$$\varepsilon = \varepsilon_{ox} \frac{C_i}{\varepsilon_{ox}\varepsilon_0 S/d - C_i d_{ox}/d}. \tag{3.2}$$

Considering that for silicon dioxide, $\varepsilon_{ox} = 3.82$, and for the studied heterostructures, $d_{ox} = 120$ nm, we obtain the dielectric constant of the LiNbO$_3$ films from equation (3.2) as $\varepsilon_{LN} = 28$. This result is in good agreement with those reported in [3, 4] for thin LiNbO$_3$ films and is close to the dielectric permittivity of bulk lithium niobate ($\varepsilon = 30$) [5].

The fact that the $C\text{–}V$ curve is shifted to the left along the voltage axis (see figure 3.1) is evidently attributed to the presence of a positive fixed charge in the films. The energy distribution of surface states at the Si–SiO$_2$ interface in the lower half of the Si bandgap is calculated according to [2] through the shift of an experimental $C\text{–}V$ curve relative to the theoretical curve (see figure 3.2).

Analysis of $C\text{–}V$ characteristics reveals that the effective density of interface states is $N_{eff} = 2.4 \times 10^{11}$ cm^{-2}. In addition, figure 3.1 shows that at zero bias, the studied heterostructures are in the deep depletion regime, i.e. close to the inversion regime, where a strong inversion layer forms at the dielectric–semiconductor interface due to a strong internal electric field in the dielectric. In this condition, even a low forward bias (negative at the metal electrode) can generate leakage currents, limiting the application of the synthesized heterostructures in nonvolatile memory units and optoelectronic devices.

The high-frequency $C\text{–}V$ characteristics of (001)Si–SiO$_2$–LiNbO$_3$–Al heterostructures, fabricated under the same regimes on p-type (001)Si wafers (without the SiO$_2$

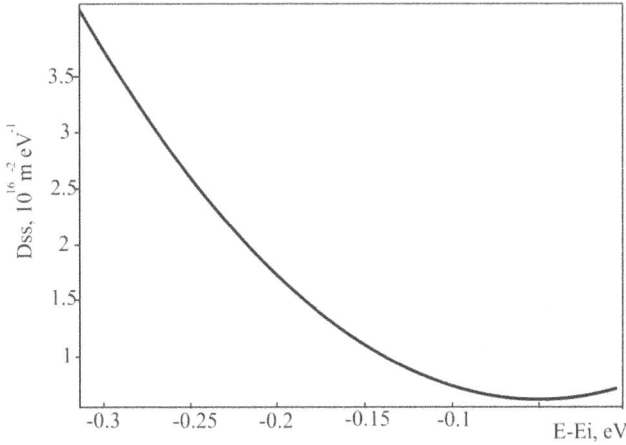

Figure 3.2. Energy distribution of surface states at the Si–SiO$_2$ interface in a (001)Si–SiO$_2$–LiNbO$_3$–Al heterostructure.

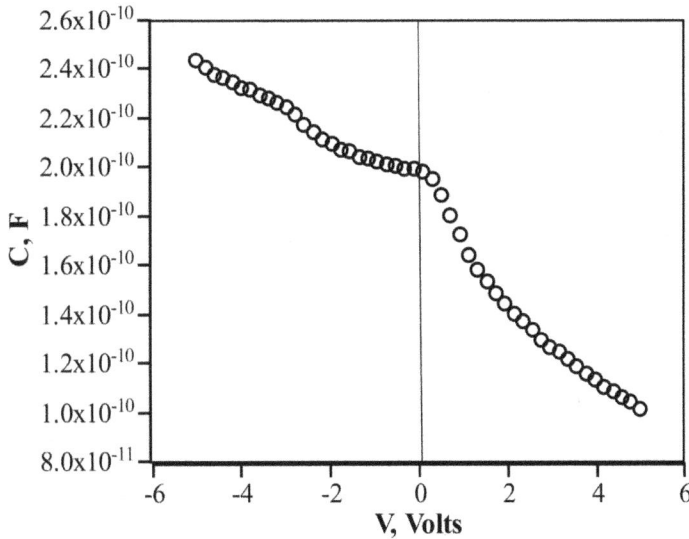

Figure 3.3. Typical high-frequency (1 MHz) C–V characteristics of a (001)Si–LiNbO$_3$–Al heterostructure fabricated by the RFMS method on p-type Si.

layer), were also similar to the C–V characteristics of an MIS capacitor with a high density of interface states, which prevents its saturation (see figure 3.3).

At higher negative bias, the leakage current in (001)Si–LiNbO$_3$–Al heterostructures fabricated on p-Si complicates proper C–V analysis. Therefore, we use I–V methods for this heterostructure. The I–V characteristics of (001)Si–LiNbO$_3$–Al heterostructures fabricated on p-type Si and measured at $T = 300$ K in ln J–V coordinates are shown in figure 3.4.

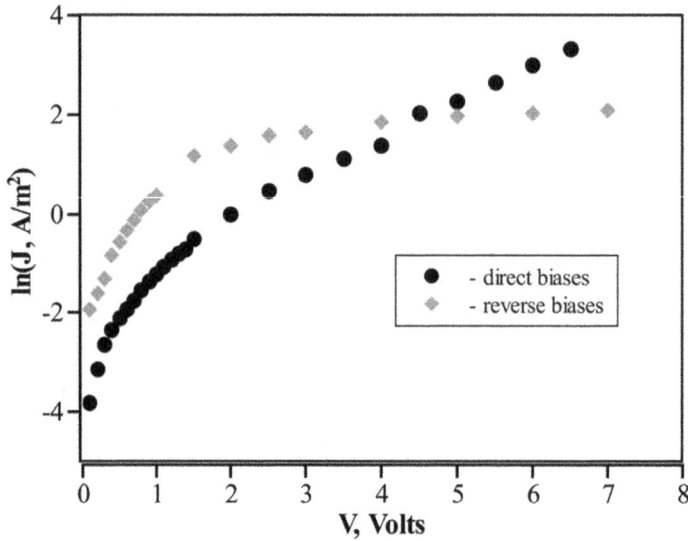

Figure 3.4. Typical *I–V* characteristics of p-type (001)Si–LiNbO$_3$–Al heterostructures fabricated by RFMS subjected to forward bias ('–' at the Al contact) and reverse bias ('+' at the Al contact).

The resistivity of the fabricated LiNbO$_3$ films is determined from the initial (ohmic) section of the *I–V* curves and equals $\rho = 1 \times 10^9\ \Omega\cdot\text{cm}$ for films deposited onto (001)Si substrates, which is in good agreement with the results of [6].

Considering the possible application of the diffusion theory for analyzing *I–V* characteristics, we estimate the Debye length $L_D = \sqrt{\varepsilon\varepsilon_0 kT/q^2 n_0}$ and the mean free path $l = V_T\langle\tau\rangle = V_T m\mu/q$ (here ε and ε_0 are the dielectric permittivity of the material and vacuum, respectively; k is the Boltzmann constant; q and m are the electronic charge and mass, respectively; n_0 is the concentration of free charges; and V_T, μ, and $\langle\tau\rangle$ are the thermal velocity, mobility, and the average effective relaxation time of carriers, respectively). For Si ($n_0 \sim 10^{15}\ \text{cm}^{-3}$, $\mu \approx 0.5 \cdot 10^3\ \text{cm V}^{-1}\text{s}^{-1}$, $T = 300\,\text{K}$, $\varepsilon = 12$, $V_T \approx 1 \cdot 10^7\ \text{cm} \cdot \text{s}^{-1}$), the Debye length and mean free path have magnitudes of $L_D \sim 10^{-5}\ \text{cm}$ and $l \sim 3 \times 10^{-2}\ \text{cm}$, respectively. Because $l \gg L_D$, the diode theory should be applied while the diffusion currents are negligible.

The *I–V* characteristics of the studied heterostructures can be described using the framework of the MIS structural model, which obeys the following expression [7]:

$$J = J_s\left(\exp\left(\frac{qV}{n_1 \cdot kT}\right) - \exp\left(-\frac{qV}{n_2 \cdot kT}\right)\right). \tag{3.3}$$

Here, J is the current density, q is the electron charge, V is the applied voltage, k is the Boltzmann constant, T is the temperature, and n_1 and n_2 are the ideality factors, which depend on the charge transport mechanism. J_s is the saturation current density, which has the following form according to thermionic emission–diffusion theory [8]:

$$J_s = A^* T^2 \exp\left(-\frac{qE_t}{kT}\right). \tag{3.4}$$

Here, A^* is the effective Richardson constant, described by the following expression [9]:

$$A^* = \frac{2q\mu}{d \cdot S} V \left(\frac{2\pi m^* k}{h^2 T^3}\right)^{1/2} \tag{3.5}$$

where d is the thickness of the dielectric, S is the contact area, E_t is the energy of a monoenergetic level in the bandgap of the dielectric, m^* is the carrier's effective mass, h is Planck's constant, and μ is the carrier's mobility.

According to equation (3.3), linear extrapolation of the I–V characteristic in $\ln(J/T^2) - 1/T$ coordinates to $1/T \to 0$ gives the effective Richardson constant A^*. The slope of this linear function allows us to determine the activation energy of conductivity.

The ideality factor is of great technical interest and can be determined by graphical differentiation of a linear section of the I–V curve in $\ln J$–V coordinates through the following formula:

$$n = \frac{q}{kT} \frac{\mathrm{d}V}{\mathrm{d}(\ln J)}. \tag{3.6}$$

Forward bias

Usually, in the bandgap and at the dielectric–semiconductor interface, there are energy levels attributed to defects and the mismatch between the crystal lattices of the two materials. For charge transport over monoenergetic levels in a dielectric, the ideality factor can be approximately written as [8]:

$$n = \frac{d}{l_t}. \tag{3.7}$$

Here, d is the thickness of the dielectric, $l_t = N_t^{-1/3}$ is the average separation between trap centers in the dielectric, and N_t is their concentration. Thus, the trap concentration in the bandgap of a dielectric can be estimated using the ideality factor obtained from the experimental I–V characteristic.

We determined the effective Richardson constant and activation energy of conductivity from the temperature dependence of current density for the studied heterostructures. A typical temperature dependence of conductivity in Arrhenius coordinates is shown in figure 3.5.

Two temperature intervals associated with different activation processes correspond to two linear sections with different slopes in figure 3.5. The activation energies for these processes are $E_{a1} = 0.23\,\text{eV}$ and $E_{a2} = 0.05\,\text{eV}$.

By extrapolating the linear part of the I–V curve in $\ln J$–V coordinates to $V \to 0$ (see figure 3.4), we obtain J_s. Using equation (3.4), we can now determine the energy of traps E_t below the bottom of the conduction band. Results obtained for a $LiNbO_3$

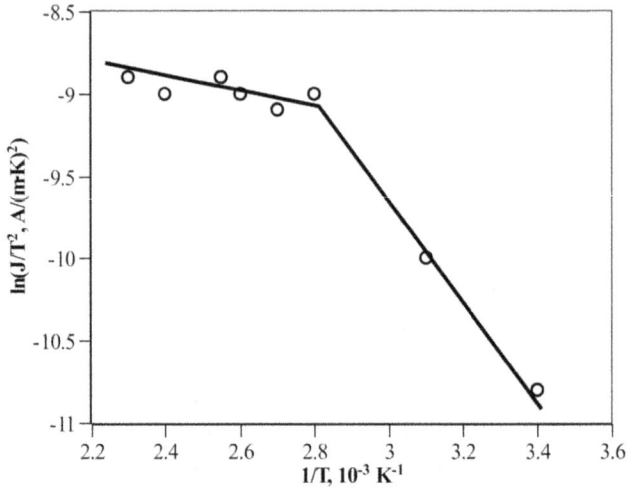

Figure 3.5. Typical temperature dependence of conductivity for (001)Si–LiNbO$_3$–Al heterostructures in $\ln(J/T^2) - 1/T$ coordinates under forward bias ('–' at the Al contact).

Table 3.1. Results of the analysis of I–V characteristics of p-type (001)Si–LiNbO$_3$–Al heterostructures fabricated by RFMS on p-type Si substrates*.

Substrate	Resistivity of LiNbO$_3$ film, ρ ($\Omega\cdot$cm)	Ideality factor, n	Effective Richardson constant A^*, $(A/m^2 \cdot K^2)$	Trap concentration in LiNbO$_3$, N_t (cm^{-3})	Energy position of traps in the bandgap of LiNbO$_3$, E_t (eV)	Activation energy of conductivity E_a (eV)	The Poole–Frenkel coefficients ratio $\beta^{ex}_{P-F}/\beta^{theor}_{P-F}$
p-Si(001)	1×10^9	51(386)	0.1	2.4×10^{17} (2.4×10^{17})	0.27	0.23(0.1)	1.08

*Parameters derived from the reverse branches of I–V characteristics are indicated in brackets.

film with a thickness of $d = 1 \times 10^{-6}$ m and dielectric permittivity $\varepsilon = 28$ are given in table 3.1.

Figure 3.6 presents both the experimental and theoretical I–V characteristics, calculated using equation (3.3) based on the parameters listed in table 3.1. As shown in figure 3.6, the theoretical model described by equation (3.3) is in good agreement with the experimental data for the (001)Si–LiNbO$_3$–Al heterostructures.

One possible conduction mechanism involving centers of localized charge (CLC) is hopping conductivity. In this scenario, the temperature dependence of current density is defined by Mott's variable-range hopping conductivity model [10]:

$$J(T) = \frac{J_0}{T^{1/2}} \exp(-(T_0/T)^{1/4}). \tag{3.8}$$

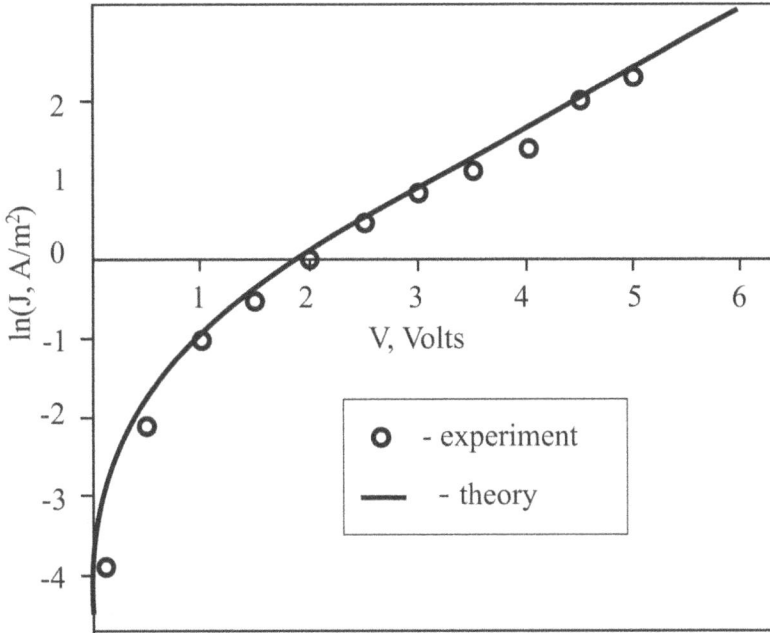

Figure 3.6. Experimental and theoretical I–V characteristics of (001)Si–LiNbO$_3$–Al heterostructures subjected to forward bias (dots—experiment, solid line—theoretical curve, calculated using equation (3.3)).

Here, T_0 is a parameter given by:

$$T_0 = \frac{\lambda}{kN(E_F)a^3} \tag{3.9}$$

where $N(E_F)$ is the energy density of localized states near the Fermi level, k is the Boltzmann constant, a is the localization radius, and λ is a dimensionless parameter (typically $\lambda \sim 16$ [10]). We have demonstrated that the experimental I–V curve is perfectly linear in Mott's coordinates [10] (see figure 3.7), allowing us to obtain the parameter T_0 from the slope of this line.

Within this model, the average hopping distance R of carriers over localized sites near the Fermi level at a temperature T is defined by [11]:

$$R = \frac{3}{8}a\left(\frac{T_0}{T}\right)^{1/4}. \tag{3.10}$$

The energy range of the localized states in this case is described by:

$$\Delta E = \frac{3}{2\pi N(E_F)R^3}. \tag{3.11}$$

The concentration of CLCs can be defined as follows:

$$N_t = N(E_F)\Delta E. \tag{3.12}$$

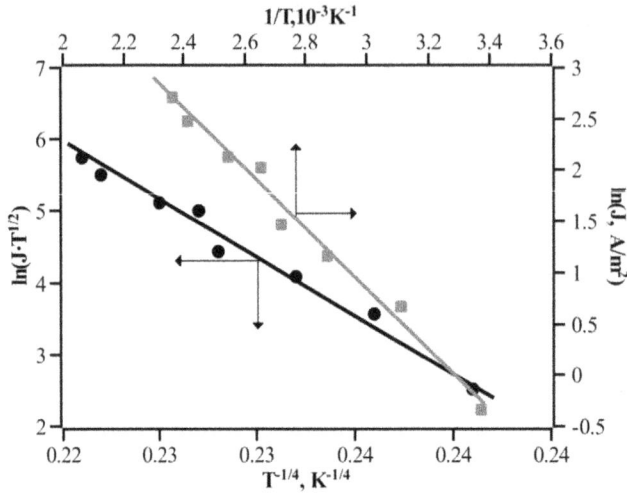

Figure 3.7. Temperature dependence of (001)Si–LiNbO$_3$–Al heterostructure conductivity in Mott's coordinates and Arrhenius coordinates at a forward bias of $V = 2$ V.

For the studied heterostructures, we obtained the following results [12]: $\Delta E = 0.5$ eV, $N(E_F) = 5 \times 10^{17}$ eV^{-1} cm^{-3}, $N_t = 2.5 \times 10^{17}$ cm^{-3}.

Reverse bias

Under reverse bias, I–V characteristics should saturate according to Schottky diode theory [2]. However, this saturation was not observed in our case. This can be explained using the Poole–Frenkel effect (thermal ionization under electric fields), which is similar to Schottky emission but occurs within the bulk of a material (see chapter 1). In this scenario, I–V characteristics can be described in terms of applied voltage by an expression similar to equation (1.15):

$$J = J_s \exp\left(\frac{\Delta\varphi}{kT}\right). \tag{3.13}$$

Here, J_s is the saturation current, and the decrease in the potential barrier (or the ionization energy of a single Coulomb potential well) is described by the Poole–Frenkel formula [13]:

$$\Delta\varphi = \beta_{P-F} V^{1/2}. \tag{3.14}$$

Here, V is an applied voltage and β_{P-F} is the Poole–Frenkel coefficient:

$$\beta_{P-F} = \left(\frac{q^3}{\pi\varepsilon\varepsilon_0 d}\right)^{1/2}. \tag{3.15}$$

It follows from equation (3.15) that I–V characteristics should be linear in $\ln(J) - \sqrt{V}$ coordinates. Indeed, the reverse I–V characteristics of p-type (001)Si–LiNbO$_3$–Al heterostructures are linear in these coordinates at relatively high

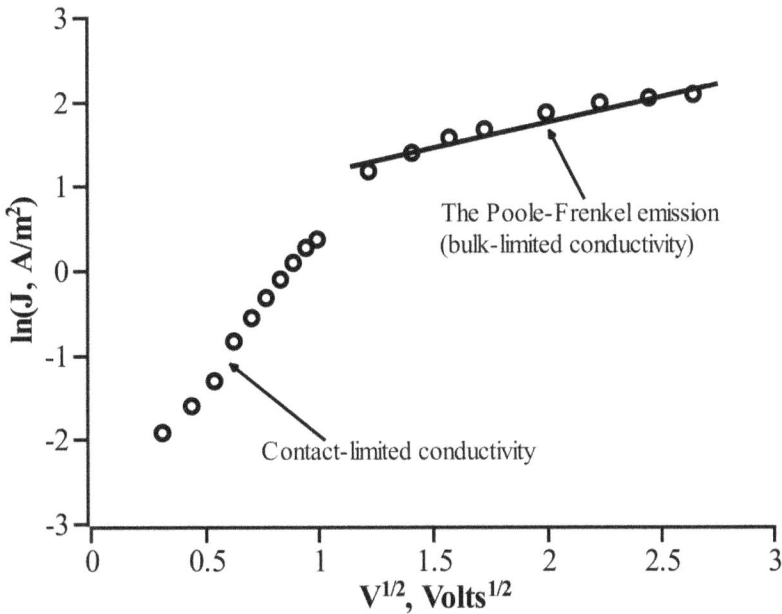

Figure 3.8. I–V characteristics of p-type (001)Si–LiNbO$_3$–Al heterostructures in $\ln(J) - \sqrt{V}$ coordinates at reverse biases ('+' at the Al contact).

voltages, as shown in figure 3.8. The low-voltage nonlinear section is likely associated with contact-limited conductivity through the Si–LiNbO$_3$ interface.

The experimental Poole–Frenkel coefficient β_{P-F}^{ex} is derived from the slope of this linear section and is very close to the theoretical coefficient β_{P-F}^{theor} calculated using equation (3.15) (see table 3.1). The good agreement between the experimental and theoretical coefficients suggests that Poole–Frenkel emission is the predominant charge transport mechanism under reverse bias in the studied heterostructures.

Our preliminary study has demonstrated that LiNbO$_3$ films fabricated by the RFMS method on p-type (001)Si substrates contain a relatively high concentration of CLCs in their bandgap. Significant leakage currents and the presence of positive oxide charge restrict the use of the C–V method and IS because the heterostructures are in the deep depletion regime and close to the inversion regime. This can be a serious limitation on their practical application.

Given these results, we will conduct all further research based on heterostructures fabricated on n-type Si substrates. Figure 3.9 shows typical C–V characteristics of n-type (001)Si–LiNbO$_3$–Al heterostructures fabricated by RFMS on n-type Si substrates, which are also similar to those of MIS structures.

As shown in figure 3.9, under zero bias, the studied heterostructures are in the accumulation regime. Thus, under forward bias ('+' at the metal electrode), these heterostructures can be analyzed as MIS structures, similar to the I–V analysis used under forward bias. It is important to note that the C–V curves are shifted to the left along the voltage axis relative to the ideal case. This shift indicates the presence of a

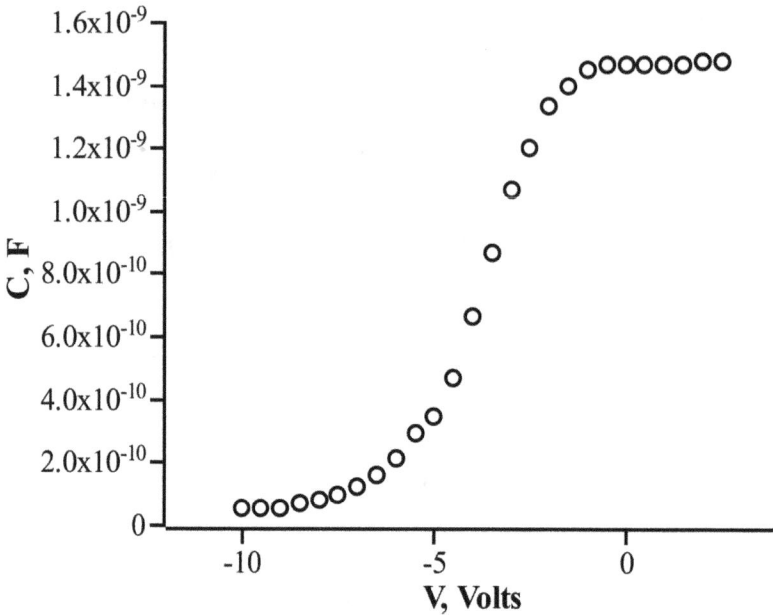

Figure 3.9. Typical high-frequency (1 MHz) C–V characteristics of (001)Si–LiNbO$_3$–Al heterostructures fabricated by the RFMS method on n-type Si substrates.

positive oxide charge in the LiNbO$_3$ film, likely due to a high defect concentration. Using standard C–V analysis [2], we determined the dielectric constant $\varepsilon = 29$ and the effective density of positive charge $Q_{ef} = 3.8 \times 10^{-8}$ C cm^{-2} in the LiNbO$_3$ film. In addition to the effective charge, it is sometimes necessary to determine the total (integral) charge in a dielectric Q_{ox} and its centroid position d_c (center of mass position). For the studied heterostructures, we derived the following parameters: $Q_{ox} = 2.2 \times 10^{-6}$ C cm^{-2}, $d_c = 134$ nm.

Typical I–V characteristics of the studied heterostructures, fabricated by the RFMS method on n-Si substrates, are shown in figure 3.10 and can be described within the framework of MIS structures using equation (3.3), similar to the analysis of heterostructures formed on p-type Si.

Following the procedure described above, we estimated the concentration of CLC in the LiNbO$_3$ films through the ideality factor of the I–V curve at 30 °C using equation (3.7), obtaining a concentration of $N_t = 7 \times 10^{17}$ cm^{-3}. Furthermore, as shown in figure 3.10, the I–V characteristics of the studied heterostructures are temperature independent in the range of $T = 56$ °C – 82 °C. One possible conductivity mechanism is non-activated hopping conductivity. In this framework, the I–V characteristic is described as [14]:

$$J = J_0 \exp(-(E_0/E)^{1/4}). \tag{3.15}$$

Here, E is the applied electric field strength, J_0 is a field-independent constant, and E_0 is the characteristic field defined by [15]:

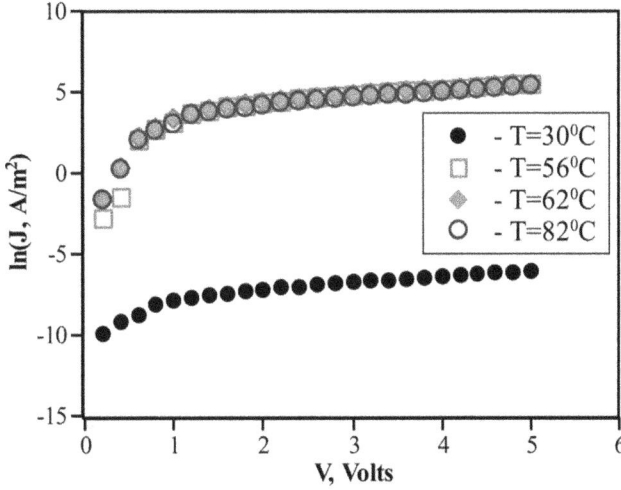

Figure 3.10. Typical *I–V* characteristics of n-type (001)Si–LiNbO$_3$–Al heterostructures fabricated by RFMS and subjected to forward bias ('+' at the Al contact) at different temperatures.

$$E_0 = \frac{\lambda}{D(E)a^4 q} \qquad (3.14)$$

where $D(E)$ is the energy density of localized states near the Fermi level, a is the localization length, and λ is a dimensionless constant, usually equal to $\sim$16 [14]. The energy dispersion within the Fermi level is described by:

$$\Delta E = \frac{3}{2\pi D(E)R^3}. \qquad (3.15)$$

The average hopping distance in the case of the non-activated mechanism is defined as [16]:

$$R = \frac{1}{(D(E) \cdot q \cdot E)^{1/4}}. \qquad (3.16)$$

Thus, the bulk concentration can be determined using the following expression:

$$N_t = D(E)\Delta E. \qquad (3.17)$$

Indeed, the *I–V* curves shown in figure 3.10 are linear in $\ln(J) - E^{-1/4}$ coordinates (figure 3.11), which is in good agreement with equation (3.15) and our previous work [17].

Using the parameter E_0 derived from the slope of the linear part of the *I–V* characteristics in figure 3.11 and equations (3.14)–(3.17), we obtain the following parameters characterizing the hopping conductivity over CLC: $R = 60\,\text{Å}$, $D(E) = 1.6 \times 10^{21}\,\text{eV}^{-1}\,\text{cm}^{-3}$, and $N_t = 2.3 \times 10^{18}\,\text{cm}^{-3}$ [17]. It is important to emphasize that, similar to heterostructures fabricated on p-Si substrates, the studied heterostructures also exhibited a relatively high concentration of CLC in LiNbO$_3$

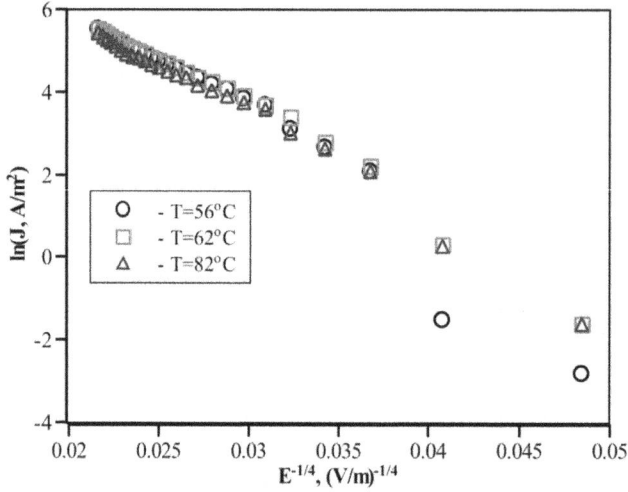

Figure 3.11. *I–V* characteristics of (001)Si–LiNbO$_3$ heterostructures fabricated by the RFMS method in ln(*J*)–*E*$^{-1/4}$ coordinates at different temperatures.

films. Therefore, the formation of this type of CLC is influenced by the sputtering conditions rather than the substrate type. A detailed study of the relationship between the sputtering conditions and the CLC parameters is presented in the following chapter.

3.1.2 Ferroelectric properties of LiNbO$_3$ thin films

As mentioned in chapter 1, the ferroelectric properties of LiNbO$_3$ are crucial for applications in memory units. Ferroelectric materials, characterized by their remnant polarization, exhibit a ferroelectric hysteresis in the dependence of their polarization on the applied electric field (*P–E* loop). This phenomenon was first observed using a method proposed by Sawyer and Tower [18]. The Sawyer–Tower circuit is illustrated in figure 3.12.

Two capacitors, C_F and C_0, connected in series represent the ferroelectric material (LiNbO$_3$ in our case) and a linear integrating capacitor, respectively. Normally, $C_0 \gg C_F$, so almost the entire applied voltage V_i is dropped across C_F. The voltage V_i is applied to the horizontal plates of an oscilloscope, so the horizontal axis on the $P(E)$ graph represents the electric field applied to the ferroelectric film:

$$E(t) = \frac{V_i(t) - V_0(t)}{d} \approx \frac{V_i(t)}{d} \tag{3.18}$$

where d is the film thickness. The voltage applied to C_0 is recorded by the vertical plates of the oscilloscope and is proportional to the polarization of the ferroelectric capacitor C_F:

$$P(t) = \sigma(t) = \frac{C_0 V_0(t)}{S} \tag{3.19}$$

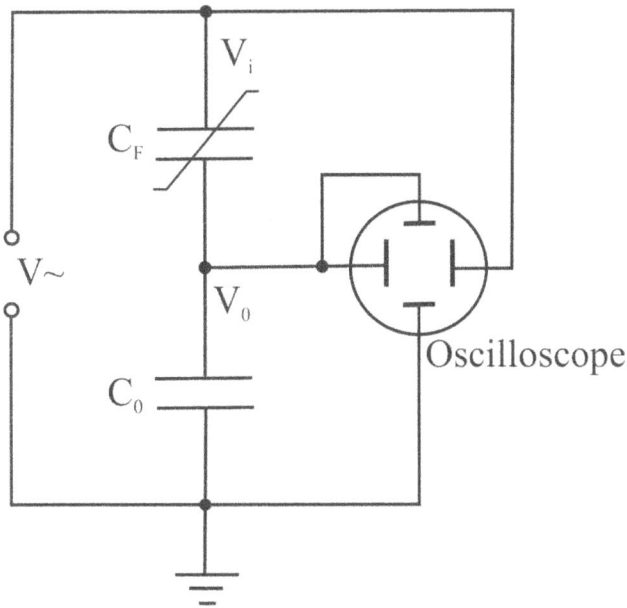

Figure 3.12. The Sawyer–Tower circuit.

where $\sigma(t)$ is the surface charge density on C_F, and S is the electrode area. Thus, the P–E loop is registered on the oscilloscope screen. In fact, equations (3.18) and (3.19) are applied only to the metal–ferroelectric–metal system. For the metal–ferroelectric–semiconductor system, equation (3.18) can be modified as follows:

$$E(t) = \frac{V_i(t) - \phi_s}{d} \tag{3.20}$$

where ϕ_s is the surface potential at the ferroelectric–semiconductor interface. Moreover, because the thin LiNbO$_3$ film is not an ideal insulator, current leakage can occur, leading to changes in $\sigma(t)$.

Figure 3.13 shows the $P(E)$ characteristics of (001)Si–LiNbO$_3$ heterostructures fabricated by the RFMS method without the ion assist effect.

As can be seen in figure 3.13, all P–E loops demonstrate saturation of the remnant polarization around $P_r = 14\,\mu\mathrm{C\,cm}^{-2}$. All loops are asymmetric (shifted to the right along the horizontal axis), and the coercive field depends on the amplitude of the applied electric field (as shown in figure 3.14). This shift can be caused by the presence of built-in fields E_b in the studied films, which also depend on the applied field magnitude E_m.

Several studies [19–21] have persuasively demonstrated that the shift of the P–E curves is triggered by various factors associated not only with the properties of the films but also with the electrical measurements. According to the Sawyer–Tower method, any shift along the horizontal axis, representing the electric field in a sample, corresponds to the presence of a built-in field (DC voltage) in the film, which is added to the applied external electric field.

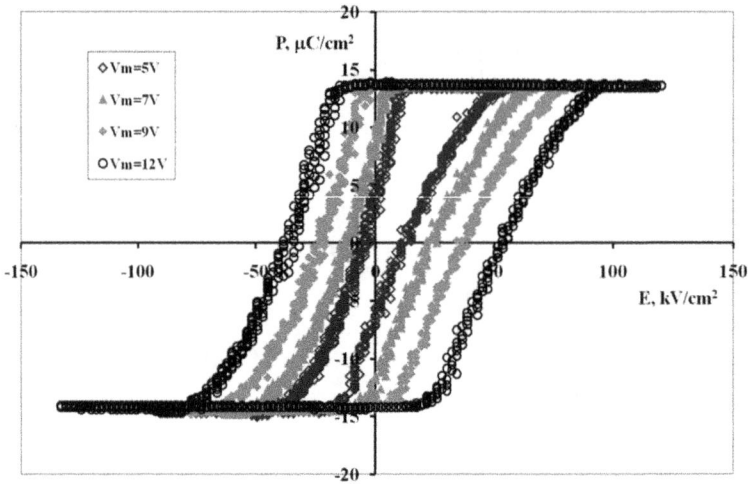

Figure 3.13. Typical P–E hysteresis loops of (001)Si–LiNbO$_3$ heterostructures fabricated by RFMS method without the plasma effect and recorded at different amplitudes of the applied voltage V_m.

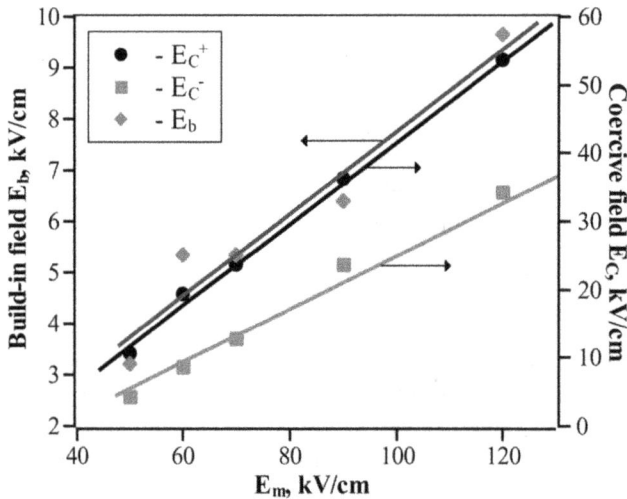

Figure 3.14. Dependence of the coercive field E_c and the built-in field E_b on the magnitude of the applied field E_m for LiNbO$_3$ films fabricated by the RFMS method without the plasma effect. E_c^+ and E_c^- are the coercive fields, measured in the positive and negative electric field range, respectively.

The built-in field can result from the formation of layers with different compositions at the top and bottom surfaces of the ferroelectric films during the growth process. This occurs for technological reasons (temperature gradients and defect concentrations), which can cause P–E loops to shift [19, 20]. A possible reason for this is the gradient of CLC concentration in a film (e.g. oxygen vacancies). To support this argument, some authors have demonstrated that the P–E hysteresis loop shift is significantly affected by inhomogeneity in the charge distribution in the film [22, 23]. By contrast, some investigators have proposed that this shift is due to the presence of dipoles in the film. These dipoles, formed by charged vacancies, contribute to the depolarization process and can block switching in ferroelectric films [23]. However, other researchers have argued that the observed shift in the *P–E* curves is not associated with dipole orientation but arises from electron trapping by vacancies [21, 24]. The distribution of dipole ensembles changes under an external electric field, modulating potential wells for the trapped electrons. In other words, polarization drives charge capture. This approach is consistent with figure 3.14, which shows a strong linear dependence of the shift in *P–E* loops on the amplitude of the applied field and a slope that agrees with [24].

Thus, it is reasonable to assume that oxygen vacancies or other charged defects at the interfaces change their distribution during the growth process, forming space charge that influences the electrical properties and, specifically, the polarization of $LiNbO_3$ films.

Earlier in this chapter, based on *C–V* analysis, we determined the total (integral) charge in a dielectric Q_{ox} and its centroid position, located at a distance d_c from the surface of the $LiNbO_3$ film (see figure 3.15).

We need to clarify the classification of charge in a dielectric. According to [25], oxide charges can be divided into four types: fixed oxide charge (Q_f), mobile ionic charge (Q_{mi}), interface-trapped charge (Q_{it}), and oxide-trapped charge (Q_{ot}). The main properties of these charges are summarized in table 3.2.

Based on the results of *C–V* analysis (see the discussion above and [26]), we can conclude that the oxide charge (Q_{ox}) in our heterostructures is positive and independent of the substrate type. Evidently, Q_{ox} is not the interface-trapped charge (Q_{it}), whose sign and value are greatly influenced by the substrate type. At this stage, we can only deduce that Q_{ox} in the studied $LiNbO_3$ films is a sum: $Q_{ox} = Q_{ot} + Q_f + Q_{mi}$. Distinguishing the contribution of each charge type is complex and requires

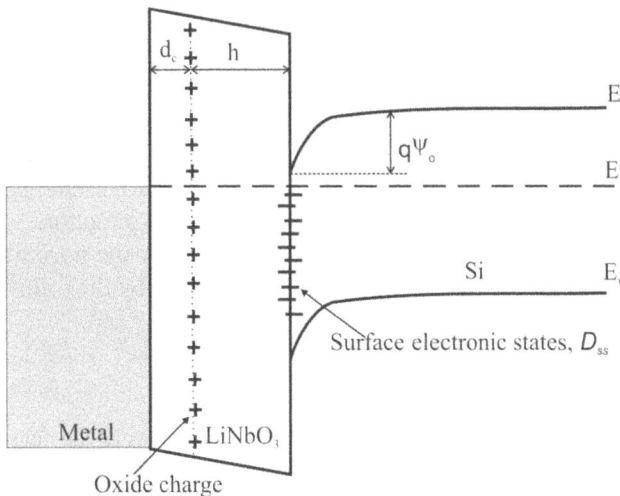

Figure 3.15. Schematic band diagram of a metal–$LiNbO_3$–Si heterostructure with a positive oxide charge distributed at a distance d_c from the film surface.

Table 3.2. Characteristics of different types of oxide charge.

Property	Types of charge			
	Interface-trapped charge (Q_{it})	Oxide-trapped charge (Q_{ot})	Fixes oxide charge (Q_f)	Mobile ionic charge (Q_{mi})
Location	At the semiconductor–oxide interface	Within the bulk of an oxide	Near the semiconductor–oxide interface	Within the bulk of an oxide
Charges	Positive/negative	Positive/negative	Positive	Positive
Sources	Structural defects, metal impurities	Ionizing radiation, injection	Structural defects	Ionic impurities (Na^+, K^+, Li^+, ...)
Dependence on applied voltage V_G	Depends on V_G	Does not depend on V_G	Does not depend on V_G	Does not depend on V_G
Charging state	Charged and discharged by applying V_G	Charged and discharged under specific conditions	Fixed	Fixed (immobilized) at temperatures below 390 °C

separate investigation. For example, thermal annealing of the studied heterostructures could help clarify this issue.

The trapped charge apparently forms the space charge in the LiNbO$_3$ film, causing the shift in *P–E* loops. A model developed in [27] introduces thin, non-switchable layers named 'passive layers' that form near the electrodes and have properties differing from those in the bulk material. The fact that these layers have finite conductivity and contain space charge is crucial in the switching of ferroelectric films. Obviously, the space charge asymmetrically trapped in LiNbO$_3$ films leads to non-symmetrical switching. Primarily, this charge is trapped by the states distributed near the electrode surface. In this case, the studied film can be modeled as an ideal ferroelectric capacitor and another capacitor (a 'surface capacitor'), whose thickness is equal to the distance from the electrode surface to the centroid of the trapped charge. The two capacitors are connected in series (see figure 3.16). The charge of the 'surface' capacitor equals the total charge sitting between the two capacitors.

The total voltage drop and the continuity equation for the normal component of the electric displacement *D* in this structure are described by the following equations:

$$V = E_f h + E_d(t)d_c$$
$$D_f - \sigma = \varepsilon_0 \varepsilon_d E_d(t).$$

(3.21)

Here, d_c and h are the thicknesses of the passive layer and the ferroelectric layer, respectively. E_f and E_d are the electric fields in the ferroelectric capacitor and passive layer, respectively; ε_d is the dielectric constant of the passive layer; and σ is the surface trapped charge density at the interface between the passive and ferroelectric layers. Combining both equations, we have:

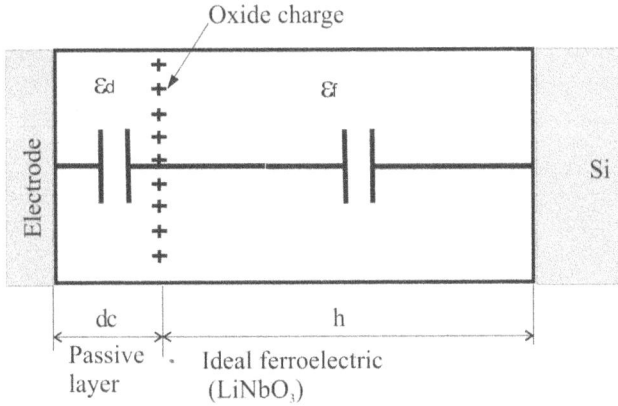

Figure 3.16. Schematic pattern of an equivalent circuit for a ferroelectric film with a passive layer, triggered by positive charge, present at a depth of d_c.

$$D_f(t) - \sigma = \frac{\varepsilon_d \varepsilon_0 (V - E_f h)}{d_c}. \tag{3.22}$$

In [27], the offset voltage V_{off} is the difference in applied voltage that produces the same switching within a structure (i.e. the same values of E_f and D_f are attained) at both charged and neutral 'passive layer/ferroelectric' interfaces. In this case, it follows from equation (3.22) that:

$$V_{\text{off}} = -\frac{\sigma d_c}{\varepsilon_d \varepsilon_0}. \tag{3.23}$$

Thus, the offset voltage directly depends on σ and d_c. In [27], the following expression for the dielectric constant of the passive layer was proposed:

$$\varepsilon_d = -\frac{\varepsilon_f d_{\text{tot}} \varepsilon_e}{h\varepsilon_e - \varepsilon_f d_{\text{tot}} - 2\varepsilon_f h}. \tag{3.24}$$

Here, ε_e and d_{tot} are the effective dielectric constant and total thickness of the film ($d_{\text{tot}} = d_c + h$), respectively, and ε_f is the dielectric constant of the ferroelectric layer. Analyzing C–V characteristics, we derive the following parameter values required for the estimation of σ: $\varepsilon_f = 28$, $\varepsilon_e = 61$, and $d_c = 134$ nm. Considering that $d_{\text{tot}} = 1$ µm and the maximum applied voltage is 12 V for the studied film, we used equations (3.24) and (3.23) to estimate the trapped charge density in the studied film, obtaining $\sigma = 4.4 \times 10^{-6}$ C cm^{-2}. This value is in good agreement with the oxide charge determined earlier from C–V analysis ($Q_{ox} = 2.2 \times 10^{-6}$ C cm^{-2}).

Therefore, in the studied LiNbO$_3$ films fabricated by RFMS, the most probable source of the shift in the P–E loops is the oxide-trapped charge. This charge has a density of $\sigma = 4.4 \times 10^{-6}$ C cm^{-2} and generates a built-in field in the film.

P–E hysteresis loops for LiNbO3 films fabricated by the IBS method were analyzed in our earlier work [28] and were analogous to those recorded for the films deposited by RFMS (see figure 3.17).

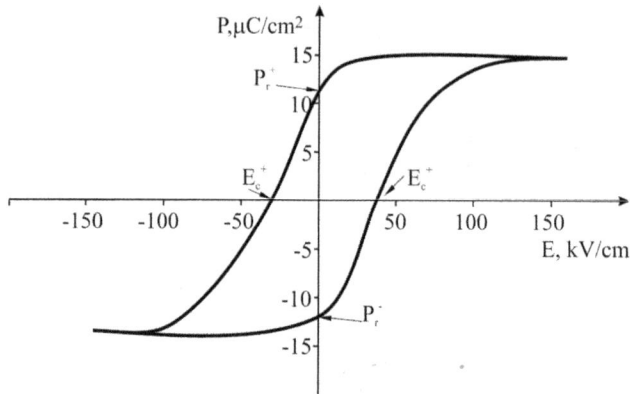

Figure 3.17. Typical hysteresis *P–E* loops of LiNbO$_3$ films, fabricated by IBS method.

Table 3.3. Parameters of the experimental *P–E* loops of LiNbO$_3$ films deposited onto (001)Si substrates by the RFMS and IBS methods.

Deposition method	Degree of crystallinity	P_r^+ (μC cm^{-2})	P_r^- (μC cm^{-2})	E_c^+ (kV cm^{-1})	E_c^- (kV cm^{-1})	Built-in field, E_b (kV cm^{-1})
IBS	Polycrystalline with	11.2	−12.4	+29	−29	0
RFMS (without the plasma effect)	random grain orientation	13.7	−14.2	+53.6	−34.3	9.7

The parameters of *P–E* loops for films deposited by both RFMS and IBS methods are listed in table 3.3.

It is worth noting that the remnant polarization P_r derived from the *P–E* loops (see table 3.3) is considerably less than that of bulk lithium niobate ($P_r = 71$ μC cm^{-2}) and is close to the values reported in [3, 29]. This lower performance can be attributed to the arbitrary orientation of polycrystalline grains in films deposited without the ion assist effect (see chapter 2), resulting in reduced domain alignment in an external electric field. We will revisit this issue in chapter 4 by examining the effects of the sputtering conditions on the ferroelectric properties of LiNbO$_3$ films.

Regarding films deposited by the IBS method, we revealed Fermi level pinning due to the high density of surface states at the Si–LiNbO$_3$ interface, which does not allow for the correct estimation of Q_{ef} in these films [12].

To sum up, based on the results discussed above, we determined the main electrical parameters of LiNbO$_3$ films fabricated by IBS and RFMS methods *without the plasma effect*, which are presented in table 3.4.

These tables provide a comprehensive summary of the electrical properties and *P–E* loop characteristics of LiNbO$_3$ films deposited by both methods, highlighting key

Table 3.4. Basic electrical properties of of LiNbO$_3$ films in Si–LiNbO3 heterostructures.

Substrate	p-Si	n-Si	n-Si
Deposition technique	RFMS (without the plasma effect)		IBS
Dielectric constant ε	28	29	29
Resistivity, ρ ($\Omega \cdot$ cm)	1×10^9	1×10^9	3×10^9
Remnant polarization, P_r (μC cm^{-2})	—	14.0	11.7
Coercive field, E_c (kV cm^{-1})	—	44.0	29.0
Effective charge in LiNbO$_3$ film, Q_{ef} (C cm^{-2})	3.8×10^{-8}	3.8×10^{-8}	—
Concentration of CLC, N_t (cm^{-3})	2.5×10^{17}	7.1×10^{17}	—
Energy of CLC E_t (eV)	0.27	—	—

differences and commonalities. The results underscore the impact of deposition methods on the ferroelectric behavior and overall electrical properties of the films, providing a basis for further investigation into ways of optimizing fabrication techniques.

3.2 Conduction mechanisms in (001)Si–LiNbO$_3$ heterostructures

As pointed out in chapter 1, the electrical properties of LiNbO$_3$-based heterostructures, which are essential components of integrated electronic devices, are significantly affected by charge transport.

It is important to note that many applications of thin films rely on the application of an external electric field to a ferroelectric capacitor, which can cause leakage currents. If these currents are significant, they can prevent the occurrence of polarization switching. Solutions to reduce these leakage currents can only be found through a deep understanding of conductivity mechanisms and the influence of current on the parameters of films and heterostructures. Specifically, leakage currents can influence the shape of the $P–E$ loop, as this curve is derived through the integration of charge released during the polarization reversal process. High leakage currents suppress hysteresis, masking the ferroelectric properties of the studied sample.

Conductivity mechanisms are influenced by microstructure, defects, composition uniformity, external fields, temperature, and charge states at the interfaces. Most perovskite ferroelectrics can be considered semiconductors with low carrier mobility [30]. When a semiconductor is connected to a metal (for example, in the formation of metallic contacts), a Schottky barrier is formed with a space-charge region or a depletion region, leading to band bending at the interface. Experimental results demonstrating a decrease in coercive field with film thickness agree with models of domain wall pinning and its formation in depletion regions [31]. Some authors argue [32, 33] that oxygen vacancies that accumulate at the interface between the electrode and the film play a crucial role in the polarization reversal process. Specifically, regions with reduced oxygen can grow toward the bulk, effectively shielding the film from the applied voltage and causing a loss of polarization.

Given these facts, there is no doubt that studying the conduction mechanisms in (001)Si–LiNbO$_3$ heterostructures is a powerful tool for the detailed examination of their electrical properties. For a detailed investigation of conduction mechanisms in (001)Si–LiNbO$_3$ heterostructures, single-phase LiNbO$_3$ films were deposited by the RFMS and IBS methods onto n-type Si substrates according to the optimal sputtering regimes developed in chapter 2. Below, we present our analysis of DC conductivity in heterostructures fabricated by the IBS method, whereas the conductivity of heterostructures synthesized by the RFMS technique will be discussed in the section on their band diagrams.

According to the results of C–V analysis reported in the previous sections of this chapter, the studied (001)Si–LiNbO$_3$–Al heterostructures are in the accumulation regime at positive biases ('+' at the Al contact). This means that a sufficient quantity of electrons is supplied to the LiNbO$_3$ layer by the silicon substrate. Considering this and the fact that the resistivity of the substrate in our case is significantly less than that of LiNbO$_3$ ($\rho = 3 \times 10^9 \, \Omega \cdot$ cm—see table 3.4 and [26]), the studied heterostructures can be considered as a metal–insulator–metal system.

The I–V characteristics of the studied heterostructures can be described by a special case of equation (3.3) [7]:

$$J = J_s\left(\exp\left(\frac{qV}{n \cdot kT}\right) - 1\right). \tag{3.25}$$

Here, J is the current density, q is the electron charge, V is the applied voltage, k is the Boltzmann constant, T is the temperature, n is the ideality factor, and J_s is the saturation current density.

I–V characteristics were studied in the temperature range of 90–300 K [34] and are shown in figure 3.18.

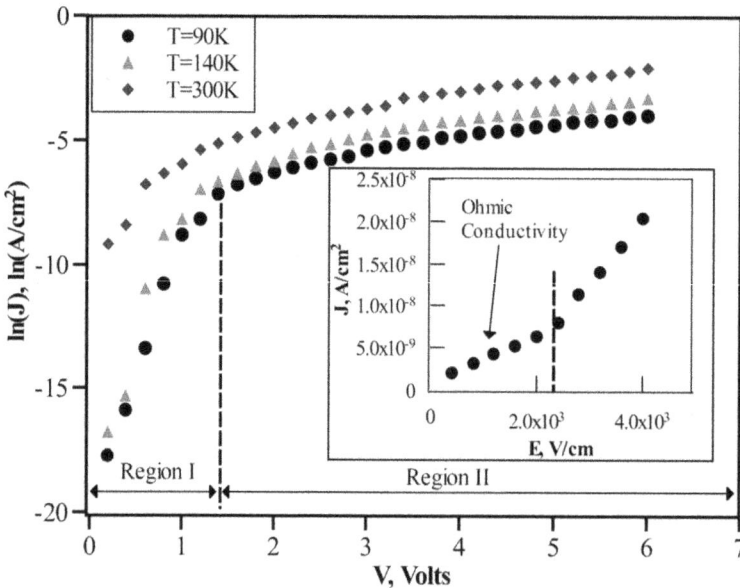

Figure 3.18. Typical I–V characteristics of (001)Si–LiNbO$_3$ heterostructures at various temperatures.

Two sections can be identified in the I–V curves (see figure 3.18): the first section, which we will call the region of low and average voltage ($0 < V < 1.5$ V) or, in terms of electric fields, $0 < E < 30\,\mathrm{kV\,cm^{-1}}$, apparently results from contact phenomena and corresponds to rapidly increasing current. The second section is associated with high voltage ($V > 1.5$ V or $E > 30\,\mathrm{kV\,cm^{-1}}$) and is influenced by the bulk of the film.

Figure 3.19 illustrates the temperature dependence of current in Arrhenius coordinates for voltages corresponding to different sections of the I–V curves. The activation energies of conductivity E_a, derived from the slopes of these graphs, are also indicated in figure 3.19.

In the low-voltage range ($0 < V < 0.1$ V), the I–V characteristics are linear ($J \propto V$), which is evidence of ohmic conductivity (see the inset in figure 3.18). The studied LiNbO$_3$ films are polycrystalline, containing, as a rule, a high concentration of CLC (traps) in the bandgap. In this case, ohmic conductivity can be realized by electron hopping over CLCs, which agrees with the results reported in [35] for hopping conductivity. The I–V characteristics within this mechanism are described by the expression [7]:

$$J = \sigma_0 \frac{V}{d} \exp\left(-\frac{E_a}{kT}\right). \tag{3.26}$$

Here, σ_0 is a constant and E_a is the activation energy of conductivity, which is derived from the slope of the graph $\ln(J)$–$1/T$ (see figure 3.19). The value of E_a is 0.14 eV, which is close to the value obtained in [35]. An activation energy of $E_a = 0.07$ eV (see figure 3.19) is also close to the value determined using the

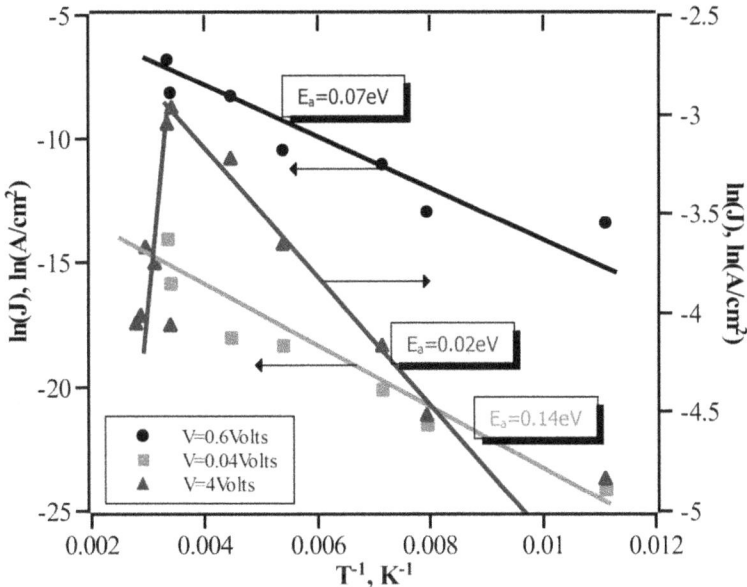

Figure 3.19. Temperature dependence of the conductivity of (001)Si–LiNbO$_3$ heterostructures at low ($V = 0.04$ V), average ($V = 0.6$ V), and high ($V = 4$ V) voltages.

framework of electronic hopping conductivity [36]. For a detailed analysis of conduction mechanisms, let us consider each voltage and temperature range on the I–V curve.

3.2.1 Region of average electric fields in the temperature interval of $T = 90$–140 K

The extremely weak temperature dependence of conductivity in the range of average electric fields ($3 < E < 30$ kV cm^{-1}) at temperatures of $T = 90$–140 K (see figure 3.19) can be attributed to Fowler–Nordheim tunneling. In this case, the current density follows equation (1.10) (see chapter 1). According to equation (1.10), the I–V characteristic should be linear in the $\ln(J/E^2)$–$1/E$ coordinates (Fowler–Nordheim coordinates), and from the slope of this linear dependence, the potential barrier height ϕ can be derived using equation (1.11) (see figure 3.20).

Tunneling through thick films is improbable, but electrons can tunnel through the potential barrier between the metal electrode and the nearest trap centers, as demonstrated in chapter 1 (see figure 1.17). Thus, if ϕ is the average potential barrier height, the energy position of traps in the bandgap E_g of LiNbO$_3$ with respect to the conduction band bottom can be found from $E_t = E_g/2 - 2\phi$. In our case, we obtained the following value: $E_t = 1.76$ eV.

3.2.2 Region of average electric fields in the temperature range of $T = 140$–300 K

I–V characteristics in this interval can be described using the framework of Richardson–Schottky emission as given by the Simmons formula (1.8) (see chapter 1). In this case, the I–V characteristics should be linear in Simmons coordinates ($\ln(J/(ET^{3/2}))$ vs. $E^{1/2}$) with a slope of β/kT (the β coefficient is given by equation (1.9) from chapter 1). Moreover, the slope of the temperature dependence of the Richardson constant in Arrhenius coordinates ($\ln(A^*)$ vs. $1/T$) provides the potential barrier height φ_0.

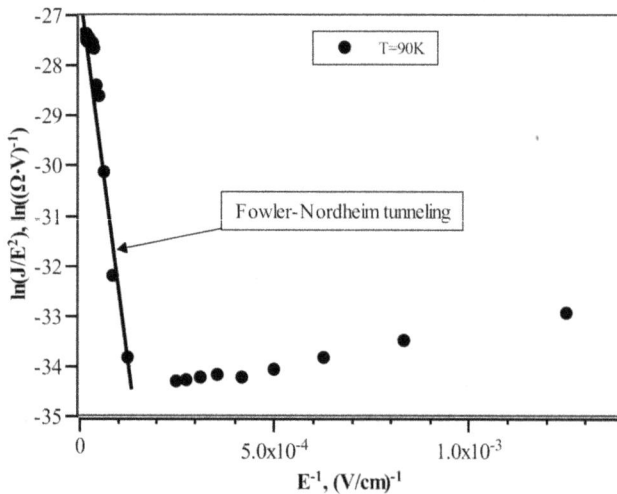

Figure 3.20. I–V characteristics of the (001)Si–LiNbO$_3$ heterostructures in Fowler–Nordheim coordinates. The rectilinear region corresponds to Fowler–Nordheim tunneling.

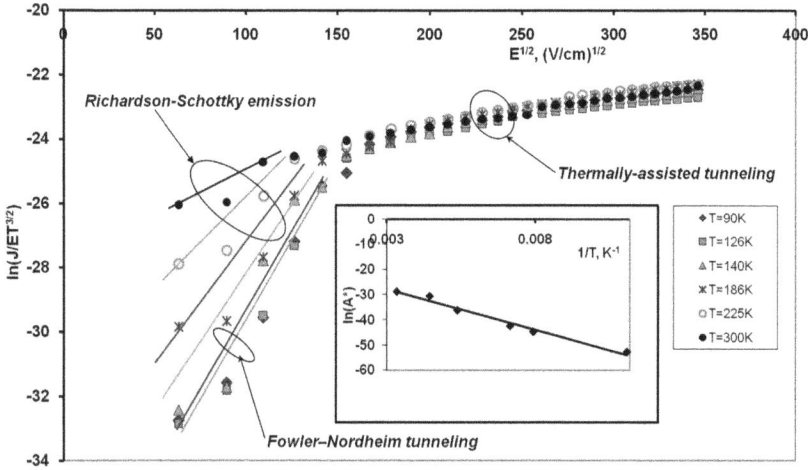

Figure 3.21. *I–V* characteristics of the (001)Si–LiNbO$_3$ heterostructures in Simmons coordinates at various temperatures. The temperature dependence of the pre-exponential factor in expression (1.8) is given in the insert.

In fact, the *I–V* characteristics of the studied heterostructures in the field range $2 < E < 30\,\text{kV cm}^{-1}$ are linear in Simmons coordinates, as shown in figure 3.21.

Analyzing these graphs yields a barrier height value of $\varphi_0 = 0.26\,\text{eV}$.

3.2.3 Region of high electric fields

As noted above, in the range of high electric fields ($E > 30\,\text{kV cm}^{-1}$), the conductivity of the studied heterostructures is influenced by the bulk properties of LiNbO$_3$ films. The conductivity of polycrystalline LiNbO$_3$ films is more affected by the properties of grain boundaries than by their volume [17]. The formation of potential barriers at the interfaces is one of the key factors influencing the electrical properties of polycrystalline films and is typically modeled by a double barrier, similar to the Schottky barrier (figure 3.22). Such heterostructures are complex to describe due to the variety of conduction mechanisms involved, some of which are shown in figure 3.22.

Within the framework of the thermally assisted tunneling mechanism through the potential barrier at the grain boundaries, the values of J_s and the factor n in equation (3.25) are given by [37]:

$$J_s = \frac{A \cdot T}{k}\sqrt{E_{00}\pi}\sqrt{\frac{q\varphi_b}{\cosh((E_{00}/kT)^2)}}\exp\left(-\frac{q\varphi_b}{E_{00}\coth(E_{00}/kT)}\right) \tag{3.27}$$

$$n = \frac{T_c}{T}\left(1 + \frac{d}{L}\right). \tag{3.28}$$

Here, A is the Richardson constant; φ_b and L represent the effective potential barrier height and the width of the depletion area at the grain boundaries, respectively; and

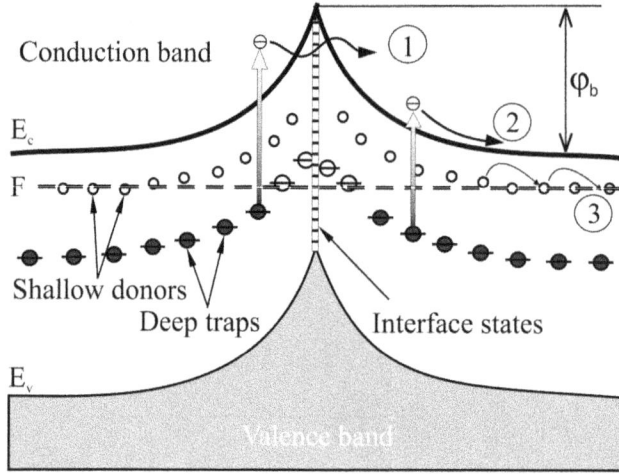

Figure 3.22. Energy diagram of the grain interface in a polycrystal and some possible charge transport mechanisms. Here, φ_b is the height of the potential barrier and F is the Fermi level. Label ① marks thermally assisted tunneling through a potential barrier, ② marks carrier emission from the deep levels, and ③ marks the hopping conduction mechanism.

d is the dielectric layer thickness. The parameters E_{oo} and T_c are defined by the following formulas:

$$E_{oo} = \frac{qh}{4\pi}\sqrt{\frac{N_d}{m^*\varepsilon\varepsilon_0}} \tag{3.29}$$

$$T_c = \frac{E_{oo}}{k}\coth\left(\frac{E_{oo}}{kT}\right). \tag{3.30}$$

Here, m^* denotes the carrier effective mass; ε and ε_0 are the dielectric permittivity of the material and the electric constant, respectively; and N_d is the concentration of ionized donors in the layer. The parameter E_{oo} can be used to estimate the contribution of thermally assisted tunneling to conductivity. Specifically, if $E_{oo}/kT \sim 1$, the tunneling component prevails, whereas if $E_{oo}/kT << 1$, thermionic emission dominates. The width of the potential barrier and the concentration of charge states N_{is} at the grain boundaries can be estimated using the following expressions [37]:

$$\varphi_b = \frac{qN_dL^2}{2\varepsilon\varepsilon_0}\left(1 - \frac{2L}{3d_g}\right) \tag{3.31}$$

$$N_{is} = \sqrt{\frac{2N_d\varepsilon\varepsilon_0\varphi_b}{q}}. \tag{3.32}$$

Table 3.5. Results of our $I–V$ analysis of (001)Si–LiNbO$_3$ heterostructures fabricated by the IBS method.

Average grain size d_g (nm)	Depletion zone's width at the intergranular interface, L (nm)	Potential barrier's effective height, φ_b (eV)	Concentration of ionized donors, N_d (cm^{-3})	Density of states, N_{is} (cm^{-2})
50	9	0.7	3×10^{19}	2.5×10^{13}

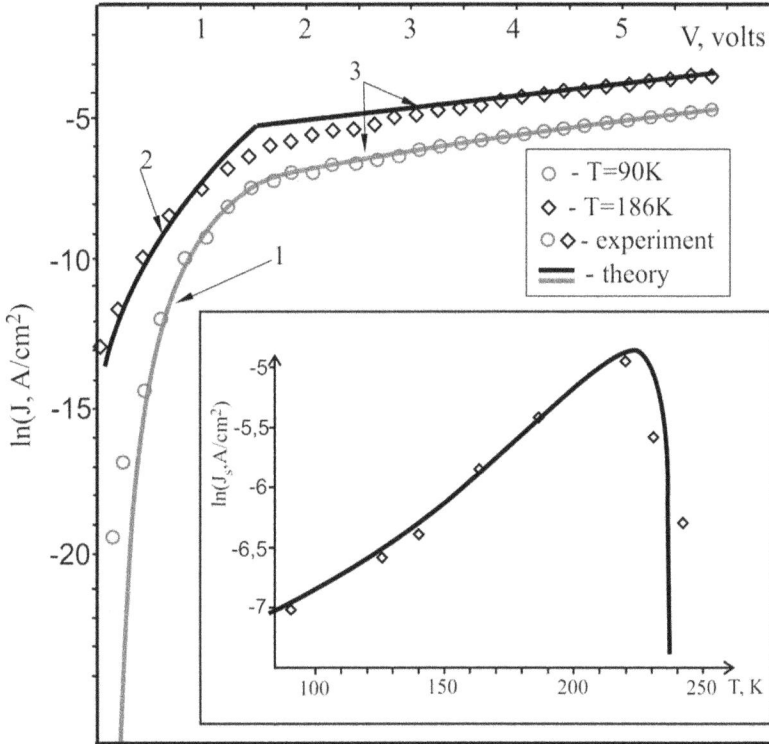

Figure 3.23. Typical $I–V$ characteristics of the (001)Si–LiNbO$_3$ heterostructures at temperatures of 90 K and 186 K. Data labeled with '1' is calculated from equation (1.10) (see chapter 1) at $T = 90$ K, the curve marked with '2' results from equation (1.8) (see chapter 1) at $T = 186$ K, and lines labeled '3' are calculated from equations (3.25) and (3.27)–(3.28). The insert shows the temperature dependence on the pre-exponential factor in equation (3.27).

Here, d_g is the average grain size. By determining J_s and n from the experimental $I–V$ characteristics, we solved the system of equations (3.27), (3.28), and (3.31) for φ_b, L, and N_d, and the results are presented in table 3.5.

The bulk concentration of traps recalculated using N_{is} is relatively high, with a value of $N_t = 9.0 \times 10^{19}$ cm^{-3}. As seen in figure 3.23, there is good agreement between the experimental $I–V$ curve and the theoretical one, calculated from equations (3.25), (1.11), and (3.27)–(3.32) [34]. The temperature dependence of the pre-exponential factor J_s in equation (3.27) is shown in the inset of figure 3.23.

The section with a negative temperature coefficient in the Arrhenius graph (see the inset in figure 3.19) is particularly interesting. This can be explained by the fact that grain boundaries significantly affect the conductivity of LiNbO$_3$ films at high electric fields. The charge states that accumulate at the grain boundaries in LiNbO$_3$ films trap electrons from the bulk of the grains, leading to the formation of depletion areas near the grain boundaries and barriers controlled by electronic traps. Electrons overcoming these barriers penetrate the intergranular layers, initially filling the traps and then forming space charges within the grain boundary areas. These space charges can overcome the potential barrier at the interface and move from the states to the conduction zone, creating current. This electronic current increases with temperature as the kinetic energy of electrons rises. Consequently, the number of space charges in the intergrain areas declines, and the electronic states are filled by carriers from deeper trap centers. This process determines the nature of the near-grain layers that carriers must overcome to maintain conductivity. Therefore, the barrier height that charge carriers must overcome increases with temperature, leading to an increase in current.

Summarizing our analysis of the I–V characteristics presented above, the conduction mechanisms in the studied heterostructures fabricated by the IBS method are schematically shown in figure 3.24 [34].

3.3 Band diagram of the Si–LiNbO$_3$ heterostructures

It is generally accepted that designing and analyzing a band diagram is a powerful way of describing and engineering the electrical properties of synthesized semiconductor heterostructures. One of the most popular approaches used to construct the band diagram is the electron affinity model [38]. Electron affinity is defined as the energy required to move an electron from the bottom of the conduction band of an uncharged crystal to infinity. Within this model, when a heterojunction is created, the conduction band offset is the difference between the electron affinities of the two materials in the heterostructure:

$$\Delta E_c = \chi_1 - \chi_2. \tag{3.33}$$

Considering that $\chi_{Si} = 4.05$ eV and $\chi_{LiNbO3} = 1.5$ [39], the conduction band offset is equal to $\Delta E_c = 2.55$ eV. However, this approach leads to an inappropriate result for ΔE_c, contradicting the experimental data, especially for two materials with significantly different bandgaps. The reason is that electron affinity depends on surface charge and dipoles, which must be taken into account when calculating ΔE_c. Moreover, when two materials are brought into contact, they exchange charge via surface states. Thus, other models that allow the estimation of the conduction band offset have been proposed.

An interesting and efficient model, based on the theory of quantum dipoles, was proposed by Tersoff [40]. This model posits the existence of 'interface-induced gap states' that are similar to those formed at the metal–semiconductor interface. From this perspective, some heterostructures can be represented by two Schottky barriers connected in series; thus, the conduction band offset can be determined as the

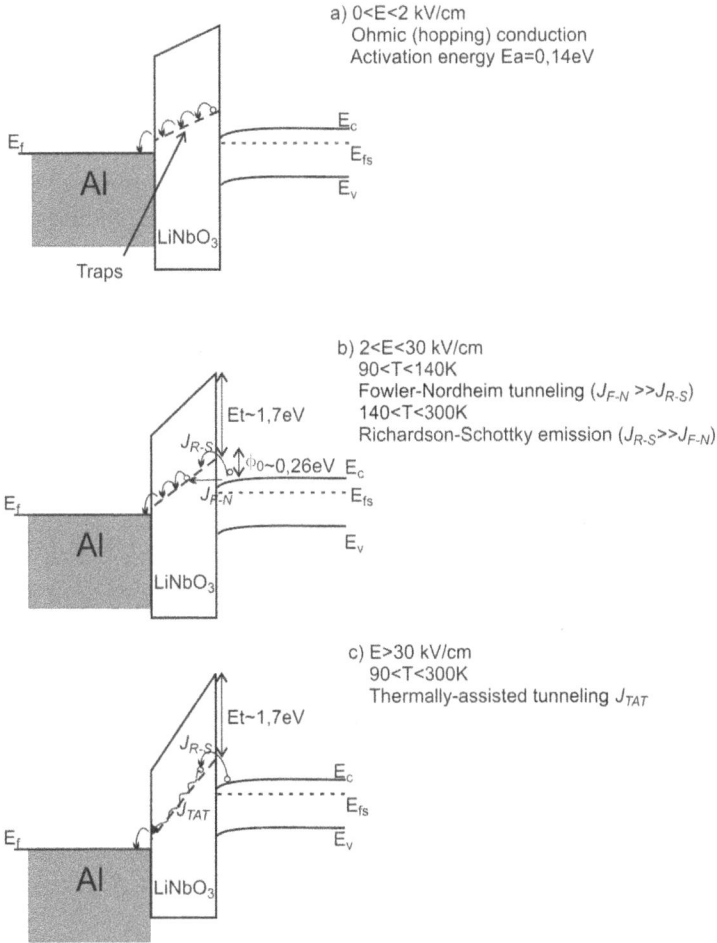

Figure 3.24. Schematic image of the conductivity mechanisms in the Si–LiNbO₃–Al heterostructures at various intervals of voltage and temperature. Reproduced from [34], with permission from Springer Nature.

difference in the Schottky barrier heights. As a result, the real band diagram of a heterostructure is influenced by a wide range of factors, such as built-in charge and interface states, which are not considered in the traditional Anderson model.

As emphasized by some authors, when surface states are presumably present at a metal–semiconductor heterojunction, the barrier height φ_b increases proportionally to the work function of the metal φ_m. It has been proposed [41] that the slope $S = \mathrm{d}\varphi_b/\mathrm{d}\varphi_m$ is affected only by the density of states and can be approximated by the following expression:

$$(1/S - 1) = 0.1(\varepsilon - 1)^2. \tag{3.34}$$

Here, ε is the dielectric constant of the material. Consequently, there are two special cases: when $S = 1$, the Schottky barrier is present; otherwise, when $S = 0$, the Bardeen barrier is formed, which is totally independent of the metal used as an electrode. Assuming $\varepsilon = 28$ [26], the slope S was estimated to be 0.01, suggesting that for Si–LiNbO$_3$ heterostructures, the Tersoff model is more realistic. A possible band diagram for Si–LiNbO$_3$ heterostructures, proposed in our recent work, is shown in figure 3.25 [42].

In the following section, we determine the parameters of the band diagram shown in figure 3.25 using I–V and C–V methods in the temperature range of 80–300 K. We have deposited films with a thickness of 0.5 μm by the RFMS method onto n-type Si substrates ($\rho = 4.5\,\Omega \cdot$ cm) under the optimal regimes with ion assist, developed in chapter 2, ensuring the formation of single-phase, c-oriented LiNbO$_3$ films (see chapter 2, table 2.6) at a growth rate of 10 nm min^{-1}. Indeed, XRD patterns show that single-phase $\langle 0001 \rangle$ textured polycrystalline LiNbO$_3$ films are formed under these conditions due to the plasma effect (figure 3.26).

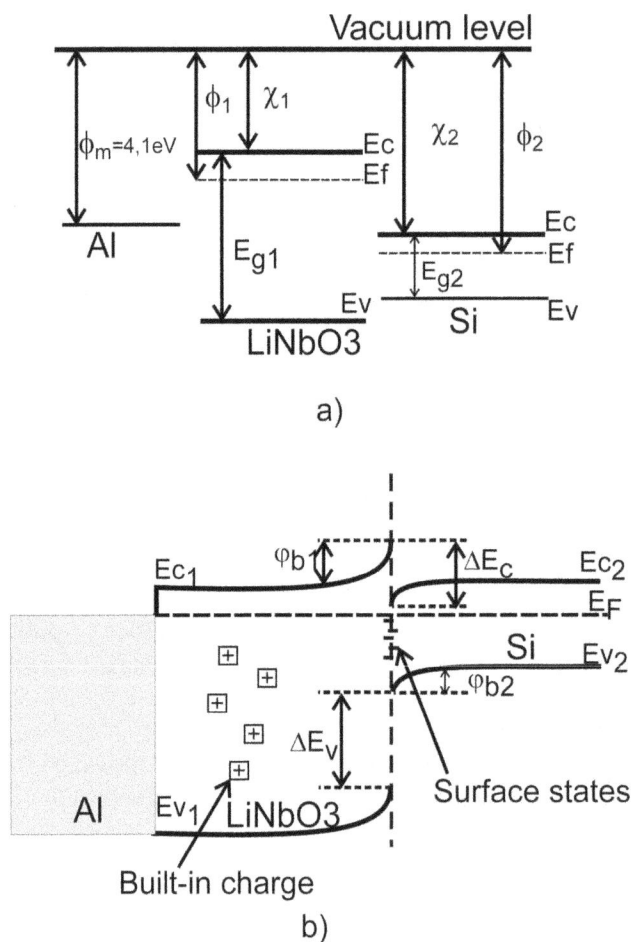

Figure 3.25. Band diagram of the (001)Si–LiNbO$_3$–Al heterostructure: (a) separate materials, (b) materials in contact. Reprinted from [42], Copyright (2013), with permission from Elsevier.

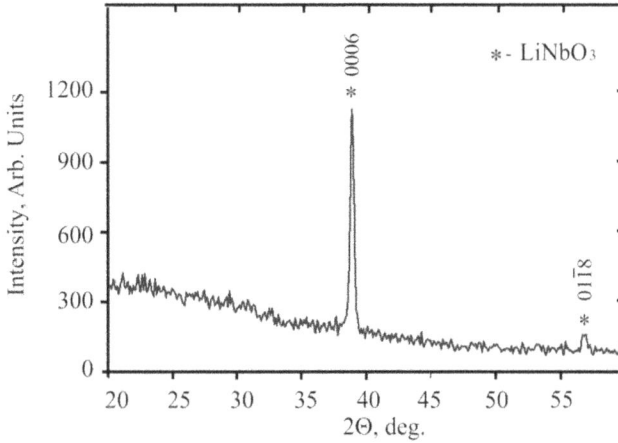

Figure 3.26. X-ray diffraction pattern of a LiNbO$_3$ film 0.5 μm thick formed on a (001)Si substrate.

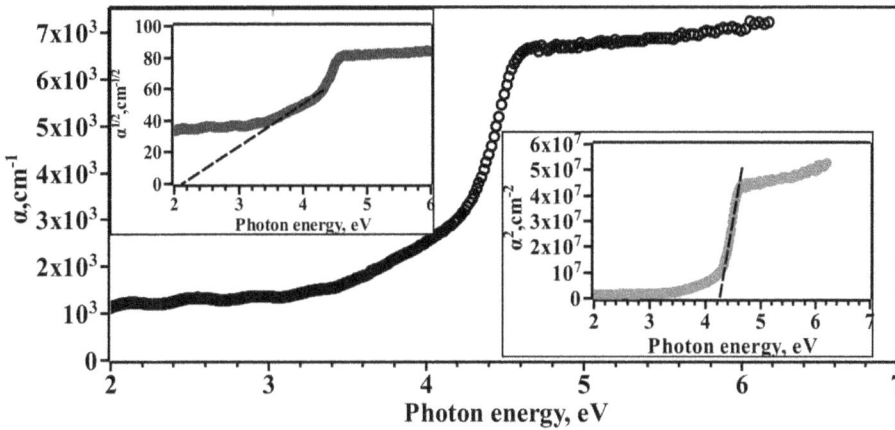

Figure 3.27. Dependence of the absorption coefficient α on the incident photon energy for the LiNbO$_3$ films. Inserts show the dependence of α^2 and $\alpha^{1/2}$ on the incident photon energy, respectively.

To successfully build the band diagram, information regarding the bandgaps of the materials in the heterostructure is needed. To determine the bandgap of our LiNbO$_3$ films, we investigated optical absorption in the region of the fundamental band edge. For this purpose, we used the same regimes to deposit LiNbO$_3$ films on fluorphlogopite substrates, ensuring the transparency of the studied heterostructures in the visible wavelength range. The absorption spectra (the absorption coefficient α vs. the incident photon energy) are shown in figure 3.27.

It can be seen from figure 3.27 that the absorption coefficient steadily increases with the photon energy and then sharply rises in the energy range of 4.2–4.6 eV. Depending on the band structure of the semiconductor, the absorption coefficient follows various laws, and the frequency dependence $\alpha(\nu)$ can be expressed as [43]:

$$\alpha(\nu) \propto B\big(h\nu - E_g\big)^r \tag{3.35}$$

Here, B is a constant, h is Planck's constant, E_g is the bandgap, $r = 1/2$ for direct optical transitions, and $r = 2$ for indirect ones. It has been demonstrated [44, 45] that the bandgap of single crystals of $LiNbO_3$ corresponds to both direct and indirect optical transitions. Thus, based on equation (3.35), we can determine the direct band energy E_g^{dir} from the intercept of the linear section of the graph $\alpha^2(h\nu)$ with the horizontal axis (see inset in figure 3.27). For the direct band energy, we obtained the following result: $E_g^{dir} = 4.2\,eV$, which is close to the value for bulk lithium niobate ($E_g^{dir} = 4.12\,eV$ [45]). The linear part with a steeper slope in figure 3.27 is attributed to the broad absorption band. The graph $\alpha^{1/2}(h\nu)$ (see the inset in figure 3.27) is linear in the range of incident photon energies of 3.5–4.2 eV, indicating indirect optical transitions [43]. The intercept of this linear graph with the horizontal axis corresponds to the edge of the indirect bandgap, which has the value $E_g^{ind} = 2.2\,eV$ in our case.

In the theory of small polarons, the absorption band edge is connected to the activation energy of conductivity W as follows: $E_{opt} = 4 \times W$ [46]. Accepting in our case $E_{opt} = 2.2\,eV$, we estimate the activation energy to be $W \approx 0.5\,eV$, which is slightly higher than the value obtained in our previous work from the AC conductivity of Si–$LiNbO_3$ heterostructures within the framework of variable range hopping conductivity [47]. To determine the band bendings, the Fermi level position, and the carrier concentrations, we applied I–V and high-frequency C–V analysis methods.

Figure 3.28 shows the typical high-frequency ($f = 10^5\,Hz$) C–V characteristic of the (001)Si–$LiNbO_3$ heterostructure. Similar to the heterostructures analyzed above, the C–V curve is shifted to the left along the voltage axis, indicating the presence of positive oxide charge in $LiNbO_3$, which is in total agreement with our previous results and those of other investigators [48]. Additionally, it follows from figure 3.28 that the studied heterostructures are operating in the accumulation regime at zero bias. According to the standard methodology [2], we replotted the C–V characteristics in

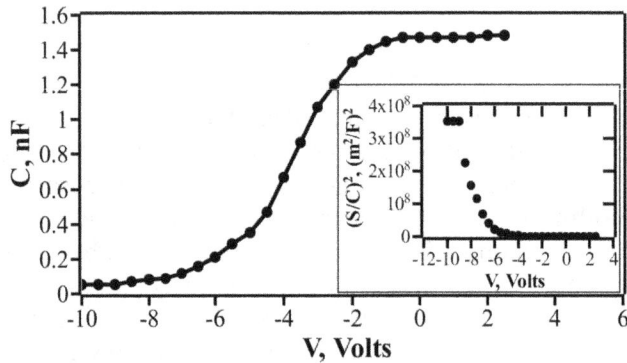

Figure 3.28. High-frequency ($f = 1 \times 10^5\,Hz$) C–V characteristics of the (001)Si–$LiNbO_3$ heterostructures. The insert displays the dependence $(S/C)^2$ vs. V for the studied heterostructures.

$(S/C)^2$ vs. V coordinates (where S is the metal contact area). This dependence is shown in figure 3.28 (see inset).

Since the graph corresponding to the uniform doping distribution is nonlinear, we have a nonuniform distribution of donors in the Si substrate of the studied heterostructures. The concentration distribution is determined by graphical differentiation of the experimental $(S/C)^2$ vs. V curve using the following well-known expression [2]:

$$N_d(x) = -\frac{2}{q \varepsilon \varepsilon_0 S^2} \left(\frac{d}{dV} \left(\frac{S}{C(V)} \right)^2 \right)^{-1} \qquad (3.36)$$

Here, ε_0 is the electric constant, ε is the dielectric constant of the semiconductor, and S is the area of the metal contact. Our calculations suggest that the concentration of ionized donors decreases from 5×10^{17} cm^{-3} at the Si–LiNbO$_3$ interface to 1×10^{17} cm^{-3} at a distance of 1 μm, which corresponds to the nominal donor concentration in the substrate (figure 3.29).

One possible origin of such a distribution could be the diffusion of Li atoms into the substrate during the deposition of LiNbO$_3$ films. Lithium, having a relatively high diffusion coefficient ($D = 2 \times 10^{-11}$ cm^2 s^{-1}), can penetrate deeply into Si, generating shallow donors there [49, 50]. Notably, shallow donors are not observed in a silicon substrate in Si–SiO$_2$–LiNbO$_3$ heterostructures. This is likely due to the presence of the SiO$_2$ layer, which prevents lithium diffusion into the substrate during RFMS. Moreover, as mentioned above, surface states influence the formation of heterojunctions. Using the C–V analysis method [2], we obtained the energy distribution of surface states $D_{ss}(E)$ at the Si–LiNbO$_3$ interface, which is presented in figure 3.30.

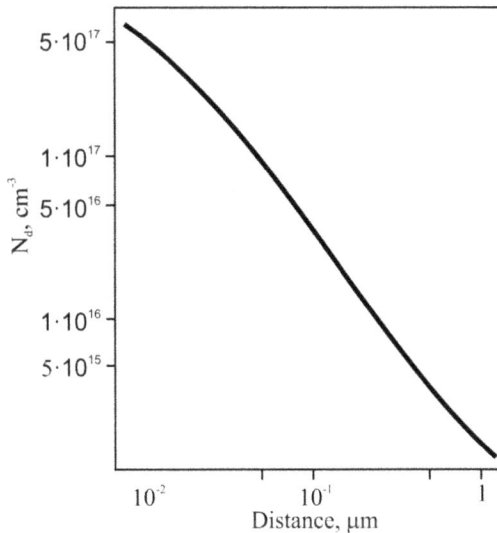

Figure 3.29. Doping profile of the (001)Si–LiNbO$_3$ heterostructures; x is the distance from the Si–LiNbO$_3$ interface toward the bulk.

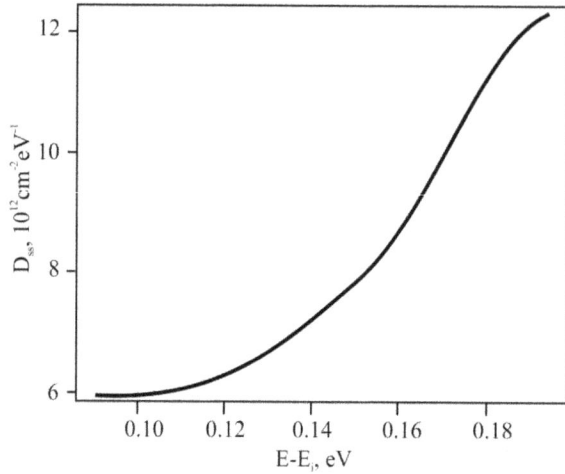

Figure 3.30. Surface state distribution for the (001)Si–LiNbO₃ heterostructure.

Table 3.6. *C–V* analysis results for the (001)Si–LiNbO₃ heterostructures.

Parameter	Value
Donor concentration in Si substrate, N_d (cm^{-3})	1×10^{15}
Flat band voltage, V_{fb} (V)	-5.6
Fermi level position in Si substrate, $E_C - E_F$ (eV)	0.31
Surface band bending, ψ_{s0} (eV)	0.36
Effective charge in LiNbO₃ layer, Q_{eff} (C cm^{-2})	$+7.8 \times 10^{-7}$
Built-in charge in LiNbO₃ layer, Q_{ox} (C cm^{-2})	$+2.2 \times 10^{-6}$

The integral charge in the LiNbO₃ layer, which is the sum of the total oxide charge Q_{ox} and the surface state charge with density D_{ss}, is determined using the following expression [2]:

$$Q_{\mathrm{sc}}(\psi_s = \psi_{s0}) - Q_{\mathrm{ox}}(\psi_s = 0) = q \int_0^{\psi_{s0}} D_{\mathrm{ss}}(\psi) \cdot f_0(\psi)d\psi. \qquad (3.37)$$

Here, ψ_{s0} is the surface potential in the flat band regime, Q_{sc} is the charge of the depletion zone in Si, D_{ss} is the density of surface states, and $f(\psi)$ is the distribution function. All results obtained from the *C–V* analysis are presented in table 3.6.

Figure 3.31 shows the *I–V* characteristics of the studied heterostructures at different temperatures. Two regions are clearly visible in the *I–V* curves: Region I, which corresponds to moderate electric fields (5–30 kV cm^{-1}), and Region II, corresponding to high electric fields (30–90 kV cm^{-1}).

It is commonly noted that, at particular voltages, the *I–V* characteristics of heterostructures are influenced by contact-limited conduction (see chapter 1). When conductivity depends on carrier mobility in a semiconductor layer, within the framework of Richardson–Schottky emission, the *I–V* curves should be linear in ln

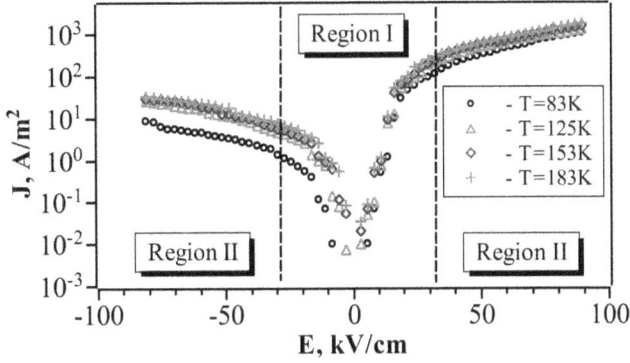

Figure 3.31. *I–V* characteristics of the Si(001)–LiNbO₃–Al heterostructures at different temperatures.

$(J/(ET^{3/2}))–E^{1/2}$ coordinates (Simmons coordinates). Such curves can be described by the Simmons–Schottky equation (1.8) from chapter 1, which is reiterated here:

$$J_{R-S} = 2q\left(\frac{2\pi m^* kT}{h^2}\right)^{3/2} \mu E \exp\left(-\frac{q\varphi}{kT}\right)\exp\left(\frac{\beta\sqrt{E}}{kT}\right). \tag{3.38}$$

On the other hand, when over-barrier emission dominates, the Richardson–Schottky formula applies [51]:

$$J = \frac{4\pi q m}{h^3}kT^2 \exp\left(-\frac{q\varphi}{kT}\right)\exp\left(\frac{\beta\sqrt{E}}{kT}\right). \tag{3.39}$$

In equations (3.38) and (3.39), φ is the potential barrier height at the heterojunction; E is the applied electric field; m^* and μ are the carrier effective mass and mobility, respectively; and β is the coefficient defined by equation (1.9). Equations (3.38) and (3.39) differ only by the pre-exponential factor, so the mobility and applied field are critical parameters in determining which formula is applicable in each case. In our experiments, the *I–V* characteristics are linear under high forward bias (Region II in figure 3.31) and are linear in Simmons coordinates (see figure 3.32).

Moreover, according to equation (3.38), the graph of $\ln(J/(ET^{3/2}))–1/T$ should be a straight line with a slope of γ:

$$\gamma = \frac{-q\varphi + \beta\sqrt{E}}{k}. \tag{3.40}$$

From equation (3.40), the barrier height φ and the factor β, which are directly related to the dielectric constant of the material, can be derived from the slope and intercept of γ vs. $E^{1/2}$ (see figure 3.33).

Regarding reverse bias (negative at the Al contact), the *I–V* characteristics are linear in Schottky coordinates (see the insert in figure 3.32), indicating that equation (3.39) should be applied. To verify the correctness of applying equations (3.38) and (3.39), we calculated the ratio of experimental to theoretical Schottky factors β_{ex}/β_{theor} for direct and reverse biases. Using equation (1.9) for β_{theor} and taking

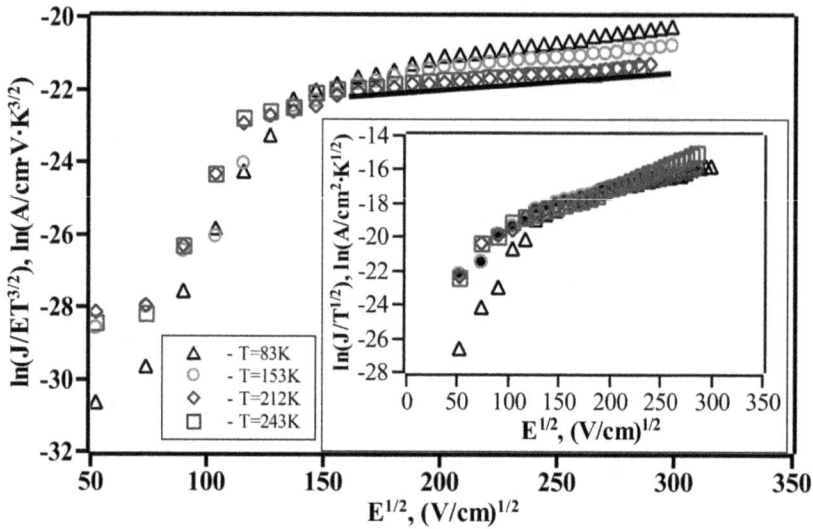

Figure 3.32. *I–V* characteristics of the (001)Si–LiNbO₃–Al heterostructures in Simmons coordinates under forward bias ('+' on the Al electrode). Insert shows the *I–V* characteristics in Schottky coordinates under reverse bias ('−' on the Al electrode).

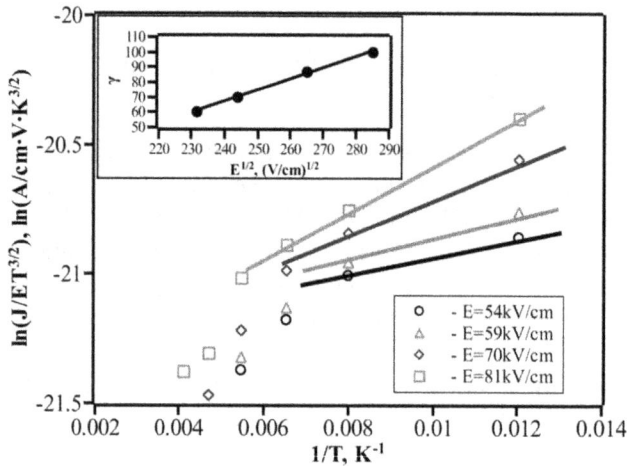

Figure 3.33. Temperature dependence of (001)-Si–LiNbO3–Al heterostructures' conductivity in Arrhenius coordinates within the framework of the Simmons–Schottky mechanism. The inset illustrates the field dependence of the slope, which is related to the linear parts of the graph $\ln(J/(ET^{3/2}))$ vs. $1/T$.

$\varepsilon = 28$, we obtained $\varphi = 0.02$ eV and $\beta_{\text{ex}}/\beta_{\text{theor}} = 1.03$ and $\beta_{\text{ex}}/\beta_{\text{theor}} = 1.09$ for direct and reverse biases, respectively.

In the range of moderate electric fields (Region I in figure 3.31), the conductivity of the studied heterostructures is influenced by Fowler–Nordheim tunneling through the barrier ϕ. In this case, the *I–V* characteristic should be linear in Fowler–Nordheim coordinates $\ln(J/E^2)–1/E$ (see chapter 1) and described by equation (1.10). The slope of this linear graph provides the barrier height ϕ. In the range of moderate electric fields, the I–V curves are linear in Fowler–Nordheim coordinates (figure 3.34) with a slope corresponding to a barrier height of $\phi = 0.03$ eV, which is close to the height obtained from I–V curves at high fields [42] and to the value observed in heterostructures fabricated by the IBS method [34].

Figure 3.35 demonstrates a good agreement between experimental data and theoretically calculated I–V curves.

It is important to emphasize that the potential barrier height derived from I–V characteristics can be a misleading value due to the significant influence of ferro-electric polarization, which can greatly reduce the barrier height in a heterojunction

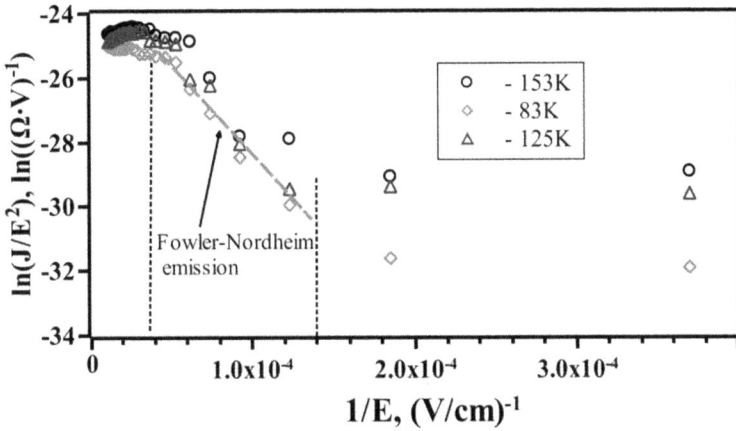

Figure 3.34. I–V characteristics of the Si(001)–LiNbO$_3$–Al heterostructures in Fowler–Nordheim coordinates under a moderate electric field.

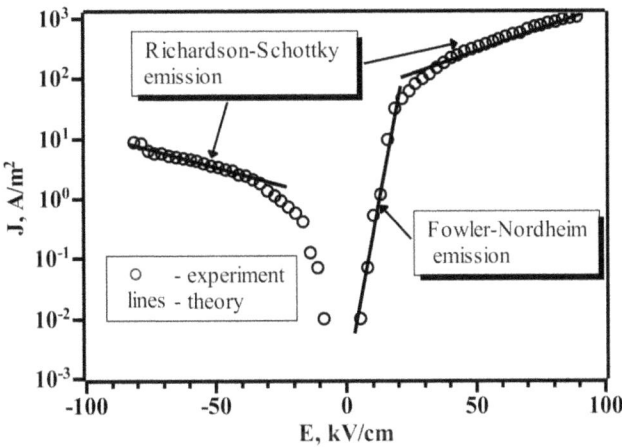

Figure 3.35. Experimental and theoretical I–V characteristics of the (001)Si–LiNbO$_3$–Al heterostructures at $T = 83$ K.

[52]. As a result, the apparent barrier height φ_{ap}, determined from $I-V$ curves, can be expressed through the actual value φ_b as follows:

$$\varphi_{app} = \varphi_b - \sqrt{\frac{qP}{4\pi(\varepsilon\varepsilon_0)^2}}. \tag{3.41}$$

Here, P is the ferroelectric polarization. Taking $P = 13.6\,\mu C\,cm^{-2}$, as obtained from the $P-E$ hysteresis loop analysis [17], and using equation (3.41), we determine the actual potential barrier height, $\varphi_b = 0.2\,eV$.

According to the Tersoff model, the band offset in a heterojunction can be found by taking the difference between two Schottky barrier heights. Thus, it is necessary to obtain information regarding the Fermi level position in the LiNbO$_3$ film of the studied heterostructures. This can be estimated using the following formula [2]:

$$E_F = \frac{kT}{q} \ln\left(\frac{N_c}{N_d}\right). \tag{3.42}$$

Here, E_F is the position of the Fermi level relative to the conduction band; N_d is the donor concentration, which in our case is $N_d = 7 \times 10^{13}\,cm^{-3}$ [17]; and N_c is the concentration of carriers in the conduction band. Using equation (3.42), we can calculate the conduction band offset as follows:

$$\Delta E_c = E_{F1} + \varphi_{b1} - \left(E_{F2} - \varphi_{b2}\right). \tag{3.43}$$

In equation (3.43), we consider that, according to $C-V$ analysis under zero bias, the bands in Si are bent down. All parameters of the proposed band diagram shown in figure 3.25 are presented in table 3.7.

It is important to note that the conduction and valence band offsets listed in table 3.7 differ significantly from those calculated within the framework of the Anderson model using the electron affinities. This indicates that interface states at the Si–LiNbO$_3$ heterojunction have a profound effect on the formation of the studied heterostructure.

Table 3.7. Parameters of the band diagram of Si–LiNbO$_3$ heterostructures (see. figure 3.25).

Band diagram parameter	Material in the heterojunction		
	LiNbO$_3$		Si
Bandgap	$E_{g1} = 4.2\,eV$		$E_{g2} = 1.1\,eV$
Fermi level position	$E_{F1} = 0.33\,eV$		$E_{F2} = 0.31\,eV$
Band bending	$\varphi_{b1} = 0.2\,eV$		$\varphi_{b2} = 0.36\,eV$
Conduction band offset ΔE_C		$0.6\,eV$	
Valence band offset ΔE_V		$2.52\,eV$	

3.4 Impedance spectroscopy and AC conductivity of thin LiNbO$_3$ films

IS is one of the most informative tools for investigating the electrical properties of materials [53, 54]. IS is widely used to study dielectric materials (solid and liquid dielectrics whose electrical properties are influenced by dipole orientation), materials with dominant electronic conductivity (crystalline and amorphous semiconductors, glasses, and polymers), and also conductive dielectrics with both ionic and electronic conductivity. AC IS is recorded over a wide frequency range, allowing different areas of a material to be characterized according to their time constants (relaxation times).

In polycrystalline materials, successfully distinguishing the grain boundaries' response from bulk phenomena depends on the appropriate equivalent circuit that describes the studied heterostructures. Generally, the response to an AC signal applied to a studied sample can be interpreted using one of the following formalisms: (1) the complex admittance $Y^* = 1/R_p + j\omega C_p$, (2) the complex impedance $Z^* = R_s + (j\omega C_s)^{-1}$, (3) the complex dielectric permittivity $\varepsilon^* = \varepsilon' - j\varepsilon''$, or (4) the complex dielectric modulus $M^* = M' + jM''$. Indices p and s in the above expressions correspond to parallel and series equivalent circuits, respectively; j is the imaginary unit, and ω is the angular frequency. The complex conductivity formalism is also used to analyze electrical properties: $\sigma^* = j\omega\varepsilon_o\varepsilon^*$ (here, ε_o is the electric constant or dielectric permittivity of free space).

Before analyzing IS data in detail, it is recommended to express the experimental data graphically to reveal their structure, reflecting the studied physical processes. For example, the material may be an insulator or a dielectric with leakages, or it may exhibit a response caused by mobile charges in a material with blocking contacts at the electrodes. For entirely blocking contacts, DC current cannot flow through the studied structure, making it difficult to analyze the dielectric response. In conductive materials, dielectric effects are usually minimal, and the most informative approach is to use a representation in terms of Z^* and M^*. When a nonconductive material is analyzed, the Y and ε formalisms are more appropriate. Nevertheless, important information can be derived from graphical data interpretation in all four formalisms, regardless of whether a material is a conductor or not.

There are different ways to graphically present experimental IS data. When capacitive effects dominate inductive ones, it is convenient to plot the dependence of the imaginary part of the impedance $-\text{Im}(Z^*)$ (or $-Z''$) vs. the real part $\text{Re}(Z^*)$ (or Z'). This type of diagram on the complex plane is called a Nyquist plot. Despite the absence of direct frequency response in this representation, the Nyquist diagram can be very useful for identifying different conductivity processes in the studied material. Different areas of a sample, associated with their resistance and capacitance, are represented by a resistor and a capacitor connected in parallel. The characteristic relaxation time (time constant) corresponding to each 'parallel RC element' is defined by the product of R and C ($\tau = RC$). Various RC elements can be derived from the impedance spectra and matched to appropriate areas in a sample. Each parallel RC element of an equivalent circuit ideally produces a semicircle in the

Nyquist plot, from which the values of the components R and C can be extracted. Values of R are obtained by the intercepts on the Z' axis, and the maximum of each semicircle is observed at the frequency $\omega_m = 1/RC$. If we define ω_m graphically and know R, it is possible to calculate the capacitance C.

Semiconducting polycrystalline ceramics such as $LiNbO_3$ are inhomogeneous materials in which the resistance of grain boundaries can dominate when IS spectra are recorded. In this case, only one semicircle is observed on the Nyquist plot. In such situations, some investigators recommend presenting experimental data in terms of complex dielectric modulus $M* = j\omega C_0 Z*$ (where $C_0 = \varepsilon_0 S/d$ is the capacitance of an 'empty' capacitor with electrode area S and an inter-electrode distance d). This allows the response from both grain boundaries and the grain bulk to be seen. Thus, a peak in the frequency dependence $Z''(\omega)$ corresponds to the element with the largest resistance, equated to the peak of $Z'' = R/2$. A peak in the frequency dependence of the imaginary part of the dielectric modulus $M''(\omega)$ corresponds to an element with minimal capacitance, its value being $M'' = C_0/2C$.

It is generally accepted that the electrical properties of various materials and structures can be entirely described using an equivalent circuit [55]. As mentioned for C–V analysis (see figure 3.9), at zero or low biases, the studied Si–$LiNbO_3$–Al heterostructures operate in the accumulation regime, allowing these heterostructures to be treated as metal–insulator-metal systems. In other words, at low bias, the electric properties of the studied heterostructures are affected by the properties of $LiNbO_3$ films and metal contacts, excluding the space-charge areas in Si from consideration. Thus, the simplified equivalent circuit of a $LiNbO_3$ film can be represented as shown in figure 1.19 (see chapter 1), where R_b and C_b represent the resistance and capacitance of the $LiNbO_3$ grain's bulk, and R_{gb} and C_{gb} represent the resistance and capacitance of the grain boundaries.

The complex impedance corresponding to the circuit shown in figure 1.19 can be written as follows:

$$Z* = \frac{R_b}{1 + j\omega\tau_1} + \frac{R_{gb}}{1 + j\omega\tau_2}. \tag{3.44}$$

The characteristic times $\tau_1 = R_b C_b$ and $\tau_2 = R_{gb} C_{gb}$ correspond to the relaxation processes, where j is the imaginary unit. Equation (3.44) can be modified as follows:

$$Z* = \frac{(R_b + R_{gb})(1 + j\omega\tau_o)}{(1 + j\omega\tau_1)(1 + j\omega\tau_2)}. \tag{3.45}$$

Here, τ_0 is given by the following expression:

$$\tau_o = \frac{R_b\tau_2 + R_{gb}\tau_1}{R_b + R_{gb}}. \tag{3.46}$$

Taking the logarithm of both sides of equation (3.45), we obtain:

$$\ln(|Z*|) = \ln(R_b + R_{gb}) + \ln(|1 + j\omega\tau_o|) - \ln(|1 + j\omega\tau_1|) - \ln(|1 + j\omega\tau_2|). \tag{3.47}$$

Therefore, each term in equation (3.47) can be determined independently from a diagram in $\ln(|Z_*|) - \ln(\omega)$ coordinates. Diagrams that present the frequency dependence of the magnitude and phase of the complex impedance in $\ln(|Z_*|) - \ln(\omega)$ and $\varphi - \ln(\omega)$ coordinates are called Bode diagrams. Each term in formula (3.47) has two frequency limits: the low-frequency case ($\omega\tau \ll 1$, when $\ln(|1 + j\omega\tau|) = 0$), and the high-frequency case ($\omega\tau \gg 1$, when $\ln(|1 + j\omega\tau|) = \ln(\omega) + \ln(\tau)$). For these two cases, the relaxation time can be derived from the break point on the Bode diagram $\ln(|Z_*|) - \ln(\omega)$. Moreover, it follows from equation (3.47) that $\lim\limits_{\omega \to 0}(|Z_*|) = R_b + R_{gb}$, which allows us to determine the total resistance ($R_b + R_{gb}$) from the intercept of the horizontal part of the graph $\ln(|Z_*|) - \ln(\omega)$ with the horizontal axis.

Additionally, in disordered materials, there is typically a distribution of characteristic times rather than a single relaxation time τ. This leads to a deformation of the Nyquist plot, resulting in a plot that is not an ideal semicircle. The complex impedance can be expressed in a manner similar to equation (3.44):

$$Z_* = \frac{R_b}{1 + (j\omega\tau_1)^{1-\alpha_1}} + \frac{R_{gb}}{1 + (j\omega\tau_2)^{1-\alpha_2}}. \tag{3.48}$$

Here, α_i is the coefficient ($0 \leqslant \alpha_i \leqslant 1$) that describes the width of the distribution spectrum of the relaxation time [56], which can be derived from the slope of a linear section in the Bode diagram $\ln(|Z_*|) - \ln(\omega)$.

For comparative analysis of IS spectra, we fabricated (001)Si–LiNbO$_3$ heterostructures using the IBS and RFMS methods under the optimal technological regimes listed in table 2.6 (see chapter 2) without the plasma effect.

Figure 3.36 shows Nyquist and Bode diagrams for (001)Si–LiNbO$_3$ heterostructures fabricated by the RFMS method and measured at 300 K. The theoretical curves in this figure were derived from equation (3.48), using parameters obtained through least-squares fitting of the experimental data. Similar diagrams for Si–LiNbO$_3$ heterostructures fabricated by the IBS method are shown in figure 3.37.

As seen from figures 3.36 and 3.37, the Nyquist diagrams are semicircles, as expected, given that the studied films are conductive. The fact that semicircles on the Nyquist diagram are elongated for both deposition methods and the presence of a break point in the Bode diagram (see figures 3.36 and 3.37) indicate a distribution of relaxation times in the films. Our analysis of the diagrams shown in figures 3.36 and 3.37 is summarized in table 3.8.

Many authors have emphasized that 'slow' relaxation processes (with relatively high characteristic times τ) correspond to processes that take place at grain boundaries or interfaces, whereas 'fast' processes occur within the bulk of the grains [54, 55]. Thus, in the studied heterostructures, processes with relaxation time τ_1 (see table 3.8) likely occur at the grain boundaries between polycrystalline grains, while the characteristic time τ_2 represents processes influenced by the bulk properties of LiNbO$_3$ grains. As shown in table 3.8, heterostructures fabricated by IBS exhibit a larger dispersion of characteristic times (larger α_1) associated with grain boundaries compared to those observed for the heterostructures obtained using RFMS. This

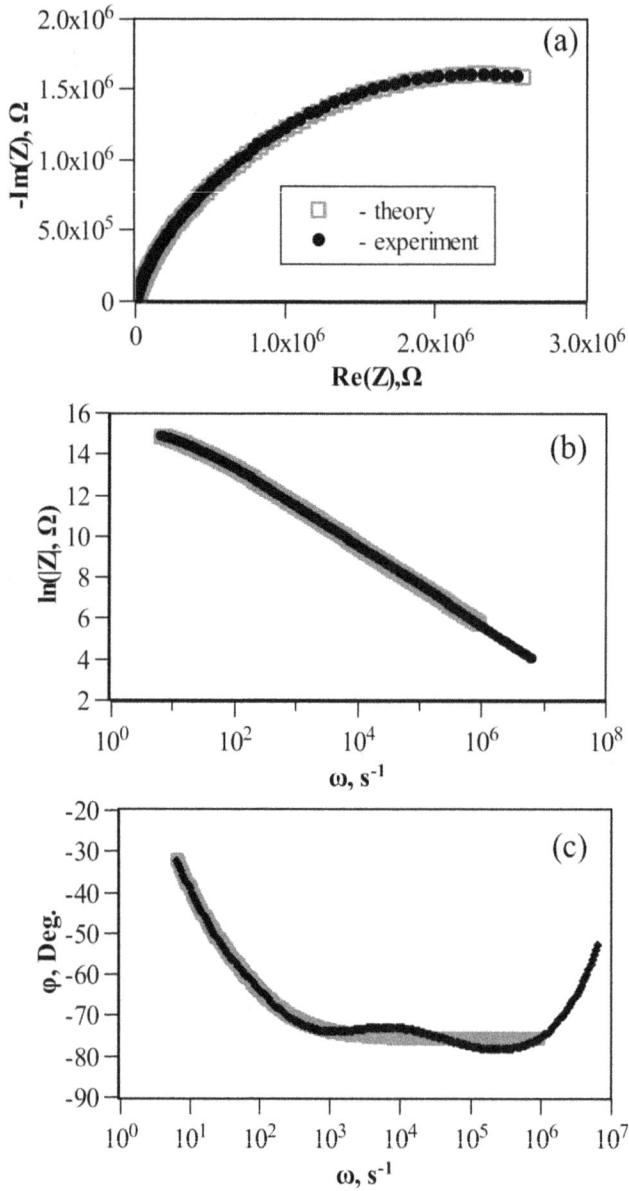

Figure 3.36. The Nyquist diagram (a), and Bode diagrams (b and c) of (001)Si–LiNbO$_3$ heterostructures grown by RFMS.

indicates that LiNbO$_3$ films deposited by the IBS method demonstrate a higher degree of structural disorder than films fabricated by the RFMS method.

According to [14], the capacitance of grain boundaries in polycrystalline materials behaves similarly to the Schottky barrier capacitance: it correlates with the concentration of charged localized centers (N_t) and the intercrystalline potential

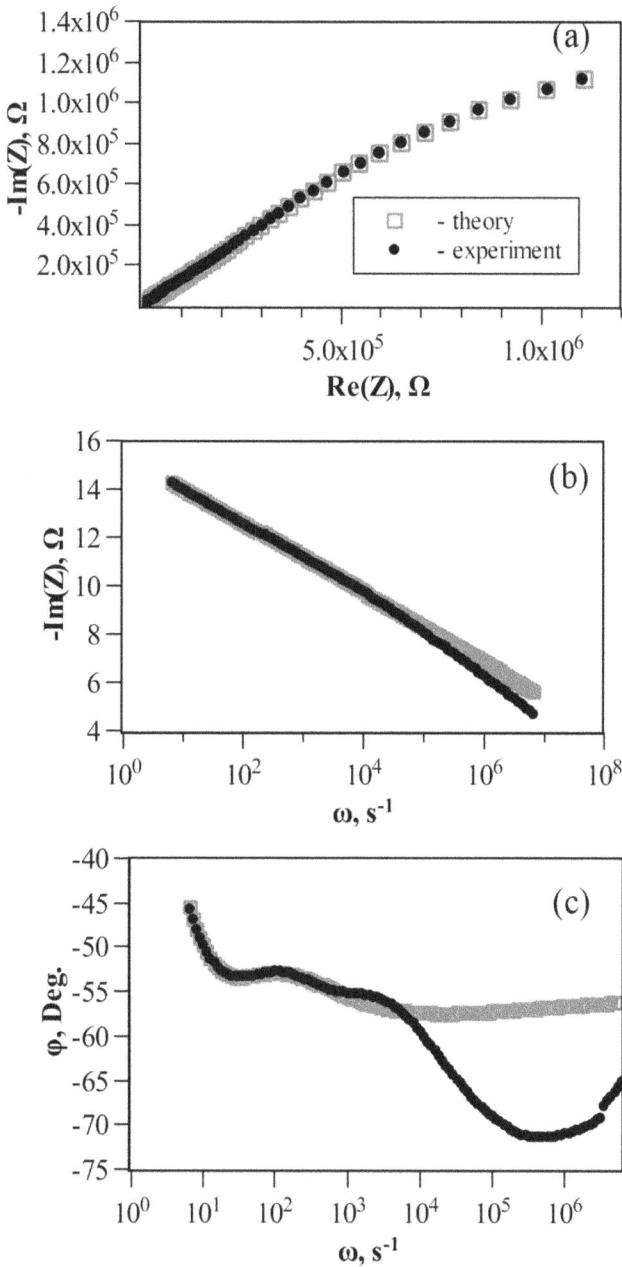

Figure 3.37. The Nyquist diagram (a), and Bode diagrams (b and c) of (001)Si–LiNbO$_3$ heterostructures grown by IBS.

barrier height φ_b according to $C_{gb} \propto (N_t / \varphi_b)^{1/2}$. Earlier in this chapter, it was shown that the concentration N_t, apparently attributed to defects, is considerably higher in LiNbO$_3$ films fabricated by IBS compared to those deposited by RFMS. This accounts for the differences in experimental relaxation times and capacitances C_{gb}

Table 3.8. Results of the impedance spectra analysis for (001)Si–LiNbO$_3$ heterostructures fabricated by two methods.

Fabrication method	R_{gb} (Ω)	R_b (Ω)	τ_1 (s)	τ_2 (s)	C_{gb} (F)	C_b (F)	α_1	α_2
IBS	5.0×10^6	4.0×10^4	0.6	3×10^{-4}	1.2×10^{-7}	7.0×10^{-8}	0.3	0.1
RFMS	4.0×10^6	3.4×10^3	0.1	1×10^{-4}	2.5×10^{-8}	2.9×10^{-8}	0.17	0.2

presented in table 3.8. For a deeper understanding of the relaxation processes in thin LiNbO$_3$ films, a detailed study of dielectric relaxation mechanisms and AC conductivity is required. When an alternating electric field is applied to a material, there is typically a delay between the field's oscillation and the polarization response of the material. Usually, a dielectric heats up in an alternating electric field. The portion of the total energy converted into heat is called the dielectric loss. This is the sum of the loss attributed to currents associated with direct voltage and the loss triggered by the active component of the displacement current. The dielectric loss tangent is an important characteristic of dielectric losses [57]:

$$\text{tg}(\delta) = \frac{\varepsilon^{//}}{\varepsilon^{/}}. \tag{3.49}$$

Here, $\varepsilon^{/}$ and $\varepsilon^{//}$ are the real and imaginary parts of the complex dielectric permittivity, respectively:

$$\varepsilon^* = \varepsilon^{/} - j\varepsilon^{//}. \tag{3.50}$$

The character of the dielectric response depends on the relaxation mechanisms occurring in the studied material. Polarization mechanisms caused by atomic polarization, electron polarization, or the polarization of an ionic lattice react almost instantly to changes in the electric field, contributing only to the real part of ε^*. On the other hand, molecular dipoles, charge defects, and carriers involved in hopping conductivity (electrons, polarons, and ions) have a delayed response to the alternating field, which contributes to the imaginary part of ε^*.

In the case of dipole (Debye) polarization, the complex dielectric permittivity can be described by the following expression [56]:

$$\varepsilon^* = \varepsilon_\infty + \frac{\varepsilon_s - \varepsilon_\infty}{1 + (j\omega\tau)^{1-\alpha}} - \frac{j\sigma_{\text{dc}}}{\varepsilon_0 \omega} \tag{3.51}$$

where ε_s and ε_∞ are the static and high-frequency dielectric permittivities, respectively; ε_0 is the electric constant; ω is the angular frequency of the testing signal; τ is the characteristic time for a given process; α is the coefficient characterizing the width of the characteristic time distribution ($0 \leqslant \alpha \leqslant 1$), σ_{dc} is the DC electrical conductivity of the sample, and j is the imaginary unit. Typically, the second term in equation (3.51) describes relaxation losses, while the third term deals with losses attributed to low-frequency conductivity through the sample.

In the case of charge polarization, when various types of charges participate in hopping conductivity, the differences in charge transport mechanisms can be revealed through an analysis of the frequency dependence of electrical conductivity, which in the most general case can be expressed as follows [14]:

$$\sigma(\omega, T) = \sigma_{dc}(T) + \sigma_{ac}(\omega, T). \tag{3.52}$$

Here, σ_{dc} and σ_{ac} are the DC and AC conductivities, respectively. In disordered systems, AC conductivity σ_{ac} is described by the 'universal law' [58]:

$$\sigma_{ac}(\omega, T) = A(T)\omega^s \tag{3.53}$$

where $A(T)$ is the frequency-independent parameter, ω is the angular frequency, and s is the exponent, which depends on the specific conductivity mechanism and usually lies in the range $0.4 < s < 1$. Various models have been proposed to explain the frequency dependence expressed by equation (3.53) and observed for a wide class of materials [59–63]. The quantum-mechanical tunneling (QMT) model [59] suggests the absence of lattice distortions associated with the motion of charges involved in AC conductivity, and within this model, the frequency exponent s in equation (3.53) does not depend on temperature. However, this model predicts the following frequency dependence for $s(\omega)$:

$$s = 1 - \frac{4}{\ln(1/\omega\tau_0)}. \tag{3.54}$$

Here, ω is the angular frequency of the testing signal, and τ_0 is the characteristic relaxation time (the inverse of the phonon frequency), typically taken to be $\tau_0 = 10^{-13}$ s. On the other hand, the correlated barrier-hopping (CBH) model proposed by Elliott [60, 61] considers only the correlated hopping of bipolarons and predicts both the frequency and the temperature dependence of the frequency exponent s. Within the CBH model, this parameter decreases with temperature as follows:

$$s = 1 - \frac{6kT}{W_H + kT \ln(\omega\tau_0)} \tag{3.55}$$

Here, k is the Boltzmann constant, τ_0 is the characteristic relaxation time corresponding to the phonon frequency, and W_H is the maximum height of the potential barrier overcome by carriers. In contrast, the small-polaron tunneling (SPT) model predicts that the exponent s rises with temperature according to the following law [61]:

$$s = 1 - \frac{4}{\ln(1/\omega\tau_0) - W_H/kT}. \tag{3.56}$$

Therefore, the study of the temperature and frequency dependence of s in equation (3.56) is crucial for understanding conduction mechanisms in the studied heterostructures.

Figures 3.38 and 3.39 show the frequency dependencies of the dielectric loss tangent for Si–LiNbO$_3$ heterostructures fabricated by the RFMS and IBS methods, respectively.

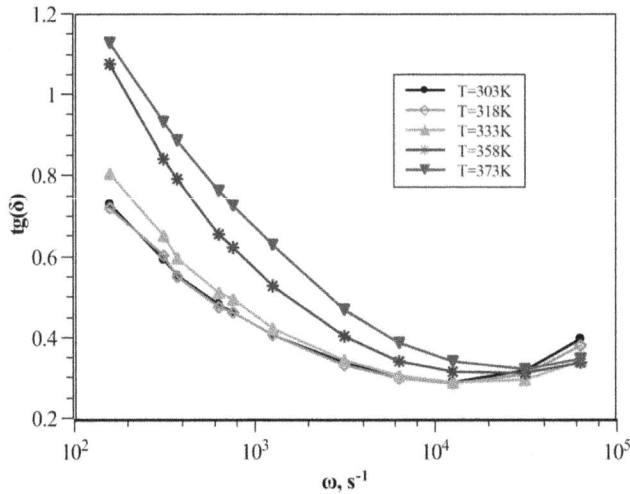

Figure 3.38. Dielectric losses as a function of frequency at different temperatures for (001)Si–LiNbO$_3$ heterostructures fabricated by RFMS.

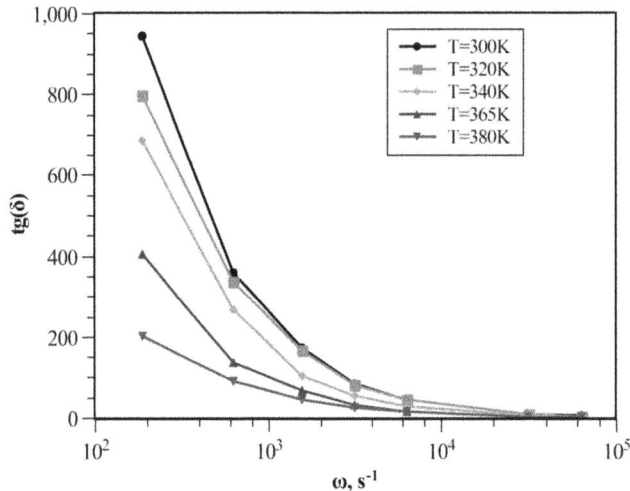

Figure 3.39. Dielectric losses as a function of frequency at different temperatures for (001)Si–LiNbO$_3$ heterostructures fabricated by IBS.

This type of frequency dependence suggests that in the studied heterostructures, the energy losses are more likely caused by electrical conductivity rather than by a Debye-type process. Furthermore, the losses in LiNbO$_3$ films deposited by IBS are significantly higher than those in films synthesized by RFMS due to their larger conductivity. For a detailed analysis of charge transport under alternating voltage, we studied the frequency dependence of conductivity in the temperature range of 290 K–420 K [47, 64]. Figure 3.40 demonstrates the typical frequency dependence of conductivity for LiNbO$_3$ films fabricated by the RFMS technique.

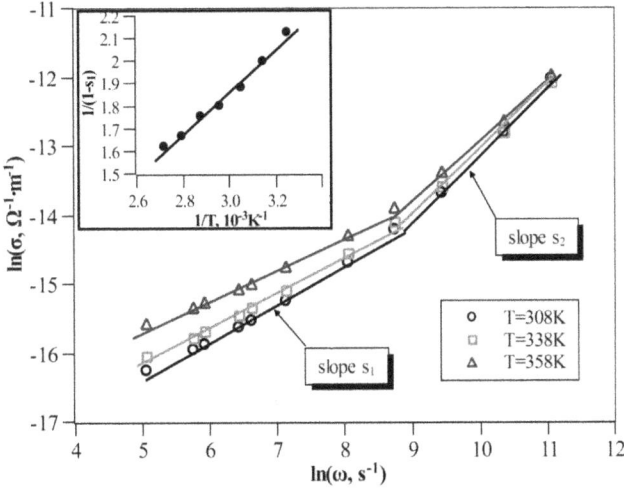

Figure 3.40. Frequency dependence of the AC conductivity of as-grown (001)Si–LiNbO$_3$ heterostructures at different temperatures. The inset shows the variation of the frequency exponent 's_1' in equation (3.53) with temperature.

In figure 3.40, two linear sections can be clearly seen with slopes s_1 and s_2, corresponding to the frequency exponent in equation (3.53), each having different temperature dependencies. The fact that s_1 decreases with temperature agrees with equation (3.55) and suggests that in the frequency range of 25–500 Hz, CBH is the prevalent conductivity mechanism. In this case, equation (3.55) can be rewritten as:

$$\frac{1}{1-s} = \frac{W_H}{6kT} + \frac{1}{6}\ln(\omega\tau_0).$$

(3.57)

In equation (3.57), the maximum barrier height for carriers W_H can be derived graphically from the slope of a linear graph in $1/(1-s)-1/T$ coordinates (see the inset in figure 3.40). For the studied films, $W_H = 0.4$ eV [47], which is in good agreement with the results of other authors [65] who studied the hopping conductivity of LiNbO$_3$ single crystals. Furthermore, as indicated by equation (3.57), the characteristic time τ_0 can be determined from the intercept of a linear graph of $1/(1-s)-1/T$ (see the inset in figure 3.40). The relaxation time obtained in this way, $\tau_0 = 1.4 \times 10^{-4}$ s, agrees with the relaxation time τ_2 determined earlier from the IS analysis (see table 3.7) and with values reported in our previous work [47]. This indicates that the relaxation processes are caused by the bulk properties of grains in LiNbO$_3$ films rather than by grain boundaries.

Another informative parameter in equation (3.53) is the coefficient A, which is expressed as follows [61]:

$$A = \frac{D^2 q^{12}}{24\pi^3(\varepsilon\varepsilon_0)^5(kT)^6}.$$

(3.58)

Here, q is the elementary charge; ε and ε_0 are the dielectric permittivity of the material and the electric constant, respectively; and D is the bulk density of CLCs in the material. The coefficient A can be determined by extrapolating the frequency dependence of conductivity in $\ln(\sigma_{ac})$ vs. $\ln(\omega)$ coordinates if $\ln(\omega) \to 0$. Using equation (3.58), it is then possible to determine D from A. Assuming the dielectric permittivity to be $\varepsilon = 28$ [26], we obtain the density of charged centers $D = 7 \times 10^{18}$ cm^{-3}, which correlates with the value determined earlier from I–V analysis and with results reported in our previous work [35]. The second linear section on the graph $\ln(\sigma_{ac}) - \ln(\omega)$ with the slope s_2 (see figure 3.40) stems from the second term in equation (3.57), which dominates at high frequencies (f > 500 Hz) when $W_H/kT < \ln(\omega\tau_0)$.

Figure 3.41 shows the frequency dependence of conductivity for (001)Si–LiNbO$_3$ heterostructures fabricated by the IBS method.

As with heterostructures grown by RFMS, the frequency dependence of conductivity is linear in $\ln(\sigma_{ac})$–$\ln(\omega)$ coordinates, in total agreement with formula (3.53). The temperature dependence of the frequency exponent s also corresponds to the CBH model (see the insert in figure 3.41). Applying the same analysis based on equation (3.57), we derive a barrier height $W_H = 1.7$ eV [64], which matches the energy of traps in the bandgap of LiNbO$_3$, determined earlier in this chapter from the analysis of conductivity mechanisms under DC voltage. Figure 3.42 demonstrates good agreement between the experimental frequency dependence of conductivity and the theoretical one calculated using the CBH model for (001)Si–LiNbO$_3$ heterostructures.

As shown in figure 3.42, CBH is the predominant charge transport mechanism in the studied LiNbO$_3$ films across the discussed frequency and temperature ranges.

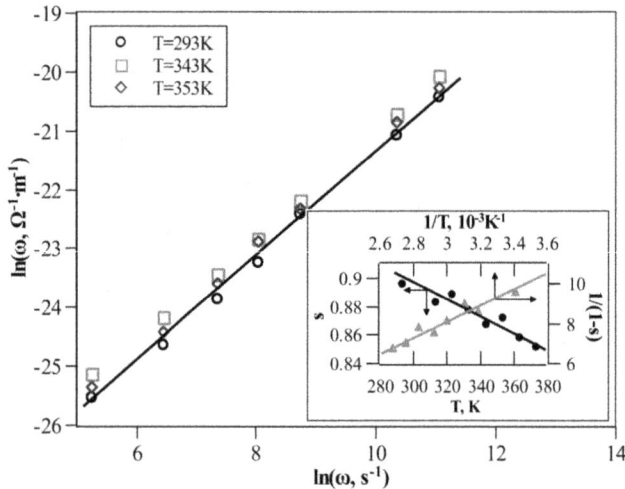

Figure 3.41. Frequency dependence of AC conductivity of (001) Si–LiNbO$_3$ heterostructures grown by the IBS method at different temperatures. The insert shows the variation of the frequency exponent 's' with temperature.

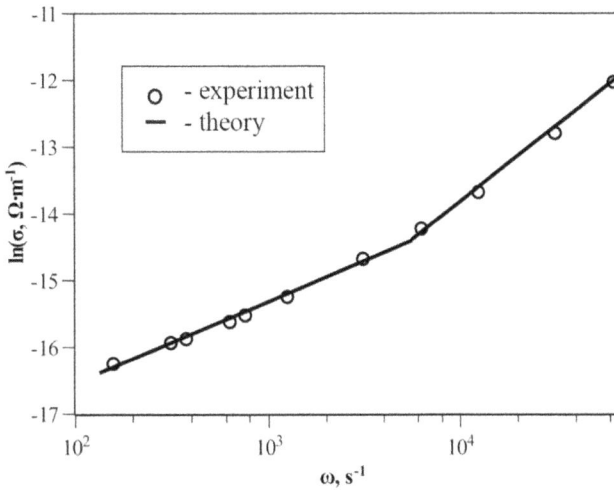

Figure 3.42. Frequency dependence of the AC conductivity of the (001)Si–LiNbO$_3$ heterostructures at $T = 300$ K (dots—experimental results, line—theoretical conductivity based on the framework of the CBH model).

To summarize, table 3.9 presents the basic electrical properties of LiNbO$_3$ films synthesized without the plasma effect, compared to the properties reported in the literature.

3.5 Summary and discussion

1. **Electrical properties and oxide charge formation**: I–V and C–V analyses revealed that the Si–LiNbO$_3$–Al heterostructures behave similarly to a metal–insulator–metal system with a conductive dielectric. The basic electrical properties of LiNbO$_3$ films deposited onto Si substrates by both RFMS and IBS methods were determined. Regardless of the conductivity type of Si wafers, a positive oxide charge with a density of $Q_{ef} = 3.8 \times 10^{-8}$ C cm^{-2} is formed, restricting the functionality of LiNbO$_3$-based heterostructures. The concentration of localized charge centers depends on the deposition method and is maximal for LiNbO$_3$ films synthesized by IBS ($N_t \sim 3 \times 10^{19}$ cm^{-3}) compared to films deposited by the RFMS technique ($N_t \sim 7 \times 10^{17}$ cm^{-3}).

2. **Ferroelectric properties**: LiNbO$_3$ films fabricated by both methods without the ion assist effect demonstrate ferroelectric properties. The remnant polarization in the synthesized films is considerably lower than in bulk lithium niobate, which is apparently caused by the arbitrary grain orientation, limiting the application of fabricated heterostructures in memory units. The built-in field in deposited LiNbO$_3$ films results from the formation of positive oxide charge during the growth process.

3. **Charge transport mechanisms**: it has been demonstrated that charge transport in Si–LiNbO$_3$–Al heterostructures is influenced by the following mechanisms:

Table 3.9. Basic electrical properties of thin LiNbO$_3$ films deposited by the RFMS and IBS techniques compared to data reported in the research literature.

Property	Result	Data reported in research literature	Notes	Source
Bandgap E_g, eV	4.2 (RFMS)	3.97		[66]
		4.43		[67]
		4.70		[68]
Resistivity ρ, $\Omega \cdot$ cm	1 × 109 (RFMS)	1 × 1010		[48, 69]
	3 × 10^9 (IBS)	2 × 10^9		[6]
Conduction mechanism	*DC conductivity:* • Low electric fields $(0<E<2\,\text{kV cm}^{-1})$ and room temperature—hopping conductivity over CLCs in the bandgap of LiNbO$_3$. Activation energy 0.14 eV. • Moderate fields $(2 < E < 30\,\text{kV cm}^{-1})$—Fowler–Nordheim tunneling via CLCs in the bandgap of LiNbO$_3$ with energy of $E_t = 1.7$ eV. • Strong fields $(E > 30\,\text{kV cm}^{-1})$—thermally assisted tunneling through the intergranular potential barrier $\varphi_b = 0.7$ eV.	Hopping electronic	Activation energy Ea = 0.3 eV	[6]
		Hopping polaronic	Ea = 0.5 eV at T > 420 K and Ea = 0.067 eV at T < 420 K	[70]
		Hopping small polaron conductivity,	Average energy of 'hops' W = 0.051 eV,	[36]
		space-charge-limited current (SCLC), Poole–Frenkel emission	'hopping' between Nb5+ and Nb4+ ionic states with an average energy of W = 0.51 eV at T = 450 K.	[71]
			W = 0.39 eV (decreases with annealing in vacuum)	[65]
		Richardson–Schottky emission	Li ions with activation energy of Ea = 0.5 eV	[72, 73]
		Ionic	Conductivity $\sigma = 5 \times 10^{-8}$ S cm−1 with activation energy of $E_a = 0.57$ eV	[48, 74, 75]
				[76]
				[77]

(Continued)

Table 3.9. (*Continued*)

AC conductivity: **CBH** conduction ($W_H = 0.4$ eV). Frequency dependence obeys the 'universal' power law $\sigma_{ac}(\omega, T) = A(T)\omega^s$.

Property	Result	Data reported in research literature	Notes	Source
Dielectric permittivity ε	28 (RFMS)	27.9		[3, 4]
	29 (IBS)	8		[70]
		29.3		[69]
		46		[3]
		50		[78]
Remnant polarization P_r, μC cm^{-2}	14.0 (RFMS)	71	Bulk LiNbO$_3$	[79]
	11.7 (IBS)	18		[3]
		2.74		[48]
		1.2		[80]
Coercive field E_c, kV cm^{-1}	44.0 (RFMS)	40	Bulk LiNbO$_3$	[81]
	29 (IBS)	115		[82]
		35.23		[29]
		170		[48]
		120		[80]

- Under low electric fields ($0 < E < 2\,\mathrm{kV\,cm^{-1}}$) and at room temperature: hopping conductivity over CLCs in the bandgap of $LiNbO_3$ with an activation energy of 0.14 eV.
- Under moderate fields ($2 < E < 30\,\mathrm{kV\,cm^{-1}}$): Schottky emission and Fowler–Nordheim tunneling, stemming from CLCs distributed in the bandgap of $LiNbO_3$ with an energy of $E_t = 1.7$ eV below the bottom of the conduction band.
- Under strong electric fields ($E > 30\,\mathrm{kV\,cm^{-1}}$): conductivity occurs due to thermally assisted tunneling through the intergranular barrier of $\varphi_b = 0.7$ eV.

4. **Band diagram**: for the first time, a detailed band diagram of the Si–$LiNbO_3$ heterostructure has been designed with all required parameters. It was established that the band offsets are affected by the high density of interface states at the Si–$LiNbO_3$ heterojunction rather than the electron affinities of the contacting materials. The bandgap of deposited $LiNbO_3$ films ($E_g = 4.2$ eV) is close to that of bulk lithium niobate.

5. **Generation of shallow donors**: for the first time, it was found that during the deposition of $LiNbO_3$ films onto Si substrates, shallow donors are generated at the Si–$LiNbO_3$ interface at a concentration of $N_d = 5 \times 10^{17}\,\mathrm{cm^{-3}}$, which declines exponentially in the substrate with depth. This phenomenon is apparently caused by the diffusion of Li atoms into the Si substrate during film deposition.

6. **Dielectric losses and relaxation processes**: using the impedance spectroscopy method, it was demonstrated that dielectric losses in the Si–$LiNbO_3$ hetero-system are caused by charge relaxation. We revealed two relaxation processes: 'fast' relaxation (in the bulk of polycrystalline grains) with a characteristic time of $\tau = 1 \times 10^{-4}$ s, and 'slow' relaxation (at grain boundaries) with a relaxation time on the order of $\tau = 0.1$ s.

7. **AC conductivity**: AC conductivity in $LiNbO_3$ films is described within the framework of the CBH model and obeys the 'universal' power law $\sigma_{ac}(\omega, T) = A(T)\omega^s$. The maximum potential barrier height overcome by carriers in $LiNbO_3$ films deposited by RFMS is found to be $W_H = 0.4$ eV, which is considerably smaller than that for films synthesized by the IBS method ($W_H = 1.7$ eV).

References

[1] Knack S, Weber J, Lemke H and Riemann H 2002 Copper–hydrogen complexes in silicon *Phys. Rev.* B **65** 165203

[2] Sze S M and Kwok K N 2006 *Physics of Semiconductor Devices* (New York: Wiley)

[3] Simões A Z, Zaghete M A, Stojanovic B D, Riccardi C S, Ries A, Gonzalez A H and Varela J A 2003 LiNbO₃ thin films prepared through polymeric precursor method *Mater. Lett.* **57** 2333–9

[4] Choi S-W, Choi Y-S, Lim D-G, Moon S-I, Kim S-H, Jang B-S and Yi J 2000 Effect of RTA treatment on LiNbO₃ MFS memory capacitors *Korean J. Ceram.* **6** 138–42 http://www.koreascience.or.kr/article/JAKO200011920780492.page

[5] Nassau K, Levinstein H J and Loiacono G M 1966 Ferroelectric lithium niobate. 2. Preparation of single domain crystals *J. Phys. Chem. Solids* **27** 989–96

[6] Shandilya S, Tomar M, Sreenivas K and Gupta V 2009 Purely hopping conduction in c-axis oriented $LiNbO_3$ thin films *J. Appl. Phys.* **105** 94105

[7] Strikha V I 1974 *Theoretical Basis of the Operation of Metal-Semiconductor Contacts* (Kiev: Naukova Dumka) (in Russian)

[8] Strikha V I and Buzanova E V 1987 *Physical Basis of the Reability of Metal-Semiconductor Cotacts in Integrated Electronics* (Moscow: Radio and Svyaz')

[9] Maissel L I and Glang R 1970 *Handbook of Thin Film Technology* (New York: McGraw-Hill)

[10] Mott N F 1969 Conduction in non-crystalline materials *Philos. Mag.* **19** 835–52

[11] Shklovskii B I and Efros A L 1984 *Electronic Properties of Doped Semiconductors* **vol 45** (Berlin: Springer)

[12] Iyevlev V, Kostyuchenko A and Sumets M 2011 Fabrication, substructure and properties of $LiNbO_3$ films *Proc. of SPIE—The Int. Society for Optical Engineering* **7747** 77471J–8

[13] Hill R M 1971 Poole–Frenkel conduction in amorphous solids *Philos. Mag.* **23** 59–86

[14] Mott N F and Davis E A 1971 *Electronic Processes in Non-Crystalline Materials* (Oxford: Clarendon))

[15] Mycielski J 1961 Mechanism of impurity conduction in semiconductors *Phys. Rev.* **123** 99–103

[16] Adalashvili D I, Adamia Z A, Lavdovskii K G, Levin E I and Shklovskii B I 1988 Negative differential resistance in the hopping conductivity region in silicon *JETP Lett.* **47** 390–2 http://jetpletters.ru/ps/1095/article_16542.shtml

[17] Iyevlev V, Sumets M and Kostyuchenko A 2012 Current–voltage characteristics and impedance spectroscopy of $LiNbO_3$ films grown by RF magnetron sputtering *J. Mater. Sci., Mater. Electron.* **23** 913–20

[18] Sawyer C B and Tower C H 1930 Rochelle salt as a dielectric *Phys. Rev.* **35** 269–73

[19] Chan H K, Lam C H and Shin F G 2004 Time-dependent space-charge-limited conduction as a possible origin of the polarization offsets observed in compositionally graded ferro-electric films *J. Appl. Phys.* **95** 2665–71

[20] Zhang J, Tang M H, Tang J X, Yang F, Xu H Y, Zhao W F, Zheng X J, Zhou Y C and He J 2007 Bilayer model of polarization offset of compositionally graded ferroelectric thin films *Appl. Phys. Lett.* **91** 162908

[21] Zheng L, Lin C, Xu W P and Okuyama M 1996 Vertical drift of *P–E* hysteresis loop in asymmetric ferroelectric capacitors *J. Appl. Phys.* **79** 8634–7

[22] Misirlioglu I B, Okatan M B and Alpay S P 2010 Asymmetric hysteresis loops and smearing of the dielectric anomaly at the transition temperature due to space charges in ferroelectric thin films *J. Appl. Phys.* **108** 34105

[23] Lohkämper R, Neumann H and Arlt G 1990 Internal bias in acceptor-doped $BaTiO_3$ ceramics: numerical evaluation of increase and decrease *J. Appl. Phys.* **68** 4220–4

[24] Pike G E, Warren W L, Dimos D, Tuttle B A, Ramesh R, Lee J, Keramidas V G and Evans J T 1995 Voltage offsets in $(Pb,La)(Zr,Ti)O_3$ thin films *Appl. Phys. Lett.* **66** 484–6

[25] Bentarzi H 2011 *Transport in Metal-Oxide-Semiconductor Structures* (Berlin: Springer)

[26] Iyevlev V, Kostyuchenko A, Sumets M and Vakhtel V 2011 Electrical and structural properties of $LiNbO_3$ films, grown by RF magnetron sputtering *J. Mater. Sci., Mater. Electron.* **22** 1258–63

[27] Tagantsev A K and Gerra G 2006 Interface-induced phenomena in polarization response of ferroelectric thin films *J. Appl. Phys.* **100** 51607

[28] Ievlev V, Shur V, Sumets M and Kostyuchenko A 2013 Electrical properties and local domain structure of LiNbO$_3$ thin film grown by ion beam sputtering method *Acta Metall. Sin.* **26** 630–4

[29] Zhao J P, Liu X R and Qiang L S 2007 Preparation and characterization of LiNbO$_3$ thin films derived from metal carboxylate gels *Key Eng. Mater.* **336–8** 213–6

[30] Waser R and Klee M 1992 Theory of conduction and breakdown in perovskite thin films *Integr. Ferroelectr.* **2** 23–40

[31] Damjanovic D 1998 Ferroelectric, dielectric and piezoelectric properties of ferroelectric thin films and ceramics *Rep. Prog. Phys.* **61** 1267–324

[32] Scott J F, Araujo C A, Melnick B M, McMillan L D and Zuleeg R 1991 Quantitative measurement of space-charge effects in lead zirconate-titanate memories *J. Appl. Phys.* **70** 382–8

[33] Duiker H M, Beale P D, Scott J F, Paz de Araujo C A, Melnick B M, Cuchiaro J D and McMillan L D 1990 Fatigue and switching in ferroelectric memories: theory and experiment *J. Appl. Phys.* **68** 5783–91

[34] Ievlev V, Sumets M and Kostyuchenko A 2013 Conduction mechanisms in Si-LiNbO$_3$ heterostructures grown by ion-beam sputtering method *J. Mater. Sci.* **48** 1562–70

[35] Ievlev V M, Sumets M P and Kostyuchenko A V 2012 Effect of thermal annealing on electrical properties of Si-LiNbO$_3$ *Mater. Sci. Forum* **700** 53–7

[36] Dhar A, Singh N, Singh R K and Singh R 2013 Low temperature dc electrical conduction in reduced lithium niobate single crystals *J. Phys. Chem. Solids* **74** 146–51

[37] Padovani F A and Stratton R 1966 Field and thermionic-field emission in Schottky barriers *Solid-State Electron.* **9** 695–707

[38] Anderson R L 1962 Experiments on Ge–GaAs heterojunctions *Solid-State Electron.* **5** 341–51

[39] Yang W-C, Rodriguez B J, Gruverman A and Nemanich R J 2004 Polarization-dependent electron affinity of LiNbO$_3$ surfaces *Appl. Phys. Lett.* **85** 2316–8

[40] Tersoff J 1984 Theory of semiconductor heterojunctions: the role of quantum dipoles *Phys. Rev. B* **30** 4874–7

[41] Mönch W 1990 Role of virtual gap states and defects in metal–semiconductor contacts *Electronic Structure of Metal–Semiconductor Contacts* ed W Mönch (Netherlands: Springer) 224–7

[42] Ievlev V, Sumets M, Kostyuchenko A, Ovchinnikov O, Vakhtel V and Kannykin S 2013 Band diagram of the Si-LiNbO$_3$ heterostructures grown by radio-frequency magnetron sputtering *Thin Solid Films* **542** 289–94

[43] Fox M 2010 *Optical Properties of Solids* (New York: Oxford University Press)

[44] Thierfelder C, Sanna S, Schindlmayr A and Schmidt W G 2010 Do we know the band gap of lithium niobate? *Phys. Status Solidi* **7** 362–5

[45] Bhatt R, Bhaumik I, Ganesamoorthy S, Karnal A K, Swami M K, Patel H S and Gupta P K 2012 Urbach tail and bandgap analysis in near stoichiometric LiNbO$_3$ crystals *Phys. Status Solidi* **209** 176–80

[46] Reik H G and Heese D 1967 Frequency dependence of the electrical conductivity of small polarons for high and low temperatures *J. Phys. Chem. Solids* **28** 581–96

[47] Ievlev V, Sumets M, Kostyuchenko A and Bezryadin N 2013 Dielectric losses and ac conductivity of Si-LiNbO$_3$ heterostructures grown by the RF magnetron sputtering method *J. Mater. Sci., Mater. Electron.* **24** 1651–7

[48] Lim D, Jang B, Moon S, Won C and Yi J 2001 Characteristics of LiNbO$_3$ memory capacitors fabricated using a low thermal budget process *Solid-State Electron.* **45** 1159–63

[49] Gosele U M 1988 Fast diffusion in semiconductors *Annu. Rev. Mater. Sci.* **18** 257–82

[50] Yoshimura K, Suzuki J, Sekine K and Takamura T 2007 Measurement of the diffusion rate of Li in silicon by the use of bipolar cells *J. Power Sources* **174** 653–7

[51] Simmons J G 1965 Richardson-Schottky effect in solids *Phys. Rev. Lett.* **15** 967–8

[52] Zubko P, Jung D J and Scott J F 2006 Space charge effects in ferroelectric thin films *J. Appl. Phys.* **100** 114112

[53] Macdonald J R 1992 Impedance spectroscopy *Ann. Biomed. Eng.* **20** 289–305

[54] Irvine J T S, Sinclair D C and West A R 1990 Electroceramics: characterization by impedance spectroscopy *Adv. Mater.* **2** 132–8

[55] Hodge I M, Ingram M D and West A R 1976 Impedance and modulus spectroscopy of polycrystalline solid electrolytes *J. Electroanal. Chem. Interfacial Electrochem.* **74** 125–43

[56] Brown W F 1956 Dielectrics *Encyclopedia of Physics* (Berlin: Springer) 1–154

[57] Zheludev I S 1971 *Physics of Crystalline Dielectrics* (New York: Plenum)

[58] Jonscher A K 1977 The 'universal' dielectric response *Nature* **267** 673–9

[59] Austin I G G and Mott N F F 1969 Polarons in crystalline and non-crystalline materials *Adv. Phys.* **18** 41–102

[60] Elliott S R 1977 A theory of a.c. conduction in chalcogenide glasses *Philos. Mag.* **36** 1291–304

[61] Elliott S R 1987 A.c. conduction in amorphous chalcogenide and pnictide semiconductors *Adv. Phys.* **36** 135–217

[62] Pollak M and Pike G E 1972 Ac conductivity of glasses *Phys. Rev. Lett.* **28** 1449–51

[63] Long A R 1982 Frequency-dependent loss in amorphous semiconductors *Adv. Phys.* **31** 553–637

[64] Ievlev V, Sumets M and Kostuchenko A 2013 Electrical conductivity of the Si-LiNbO$_3$ heterostructures grown by ion sputtering method *Proc. of SPIE—The Int. Society for Optical Engineering* **8770** 87701M

[65] Akhmadullin I S, Golenishchev-Kutuzov V A, Migachev S A and Mironov S P 1998 Low-temperature electrical conductivity of congruent lithium niobate crystals *Phys. Solid State* **40** 1190–2

[66] Fakhri M A, Al-Douri Y, Hashim U, Salim E T, Prakash D and Verma K D 2015 Optical investigation of nanophotonic lithium niobate-based optical waveguide *Appl. Phys.* B **121** 107–16

[67] Shandilya S, Sharma A, Tomar M and Gupta V 2012 Optical properties of the c-axis oriented LiNbO$_3$ thin film *Thin Solid Films* **520** 2142–6

[68] Satapathy S, Mukherjee C, Shaktawat T, Gupta P K and Sathe V G 2012 Blue shift of optical band-gap in LiNbO$_3$ thin films deposited by sol–gel technique *Thin Solid Films* **520** 6510–4

[69] Gupta V, Bhattacharya P, Yuzyuk Y I, Katiyar R S, Tomar M and Sreenivas K 2004 Growth and characterization of c-axis oriented LiNbO$_3$ film on a transparent conducting Al: ZnO inter-layer on Si *J. Mater. Res.* **19** 2235–9

[70] Easwaran N, Balasubramanian C, Narayandass S A K and Mangalaraj D 1992 Dielectric and AC conduction properties of thermally evaporated lithium niobate thin films *Phys. Status Solidi* **129** 443–51

[71] Dhar A and Mansingh A 1990 Polaronic hopping conduction in reduced lithium niobate single crystals *Philos. Mag.* B **61** 1033–42

[72] Hao L Z, Zhu J and Li Y R 2011 Integration between $LiNbO_3$ ferroelectric film and AlGaN/GaN system *Mater. Sci. Forum* **687** 303–8

[73] Joshi V, Roy D and Mecartney M L 1995 Nonlinear conduction in textured and non textured lithium niobate thin films *Integr. Ferroelectr.* **6** 321–7

[74] Hao L-Z, Liu Y-J, Zhu J, Lei H-W, Liu Y-Y, Tang Z-Y, Zhang Y, Zhang W-L and Li Y-R 2011 Rectifying the current–voltage characteristics of a $LiNbO_3$ film/GaN heterojunction *Chin. Phys. Lett.* **28** 107703

[75] Guo S M, Zhao Y G, Xiong C M and Lang P L 2006 Rectifying *I–V* characteristic of $LiNbO_3$/Nb-doped $SrTiO_3$ heterojunction *Appl. Phys. Lett.* **89** 223506

[76] Kim S H, Lee S J, Kim J P, Chae B G and Yang M J Y S 1998 Low-frequency dielectric dispersion and Raman spectroscopy of amorphous $LiNbO_3$ *J. Korean Phys. Soc.* **32** 830–3

[77] Can N, Ashrit P V, Bader G, Girouard F and Truong V 1994 Electrical and optical properties of Li-doped $LiBO_2$ and $LiNbO_3$ films *J. Appl. Phys.* **76** 4327–31

[78] Edon V, Rèmiens D and Saada S 2009 Structural, electrical and piezoelectric properties of $LiNbO_3$ thin films for surface acoustic wave resonators applications *Appl. Surf. Sci.* **256** 1455–60

[79] Wemple S H, DiDomenico M and Camlibel I 1968 Relationship between linear and quadratic electro-optic coefficiens in $LiNbO_3$, $LiTaO_3$, and other oxygen-octahedra ferroelectrics based on direct measurement of spontaneous polarization *Appl. Phys. Lett.* **12** 209–11

[80] Kim K-H, Lee S-W, Lyu J-S and Kim H-J Y B-W 1998 Properties of lithium niobate thin films by RF magnetron sputtering with wafer target *J. Korean Phys. Soc.* **32** 1508–12

[81] Gopalan V, Mitchell T E, Furukawa Y and Kitamura K 1998 The role of nonstoichiometry in 180° domain switching of $LiNbO_3$ crystals *Appl. Phys. Lett.* **72** 1981

[82] Kim Y-S, Jung S-W, Jeong S-H, In Y-I, Kim K-H and No K 2003 Properties of $LiNbO_3$ thin films fabricated by CSD (chemical solution decomposition) method (AWAD2003: Asia-Pacific Workshop on Fundamental and Application of Advanced Semiconductor Devices) *Tech. Rep. IEICE. SDM* **103** 33–6

IOP Publishing

Lithium Niobate-Based Heterostructures (Second Edition)
Synthesis, properties, and electron phenomena
Maxim Sumets

Chapter 4

Effect of sputtering conditions and post-growth treatment on electron phenomena in Si–LiNbO$_3$ heterostructures

In this chapter, we demonstrate that the plasma composition, its spatial inhomogeneity, and the relative target–substrate position significantly influence the electrical properties of Si–LiNbO$_3$ heterostructures through the charge centers in the bandgap of LiNbO$_3$ (LN) films and at the Si–LN interface. Using an Ar + O$_2$ reactive gas mixture results in a reduction of positive oxide charge and coercive field. The barrier properties of the Si–LN heterojunction are affected by the plasma composition. Thermal annealing (TA) of fabricated films leads to an increase in the amount of the LiNb$_3$O$_8$ phase due to the diffusion of oxygen atoms toward grain boundaries. Post-growth treatments, such as TA and pulsed photon treatment (PPT), have a profound effect on the optical properties of LN films.

It was shown in chapter 3 that the relatively high concentration of centers of localized charge (CLCs) and positive oxide charge, as well as the low remnant polarization in LN films fabricated without the ion assist effect, restrict the applications of LN-based heterostructures. Given that the technological regimes of radio-frequency magnetron sputtering (RFMS) affect the structure and surface morphology of LN films, chapter 2 proposed optimal deposition regimes for the formation single-phase oriented films with a high degree of crystallinity and minimal surface roughness.

Motivated by the lack of systematic investigations into the influence of synthesis conditions on the electrical properties of fabricated LN-based heterostructures, this chapter focuses on methods for changing these properties through variations in the technological regimes of RFMS and subsequent TA. It was demonstrated in chapter 2 and in our works [1, 2] that fundamental sputtering parameters, such as the reactive gas pressure and composition, the substrate temperature, and its relative position to the target, have critical effects on the structure, composition, and surface morphology of deposited LN films. Furthermore, TA leads to the recrystallization of the films, the disappearance of texture, and the formation of 'parasitic' phases such as $LiNb_3O_8$. Undoubtedly, these factors should be reflected in changes in the electrical properties of LN-based heterostructures.

For electrical measurements, we fabricated films by the RFMS and ion-beam sputtering (IBS) methods onto heated ($T = 550$ °C) silicon wafers of n-type conductivity ($\rho = 4.5\,\Omega\cdot$cm) under optimal conditions (see chapter 2) with the plasma effect to improve polarization reversal. As described in chapter 3, the electrical properties of the fabricated heterostructures were studied using techniques based on measuring current–voltage (I–V) and high-frequency (1 MHz) capacitance–voltage (C–V) characteristics, the Sawyer–Tower method, tangent loss frequency dependence, and impedance spectroscopy (IS). The local domain structure was studied using the piezoresponse force microscopy (PFM) method, the fundamentals of which can be found in [3].

4.1 Effects of spatial plasma inhomogeneity and composition and the relative target–substrate position

Films were synthesized using the RFMS and IBS methods with the ion assist effect according to the regimes listed in table 4.1 to study the influence of reactive gas pressure on the films' electrical properties.

The structural properties of heterostructures similar to those listed in table 4.1 were described in detail in chapter 2 and in our works [1, 4, 5], so they are not analyzed here.

4.1.1 Capacitance–voltage characteristics and ferroelectric properties of Si–LiNbO₃–Al heterostructures

Figure 4.1 shows typical normalized high-frequency C–V characteristics of the studied Si–LN–Al heterostructures, which are also similar to those of metal–insulator–semiconductor (MIS) structures (see chapter 3).

Table 4.1. Technological regimes used in the fabrication of Si–LN–Al heterostructures by the RFMS method.

Sample #	Film thickness, d (μm)	Magnetron power (W)	Reactive gas composition	Reactive gas pressure, P (Pa)	Substrate–target distance (cm)	Film composition and structure
LN133	1	100	Ar	5.0×10^{-1}	5	Polycrystalline two-phase (LN and $LiNb_3O_8$) films with arbitrary grain orientation
LN134	0.37	100	Ar	1.5×10^{-1}	5	Single-phase polycrystalline $\langle0001\rangle$ textured LN films
LN135	0.60	100	Ar(60%) + O₂(40%)	1.5×10^{-1}	5	Single-phase polycrystalline $\langle0001\rangle$ textured LN films

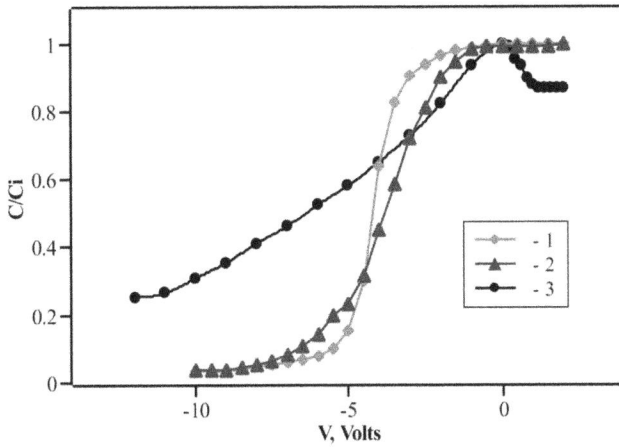

Figure 4.1. Typical C–V characteristics of studied Si–LN–Al heterostructures at a temperature of $T = 293$ K. 1—sample LN133, 2—sample LN134, 3—sample LN135.

Table 4.2. Results of C–V analysis for Si–LN–Al heterostructures fabricated by RFMS under different regimes.

Sample #	Dielectric constant of the film, ε	Flat band voltage, V_{FB} (V)	Effective charge in the film, Q_{eff} (C cm^{-2})	Effective density of states, N_{eff} (cm^{-2})	Position of a charge centroid (relative to the film surface) in thickness units, d_c/d
LN133	32	−4.8	$+9 \times 10^{-7}$	5.7×10^{12}	0.96
LN134	28	−5.6	$+8 \times 10^{-7}$	5.0×10^{12}	0.36
LN135	29	−3	$+5.1 \times 10^{-8}$	3.2×10^{11}	—

As shown in figure 4.1, the C–V characteristics of samples LN133 and LN134 are shifted to the left along the horizontal axis, indicating the presence of a positive oxide charge in the LN films of both samples. By applying standard methods of C–V analysis, we determined the effective density of positive charge and the effective density of states in the studied Si–LN–Al heterostructures [5]. Table 4.2 summarizes these results.

As shown in table 4.2, similar to previous cases where films were deposited without the ion assist effect, the studied heterostructures operate in the accumulation regime under zero or low positive bias ('+' at the Al contact), allowing them to be analyzed as metal–insulator–metal systems. It is important to note that the effective density of CLCs in the film deposited at higher reactive gas pressure (sample LN133) is comparable to that fabricated under nearly optimal conditions (sample LN134). In chapter 2 and in our work [1], we demonstrated that increasing the reactive gas pressure fivefold compared to the optimal value (0.1 Pa) leads to intensive bombardment of the film surface by plasma particles, resulting in the formation of defective layers, a lack of texture, and the creation of a 'parasitic' $LiNb_3O_8$ phase. Consequently, the effective positive charge in the deposited films is not caused by the formation of the $LiNb_3O_8$ phase or by defects produced by bombardment. Almost the same value of positive charge was observed by other researchers in LN films fabricated by the laser ablation method [6], and it was attributed to the formation of lithium vacancies. On the other hand, films deposited in an $Ar + O_2$ reactive gas environment (sample LN135) demonstrate the lowest effective density of positive charge. Thus, the presence of oxygen in the reaction chamber plays a key role in reducing built-in charge.

It is generally accepted that mechanical strain affects the structural and electrical properties of deposited coatings. Strains in a film can be divided into two types: internal and external. Internal strains are triggered by point defects and lattice distortions in a deposited layer and occur inevitably during the RFMS process due to relatively intensive bombardment of the substrate surface by high-energy plasma particles. The crystal lattice mismatch and the difference in linear expansion coefficients at the film–substrate interface are the main sources of external strain. According to the classification given in chapter 3, oxide charge is the sum of the interface-trapped charge (Q_{it}), oxide-trapped charge (Q_{ot}), fixed oxide charge (Q_f), and the mobile ionic charge (Q_{mi}), which is immobile at temperatures below 390 °C. Since, in our case, Q_{eff} does not depend on the substrate type and orientation or the intensity of bombardment of the film surface, we can exclude Q_{it} from consideration and assume that $Q_{eff} = Q_{ot} + Q_f$. Thus, internal strain is the main source of positive oxide charge observed in the studied LN films.

It was shown in [7] that the technological parameters of RFMS greatly affect the degree of mechanical strain in LN films. Specifically, this paper recommended that sputtering should be conduced at a pressure of 1.3 Pa in a reaction chamber using an $Ar(80\%) + O_2(20\%)$ gas mixture as the reactive environment. The authors indicated that these conditions enable the formation of LN films with minimal internal strains associated with oxygen vacancies. On the other hand, there is strong evidence that the presence of O_2 in the reaction chamber influences plasma properties, increasing the concentration of Li ions [8]. Analysis of various models of defect formation suggests that $Nb_{Li}^{4\cdot}$ antisite positively charged defects (Nb at a Li site) are the most likely source of positive charge in LN single crystals [9]. If so, the presence of oxygen reduces the concentration of lithium vacancies (V_{Li}) and consequently the formation of positively charged complexes $Nb_{Li}^{4\cdot}$.

Before we determine the distribution of electrically active defects and impurities in the Si substrate and in the LN films, it is interesting to analyze the $C–V$ characteristics of sample LN135. As clearly seen in figure 4.1, a peak and further modulation of capacitance are observed in the range of positive biases, which radically differs from samples LN133 and LN134. This unusual behavior of the $C–V$ curves can be explained using the framework of the electrical characteristics of isotype heterojunctions. The authors of [10] explained a similar $C–V$ curve for n–n heterojunctions based on the double depletion layer model. The heterojunction in

Direct over-barrier transition

ΔE_c

ϕ_1

E_{c1}

Two-stage charge transition

ϕ_2

E_{c2}

Applied voltage

Fermi level

E_{v2} Si

Interface states

LiNbO$_3$

E_{v1}

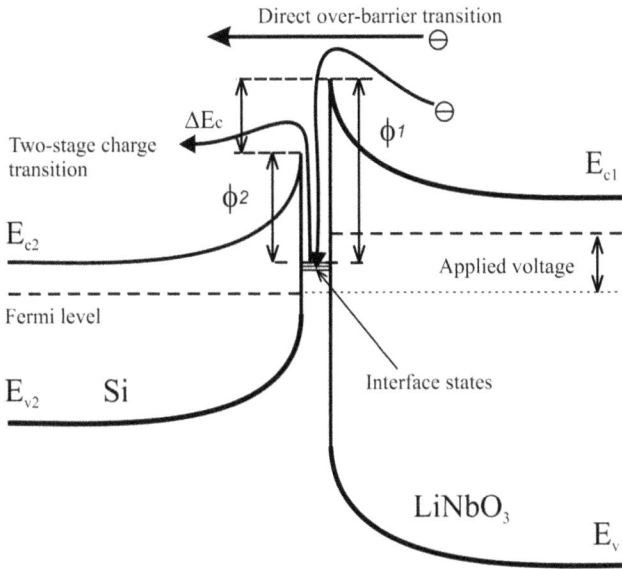

Figure 4.2. Schematic band diagram of an n–n isotype heterojunction and applied voltage using the model of two series-connected back-to-back Schottky diodes, forming a double depletion layer. The arrows indicate two possible charge transport mechanisms.

this model is represented as two Schottky diodes connected in series back-to-back, provided a high density of electronic states exists at the interface, which is equivalent to the presence of a metal contact between two barriers (see figure 4.2).

In heterostructures fabricated in an Ar(60%) + O$_2$(40%) environment by RFMS, an extended intermediate layer with variable composition is formed (see chapter 3 and [4]). Assuming this layer has high conductivity, which is supported by *I–V* analysis, it should have a large concentration of CLCs. Thus, the presence of this layer can be modeled as a metal contact between two semiconducting materials, as shown in figure 4.2.

Figure 4.2 indicates that regardless of the polarity of the applied voltage, the *C–V* characteristic decreases with bias and is affected by the capacitance of the respective reverse-biased Schottky diode. Regarding electrical conductivity, which will be analyzed in the next section, it is limited by the direct over-barrier charge transition as well as by two-stage charge transport, indicated by arrows in figure 4.2. Let us determine the spatial distribution of electrically active defects and impurities in the Si substrates and LN films.

Within the framework of the depletion layer approximation, the concentration of doping impurities in a semiconductor $N(x)$ can be expressed through the measured high-frequency capacitance C by the following expression [11]:

$$N(x) = \pm \frac{2}{q\varepsilon_0\varepsilon_s}\left[\frac{\mathrm{d}}{\mathrm{d}V}\left(\frac{S}{C}\right)^2\right] \qquad (4.1)$$

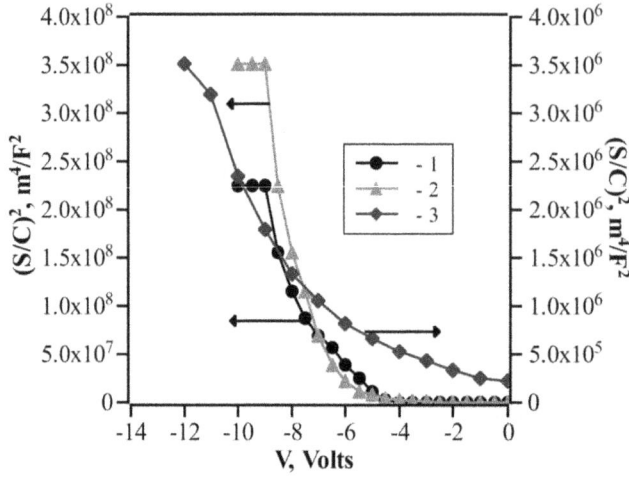

Figure 4.3. $C-V$ characteristics of the studied Si–LN–Al heterostructures in $(S/C)^2-V$ coordinates in the depletion regime ('−' at the Al contact).

where '−' represents an n-type semiconductor and '+' corresponds to a p-type semiconductor. In equation (4.1), q is the electron charge, $\varepsilon_0 = 8.85 \times 10^{-12}\,\mathrm{F\,m^{-1}}$ is the dielectric permittivity of vacuum, ε_s is the dielectric permittivity of the material, S is the contact area, and V is the applied voltage. Thus, the slope of the experimental $C-V$ characteristic in $(S/C)^2-V$ coordinates is determined by the impurity concentration $N(x)$ in the depletion layer. The coordinate x in equation (4.1) is calculated using the measured capacitance of an MIS structure [11]:

$$x = \varepsilon_0 \varepsilon_s S \left(\frac{1}{C} - \frac{1}{C_i} \right).\qquad(4.2)$$

Here, C_i is the capacitance of the dielectric layer (the capacitance of the MIS structure in the accumulation regime). As for Schottky diodes, a doping profile in a semiconductor can also be determined using equations similar to (4.1) and (4.2) using the approximation $1/C_i \to 0$. Figure 4.3 shows $C-V$ characteristics of Si–LN–Al heterostructures in the depletion regime ('−' at the Al contact) in $(S/C)^2-V$ coordinates.

Donor distributions in the Si substrates of the studied heterostructures, obtained by both graphical integration of experimental $C-V$ curves and the use of equations (4.1) and (4.2), are shown in figure 4.4.

As seen in figure 4.4, donors are distributed at nonuniform depths in the substrates of all samples. In samples LN133 and LN134, donor concentration declines with depth, reaching the nominal concentration attributed to the n-type Si substrates used in our experiments at a depth of 500 nm [5]. It is worth noting that in heterostructures fabricated in an Ar(60%) + O$_2$(40%) gas mixture (sample LN135), the donor concentration is considerably higher than in those formed in a pure Ar atmosphere, extending to a depth of up to 1 μm. This result can be explained by the

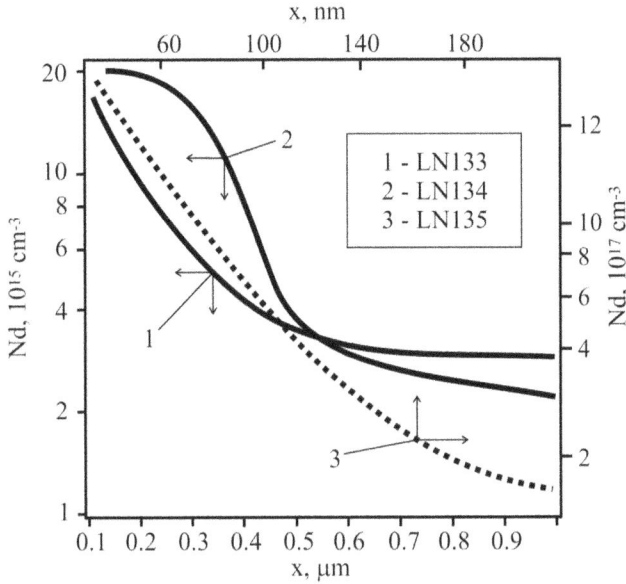

Figure 4.4. Donor concentration profiles in a silicon substrate for the studied Si–LN–Al heterostructures.

fact that diffusive molecular oxygen leads to the formation of donor centers in silicon [12]. This process is very sensitive to substrate temperature and oxygen pressure. Molecular oxygen has a diffusion coefficient of $D = 10^{-9}\,\mathrm{cm^2\,s^{-1}}$ at 450 °C, which is tenfold higher than that of atomic oxygen. On the other hand, it was shown in [8] that the presence of oxygen in the reaction chamber affects the reactive gas composition, increasing the concentration of Li atoms in the plasma. As a result, Li atoms, which are shallow donors in Si [13] with a diffusion coefficient of $D = 10^{-11}\,\mathrm{cm^2\,s^{-1}}$, penetrate the substrate and do not form a uniform impurity distribution, which fully agrees with our experimental results.

According to the Schottky diode theory, in the case of uniform donor distribution in a semiconductor with concentration N_d, the capacitance of the Schottky barrier is a linear function of the applied depleting voltage in $(S/C)^2$–V coordinates, and it is described by the following equation [11]:

$$\left(\frac{S}{C}\right)^2 = \frac{2(\varphi_D - V)}{q\varepsilon\varepsilon_0 N_d} \tag{4.3}$$

where φ_D is the diffusion potential, ε is the dielectric constant of the semiconductor, and V is the applied voltage. Thus, the donor concentration can be determined through the slope of the $(S/C)^2$—V graph, and φ_D is defined by the intercept with the horizontal axis.

Figure 4.5 shows the C–V characteristic of the LN135 sample under positive bias, corresponding to the depletion zone of the LN layer (attributed to the capacitance of the Schottky barrier in LN) in $(S/C)^2$–V coordinates.

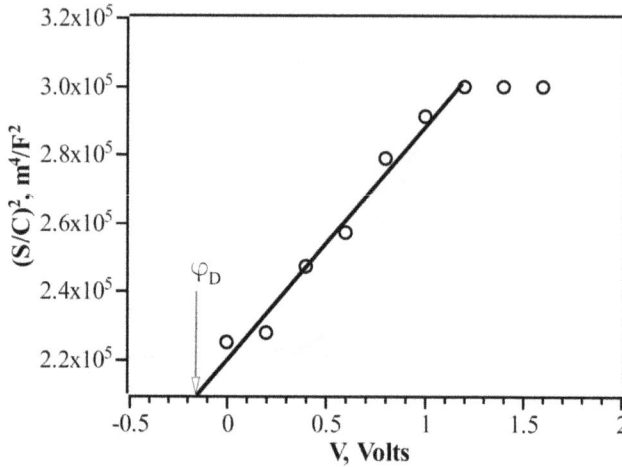

Figure 4.5. Plot of $(S/C)^2$ vs. V for sample LN135, associated with the depletion zone in LN at $T = 295$ K.

From figure 4.5, it follows that there is a uniform donor distribution in the LN film with a concentration of $N_d = 7 \times 10^{17}$ cm^{-3}, which is derived from the slope of the linear part in figure 4.5 using equation (4.3) and taking $\varepsilon = 28$ [14]. Furthermore, the diffusion potential, determined from the same graph, is $\varphi_D = 0.3$ V.

We also determined the energy distribution of surface states D_{ss} at the Si–LN interface through the shift of experimental C–V curves relative to the theoretical ones. These distributions in the upper half of the Si bandgap for the studied samples are shown in figure 4.6.

Figure 4.6 shows that the energy distribution does not change significantly when the Ar pressure in the reaction chamber is increased, and D_{ss} is one order of magnitude lower than the values reported by other authors [15]. The density of states in heterostructures fabricated in an Ar(60%) + O$_2$(40%) reactive gas mixture (sample LN135) greatly exceeds the D_{ss} of those grown in pure Ar (samples LN133 and LN134). This agrees with the model used to describe the electrical properties of sample LN135, i.e. a double depletion layer with a high density of states between two Schottky diodes (see figure 4.2).

Films fabricated in all three regimes exhibited ferroelectric properties [5], as reflected in the ferroelectric hysteresis loops shown in figure 4.7.

The results of the analysis of P–E loops shown in figure 4.7 are presented in table 4.3. These results indicate that the remnant polarization does not depend on the deposition regime used and is close to that of single-crystal lithium niobate, $P_r = 71$ µC cm^{-2} [16].

It is important to note that the remnant polarization for all samples is considerably higher than that for LN films with arbitrary grain orientation grown without the ion assist effect, as studied in chapter 3. Apparently, a single-axis ⟨0001⟩ texture in LN films deposited under the ion assist regime (see table 4.1) promotes effective polarization reversal in the studied films, with the result that their ferroelectric properties approach those of single-crystal lithium niobate. The coercive field is

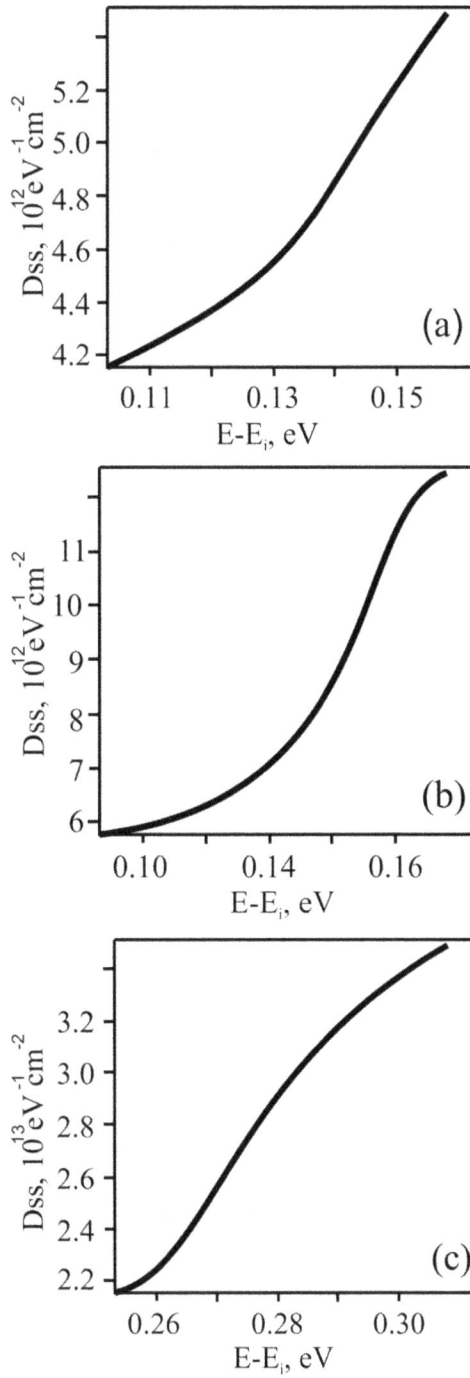

Figure 4.6. Energy distributions of surface states in the upper half of the bandgap of Si for the studied Si–LN–Al heterostructures: (a)—sample LN133, (b)—sample LN134, (c)—sample LN135.

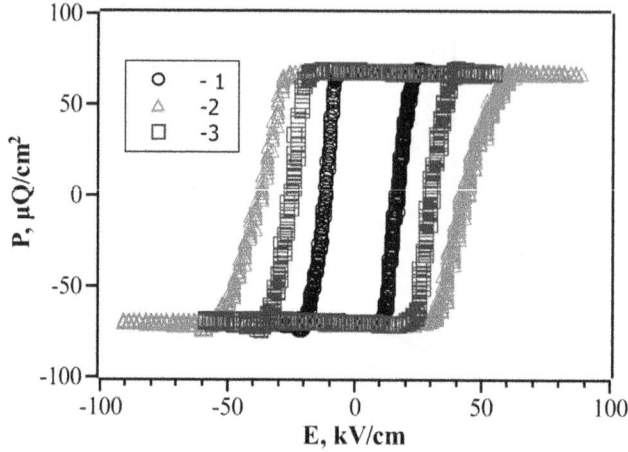

Figure 4.7. *P–E* loops for the studied heterostructures: 1—sample LN133, 2—sample LN134, 3—sample LN135.

Table 4.3. Parameters of *P–E* loops of the studied heterostructures.

Sample #	Remnant polarization, P_r ($\mu C\,cm^{-2}$)	Coercive field, E_c ($kV\,cm^{-1}$)	Vertical shift of *P–E* loop, ΔP_r ($\mu C\,cm^{-2}$)	Built-in field, E_b ($kV\,cm^{-1}$)
LN133	69	14.0	−1	2.8
LN134	69	39.5	−1	2.8
LN135	69	27.3	−1	2.0

minimal for films fabricated at higher working pressures and apparently depends on ferroelectric domain dynamics; however, this issue is beyond the scope of this monograph and requires additional study. Moreover, like the LN films with arbitrary grain orientation (see chapter 3), the textured films studied here also demonstrate a built-in field and a slight shift of *P–E* loops along the vertical axis, which can be caused by preferential domain orientation. As discussed in chapter 3, the most likely origin of this built-in field in LN films is oxide- or interface-trapped charge. Indeed, as follows from the *C–V* analysis, the positive oxide charge present in LN films does not depend on the reactive gas pressure (samples LN133 and LN134 in table 4.2) and declines when oxygen is present in the reaction chamber (sample LN135). This fact correlates with data related to the built-in field in table 4.3.

There is evidence that various defects and space charge effects greatly influence *P–E* loops [17, 18]. The coercive field in this case can be expressed as:

$$E_c = E_c^{/} - E_{sc} + E_{\text{defect}} \tag{4.3}$$

where E_c is the coercive field, $E_c^{\prime}$ is the coercive field attributed to domain motion, E_{sc} is the space charge field, and E_{defect} is the field generated by defects in the ferroelectric film. The minimal coercive field (see table 4.3) for films that are fabricated at higher working gas pressure (sample LN133) and have arbitrary grain orientation is caused by the terms $E_c^{\prime}$ and E_{sc} in equation (4.3). In this case, a lower electric field is needed for ferroelectric domain reversal compared to the field required for the textured $\langle 0001 \rangle$ LN films (samples LN134 and LN135 in table 4.1). Further study is needed to analyze the I–V characteristics and frequency dependence of the AC conductivity of Si–LN–Al heterostructures fabricated by RFMS under different regimes.

4.1.2 Current–voltage characteristics of Si–LiNbO$_3$–Al heterostructures

The I–V characteristics of Si–LN–Al heterostructures in ln J–V coordinates under forward bias ('+' at the Al contact) are shown in figure 4.8.

The rapidly increasing sections of the I–V curves with applied voltage, observed in figure 4.8, are influenced by the barrier properties of the Si–LN heterojunction, whereas the later, sloping sections are attributed to the bulk properties of the films (see discussion in chapter 3). It follows from figure 4.8 that the barrier properties of structures fabricated in an Ar environment are not influenced by the reactive gas pressure (samples LN133 and LN134), but these properties depend on the reactive gas composition (sample LN135 in figure 4.8).

First, let us analyze possible conduction mechanisms in sample LN133 based on the band diagram of the studied heterostructures proposed in chapter 3. Under a relatively low forward bias, the most probable barrier-limited conduction mechanisms are Fowler–Nordheim tunneling (at low temperatures) and Richardson–

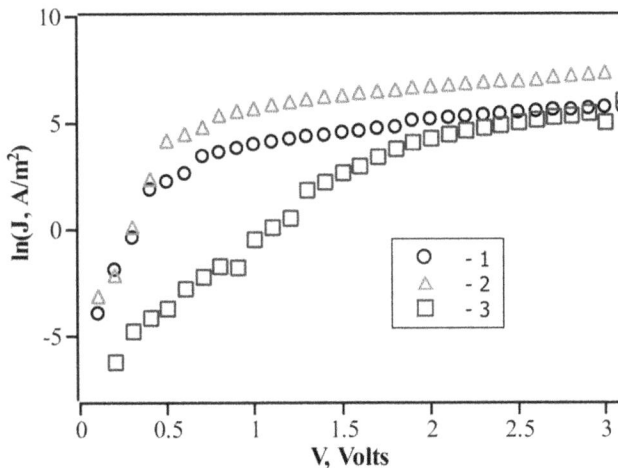

Figure 4.8. Typical I–V characteristics of Si–LN–Al heterostructures fabricated using different regimes of the RFMS method under forward bias ('+' at the Al contact) at a temperature of $T = 300$ K. 1—sample LN133, 2 —sample LN134, 3—sample LN135.

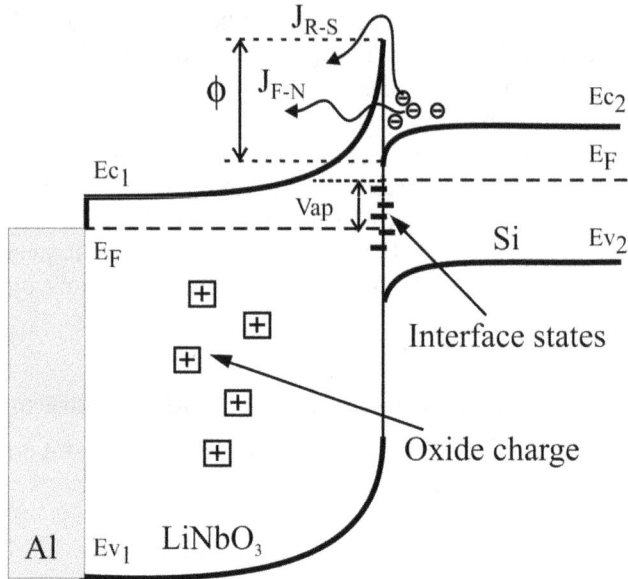

Figure 4.9. Band diagram of sample LN133 under low forward bias V_{ap}. The curved arrows indicate two conduction mechanisms, limited by the potential barrier ϕ.

Schottky emission (at higher voltages). The electronic currents corresponding to these mechanisms are labeled J_{F-N} and J_{R-S} in figure 4.9.

As shown in chapter 1, in the case of Fowler–Nordheim tunneling, the $I–V$ characteristics should be linear in $\ln(J/E^2)–1/E$ coordinates, where J is the current density and E is the electric field strength. The slope of this linear graph is determined by the potential barrier height ϕ and can be described by the following expression [19]:

$$B = \frac{8\pi\sqrt{2m^*}\,\phi^{3/2}}{3qh}.$$ (4.4)

Here, h is Planck's constant, m^* is the effective mass of the electron, and ϕ is the average potential barrier height. Indeed, the $I–V$ characteristics of Si–LN–Al heterostructures fabricated at relatively high working pressure (sample LN133) and recorded at low temperatures have a linear section in Fowler–Nordheim coordinates (see figure 4.10), which suggests that tunneling takes place in this temperature range.

Taking into account corrections caused by polarization effects in a ferroelectric film (see equation (3.41) in chapter 3), the potential barrier height derived from the slope of the linear part of the $I–V$ curve in figure 4.10 is $\phi = 0.4\,\text{eV}$. This value is less than those of the heterostructures similar to LN134 determined in chapter 3 (the value of the conduction band offset ΔE_c). Additionally, it is evident from figure 4.10 that the tunneling component of conductivity declines with temperature, which is reflected in the shortening of the linear section of the $I–V$ curves.

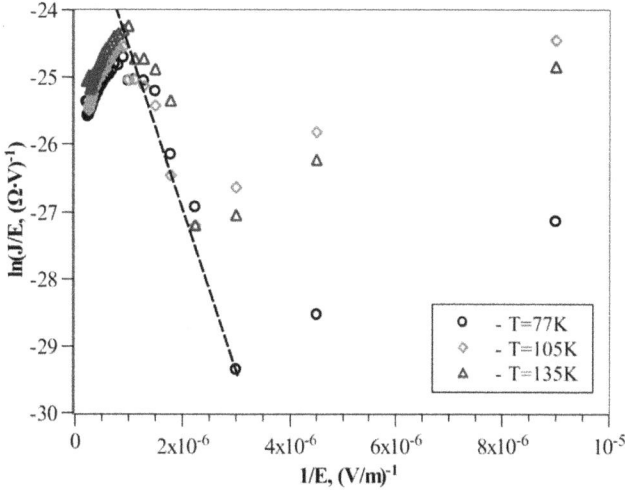

Figure 4.10. *I–V* characteristics of sample LN133 in Fowler–Nordheim coordinates. The dashed line indicates Fowler–Nordheim tunneling through the potential barrier ϕ (current J_{F-N} in figure 4.9).

Next, let us analyze the *I–V* characteristics of sample LN135 under a relatively low bias (low applied electric field). According to [10], within the framework of the double depletion model (two Schottky barriers connected in series back-to-back), the *I–V* characteristics are limited by a two-stage charge transport mechanism (see figure 4.2) and depend greatly on the Schottky barrier heights ϕ_1 and ϕ_2. The current densities within the two diodes are described by the following equations:

$$J_1 = Js_1(\exp(V_1/V_{o1}) - 1); \quad J_2 = Js_2(\exp(V_2/V_{o2}) - 1). \tag{4.5}$$

Here, V_1 and V_2 are the voltage drops across barriers ϕ_1 and ϕ_2 respectively, V_0 is a coefficient dependent on the conduction mechanism through the barrier, and J_{s1} and J_{s2} are the saturation currents, given by the following equations:

$$Js_1 = A*T^2 \exp\left(-\frac{q\phi_1}{kT}\right); \quad Js_2 = A*T^2 \exp\left(-\frac{q\phi_2}{kT}\right). \tag{4.6}$$

Here, $A*$ is the effective Richardson constant, k is the Boltzmann constant, and T is the temperature. Since $J_{s1} = J_{s2}$ and $V = V_1 + V_2$, we can obtain the following expression for total current density:

$$J_{tot} = \frac{Js_1 Js_2(\exp(qV/kT) - 1)}{Js_2 + Js_1 \exp(qV/kT)}. \tag{4.7}$$

The *I–V* characteristics described by equation (4.7) have an inflection point, corresponding to the following voltage and current density:

$$V_{inf} = \frac{kT \ln(Js_2/Js_1)}{q}; \quad J_{inf} = \frac{1}{2}Js_1(Js_2/Js_1 - 1). \tag{4.8}$$

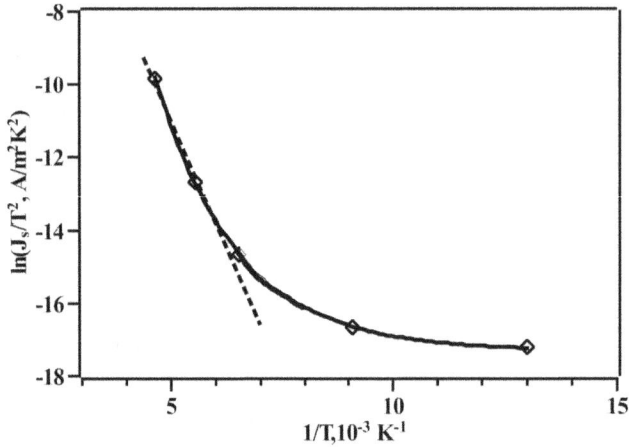

Figure 4.11. Temperature dependence of the saturation current J_s for sample LN135 in $\ln(J_s/T^2)$–$1/T$ coordinates.

In our case, $J_{s2} \gg J_{s1}$ and $J_{inf} = J_{s2}/2$. Furthermore, as follows from equation (4.6), the potential barrier height ϕ can be found from the linear section of the $\ln(J_s/T^2)$–$1/T$ graph [4]. This graph for sample LN135 is shown in figure 4.11.

Figure 4.11 shows that the temperature dependence of the saturation current is approximated by a linear law $\ln(J_s/T^2)$–$1/T$ only at relatively high temperatures (140 K–300 K). The deviation from a straight line observed at low temperatures results from the fact that within this temperature range, Fowler–Nordheim tunneling rather than Richardson–Schottky emission affects the I–V curves. The Schottky barrier height (at the Si site), determined from the slope of this linear section in figure 4.11, equals $\phi_2 = 0.27$ eV.

On the other hand, by substituting equations (4.6) into (4.8) for V_{inf}, we can obtain the following expression:

$$V_{inf} = \Delta E_c + \frac{kT}{q} \ln((m_2/m_1)^{3/2}). \tag{4.9}$$

Here, we take into account that, as shown in figure 4.2, the conduction band offset equals $\Delta E_c = \phi_1 - \phi_2$. In addition, we use the Richardson constant described by equation (1.8) from chapter 1. Equation (4.9) suggests that the temperature dependence of V_{inf} should be a linear function with a slope that depends on the ratio m_2/m_1 between the effective masses of carriers in the substrate and the LN film. The intercept of this linear graph with the vertical axis gives the conduction band offset ΔE_c in the heterojunction. Figure 4.12 demonstrates the temperature dependence $V_{inf}(T)$ for sample LN135.

Analyzing the temperature dependence $V_{inf}(T)$ shown in figure 4.12, we obtain the following parameters: $\Delta E_c = 0.16$ eV and $m_2/m_1 = 177$ [4]. Furthermore, using the barrier height ϕ_2 obtained earlier, we calculate the Schottky barrier at the LN site as follows: $\phi_1 = \phi_2 + \Delta E_c = 0.43$ eV.

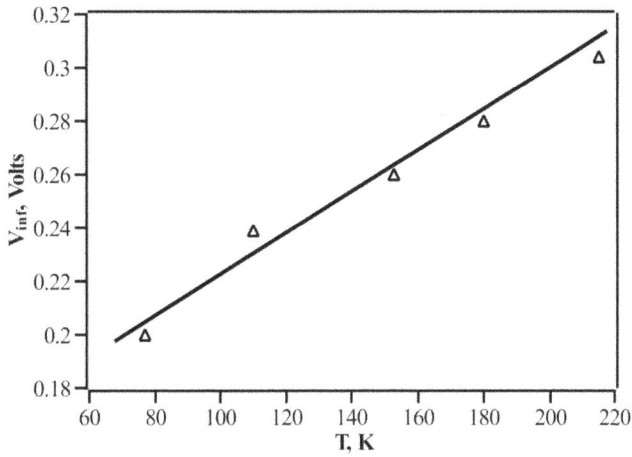

Figure 4.12. Temperature dependence of the voltage associated with the inflection point of the I–V curves for sample LN135.

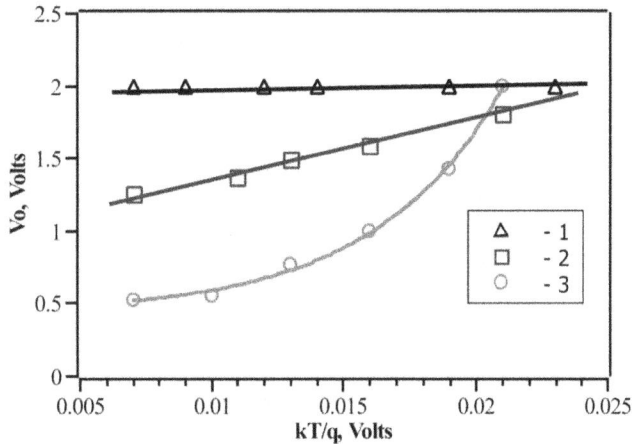

Figure 4.13. Temperature dependence of $V_0(T)$ (V_0 vs. kT/q plots) for the studied samples under high applied electric fields. 1—sample LN133, 2—sample LN134, 3—sample LN135.

Next, let us analyze the I–V characteristics of the studied samples under high electric fields when the current is limited by the bulk properties of the LN films. These I–V curves can be described by an equation similar to (4.5). An original method of determining the most probable charge transport mechanism in diode-like heterostructures was proposed in work [20]. This approach is based on an analysis of the temperature dependence of the parameter V_0 in equation (4.5), which is determined by the slope of the I–V characteristics in ln J–V coordinates. Figure 4.13 shows the temperature dependence of V_0 (T) attributed to the I–V characteristics of the studied samples and analyzed in our work [5].

According to the classification proposed in [20], curve 1 in figure 4.13 corresponds to the field emission mechanism that occurs when the parameter V_0 does not depend on temperature. Curve 2 corresponds to Richardson–Schottky emission, which agrees with the results for heterostructures similar to sample LN134, as described in chapter 3. Curve 3 is observed if thermally assisted tunneling dominates, and the intercept of the corresponding horizontal section of the graph with the vertical axis gives the value of V_{00}, which is defined as (see chapter 1):

$$V_{00} = \frac{h}{4\pi} \sqrt{\frac{N_d}{m^* \varepsilon \varepsilon_0}}. \qquad (4.10)$$

Here, N_d is the concentration of ionized donors in the dielectric layer, h is Planck's constant, m^* is the effective mass of the carrier, and ε is the dielectric constant of the material.

As shown in chapter 1 (see equation (1.21)), when the field emission mechanism corresponds to non-activated hopping conductivity over CLCs (defects), the current density can be described as follows:

$$J = J_0 \exp(-(E_o/E)^{1/4}) \qquad (4.11)$$

Here, E is the applied electric field, J_0 is a field-independent constant, and E_0 is the characteristic field, denoted by the following formula:

$$E_0 = \frac{16}{D(E)a^4 q}. \qquad (4.12)$$

Here, $D(E)$ is the energy density of localized states near the Fermi level, q is the electron charge, and a is the localization length, which is taken to be $a \approx 3a_0$ [21], where a_0 is the crystal lattice parameter.

As follows from equation (4.11), the I–V characteristics are straight lines in $\ln(J)$ vs. $E^{-1/4}$ coordinates with a slope giving E_0. Figure 4.14 shows the I–V characteristic of sample LN133 under a high electric field in $\ln(J)$ vs. $E^{-1/4}$ coordinates.

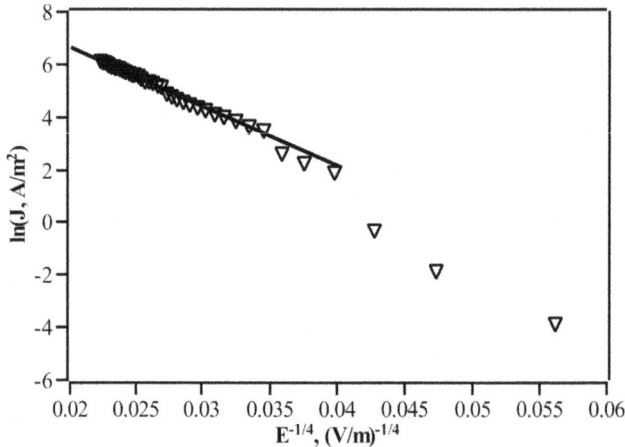

Figure 4.14. Electric field dependence of current density for sample LN133.

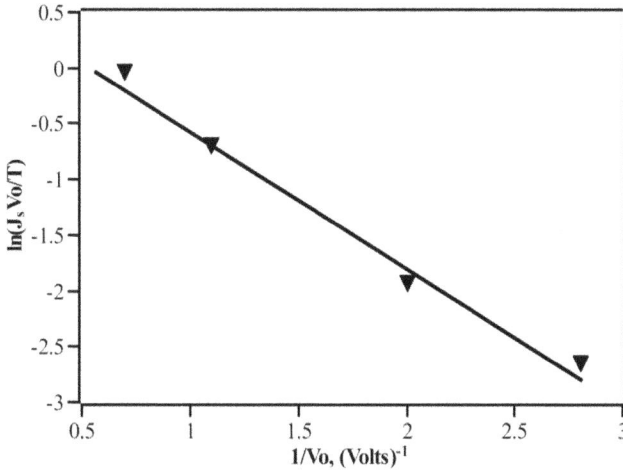

Figure 4.15. The $\ln(J_s V_0/T) - 1/V_0$ plot for sample LN135.

Figure 4.14 clearly exhibits a linear section, in full agreement with equation (4.11). Taking the lattice parameter for LN to be $a_0 = 5.1$ Å, we estimated the energy density of states near the Fermi level in our work [5]. On the other hand, the average hopping range and effective concentration of CLCs in the framework of this mechanism are described by:

$$R = (q \cdot D(E) \cdot E)^{-1/4} \tag{4.13}$$

$$N_t = \frac{2}{3\pi R^3}. \tag{4.14}$$

Applying equations (4.13) and (4.14) for sample LN133, we obtain the parameters $R = 9.0$ Å and $N_t = 6 \times 10^{17}$ cm^{-3}.

As regards sample LN135, within the framework of thermally assisted tunneling, the parameter V_{00} obtained by extrapolation of the $V_0(T)$ dependence to $T \rightarrow 0$ gives the donor concentration using equation (4.10): $N_d = 4.0 \times 10^{18}$ cm^{-3} [5]. Moreover, analysis of equation (1.14) suggests that the donor concentration should be a straight line in $\ln(J_s V_0/T) - 1/V_0$ coordinates with a slope equal to the barrier height φ_b. Since, in this case, we deal with bulk-limited conductivity, φ_b can be interpreted as an intergranular barrier. The $\ln(J_s V_0/T) - 1/V_0$ plot for sample LN135 is shown in figure 4.15.

The intergranular barrier derived from the slope of this graph is $\varphi_b = 0.7$ eV.

4.1.3 Impedance spectroscopy and AC conductivity of Si–LiNbO$_3$–Al heterostructures

Nyquist diagrams and the frequency dependencies of the imaginary parts of complex impedance Z'' and dielectric modulus M'' are shown in figure 4.16 for sample LN133 at different temperatures.

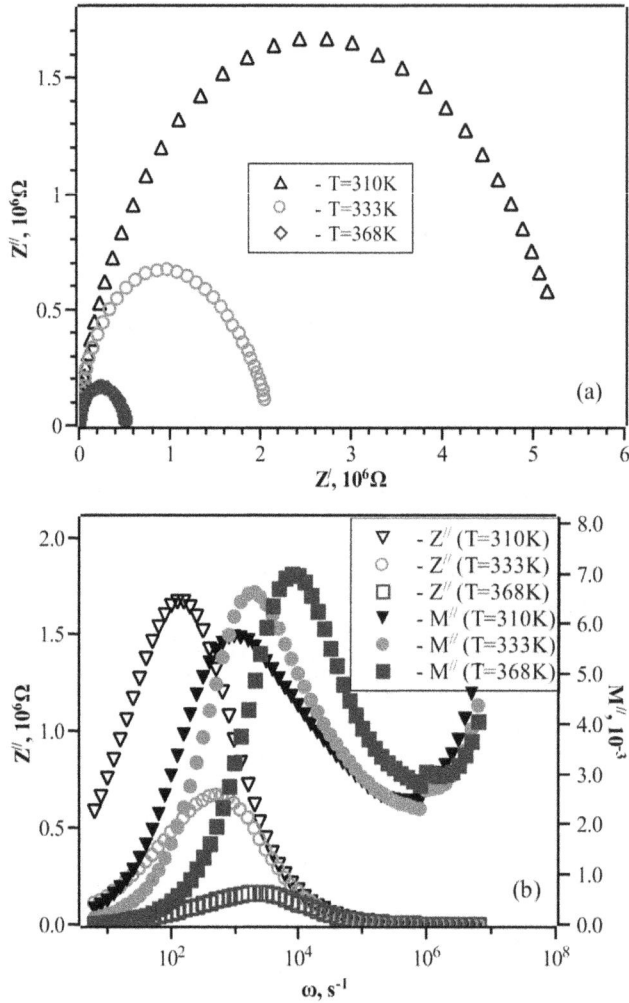

Figure 4.16. Nyquist diagrams (a) and the frequency dependencies of the imaginary parts of complex impedance Z'' and dielectric modulus M'' (b) for sample LN133.

Figure 4.16(a) indicates that we are dealing with conducting films where dielectric effects are minimal. As in chapter 3, we will analyze the studied sample using the equivalent circuit shown in figure 1.19 (see chapter 1), where R_b and C_b represent the resistance and capacitance of the bulk of LN grains, respectively, while R_{gb} and C_{gb} represent the resistance and capacitance of the grain boundaries. Considering the possible distribution of relaxation times, the complex impedance of this equivalent circuit can be written as:

$$Z^* = \frac{R_b}{1 + (j\omega\tau_1)^{1-\alpha_1}} + \frac{R_{gb}}{1 + (j\omega\tau_2)^{1-\alpha_2}}. \tag{4.15}$$

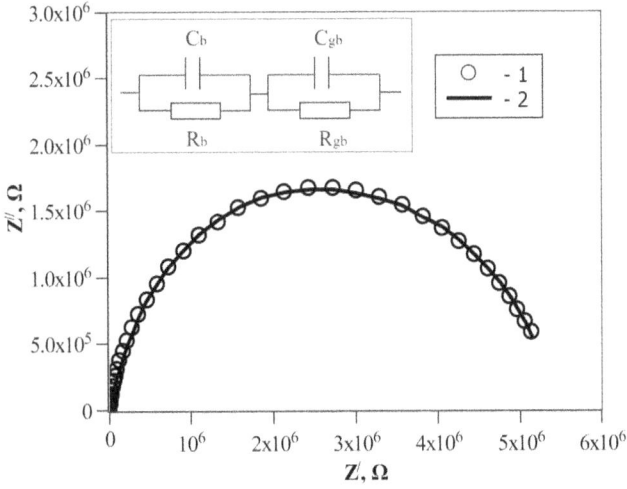

Figure 4.17. Nyquist plots for sample LN133. 1—experimental results, 2—theoretical curve calculated through equation (4.15) using the equivalent circuit shown in the inset.

Here, $\tau_1 = R_b C_b$ and $\tau_2 = R_{gb} C_{gb}$ are the relaxation times, and j is the imaginary unit. The coefficient α_i ($0 \leqslant \alpha_i \leqslant 1$) describes the width of the distribution spectrum of the relaxation time. The distribution of relaxation times is inferred from the fact that Nyquist diagrams are deformed and deviate from ideal semicircles (see figure 4.16). Using both spectra ($Z''(\omega)$ and $M''(\omega)$) shown in figure 4.16 and the methods of IS described in chapter 3, we obtained the following parameters for sample LN133: $R_{gb} = 3.3 \times 10^6$ ohms, $C_b = 7.7 \times 10^{-10}$ F [22]. Additionally, using the peak frequencies $\omega_m = \tau^{-1} = (RC)^{-1}$ attributed to the same spectra, two other parameters of the equivalent circuit were determined: $C_{gb} = 2.2 \times 10^{-9}$ F and $R_b = 1.3 \times 10^6$ ohms. The characteristic times attributed to the responses of the grain bulk and the grain boundaries in sample LN133 were $\tau_1 = 7 \times 10^{-3}$ s and $\tau_2 = 4 \times 10^{-4}$ s, respectively. By fitting the theoretical equation to the experimental results, we obtained coefficients $\alpha_1 = 0.16$ and $\alpha_2 = 0.26$. Experimental and theoretical Nyquist plots calculated using the parameters mentioned above are presented in figure 4.17.

As shown in chapter 2 and our work [1], at higher reactive gas pressures in a reaction chamber (sample LN133), polycrystalline films with arbitrary grain orientation and a two-phase composition (LN and $LiNb_3O_8$) are formed in the RFMS process. The authors of [23] demonstrated that under certain synthesis conditions, $LiNb_3O_8$ crystallites precipitate while retaining the epitaxial ratio of the initial LN crystallites. This is caused by the desorption of Li_2O and oxygen atoms from the inner space of LN grains, 'covered' by shells of the $LiNb_3O_8$ phase. In the first approximation, two-phase films can be represented as grains consisting of a LN core and grain boundaries composed of the $LiNb_3O_8$ phase. If d_b and d_{gb} denote the widths of the bulk core and grain boundaries, respectively, and we use the expression for the capacitance of a parallel plate capacitor, we can estimate the average width of the grain boundaries d_{gb} using the following system of equations:

$$\begin{cases} d_g = d_{gb} + d_b \\ \dfrac{d_{gb}}{d_b} = \dfrac{\varepsilon_{gb} C_b}{\varepsilon_b C_{gb}} \end{cases}. \tag{4.16a}$$

Here, d_g is the average grain size. From these equations, the width of the grain boundaries can be expressed as:

$$d_{gb} = \frac{d_g}{1 + \varepsilon_b C_{gb} / \varepsilon_{gb} C_b}. \tag{4.16b}$$

Using the values for C_{gb} and C_b calculated above and considering that $d_g \simeq 50$ nm [1], $\varepsilon_{gb}(\mathrm{LiNb_3O_8}) = 34$ [24], and $\varepsilon_b(\mathrm{LN}) = 28$ [14], we obtain the following results: $d_{gb} = 15$ nm, $d_b = 35$ nm, and $d_{gb}/d_b \approx 0.4$ [22].

Grain-boundary-limited conductivity is affected by charge transport mechanisms under an alternating signal. To investigate AC conductivity, resistances R_b and R_{gb}, derived from IS, were recalculated into conductivity σ via a simple relationship: $R = d/(\sigma S)$ (here, d is the thickness of d_b or d_{gb}, and S is the contact area). The conductivities σ_{gb} and σ_b of sample LN133 at different temperatures are shown in figure 4.18 in Arrhenius coordinates.

As shown in figure 4.18, grain boundary conductivity is an activated process obeying the following law:

$$\sigma_{gb} = \sigma_0 \cdot \exp\left(-\frac{qE_a}{kT}\right). \tag{4.17}$$

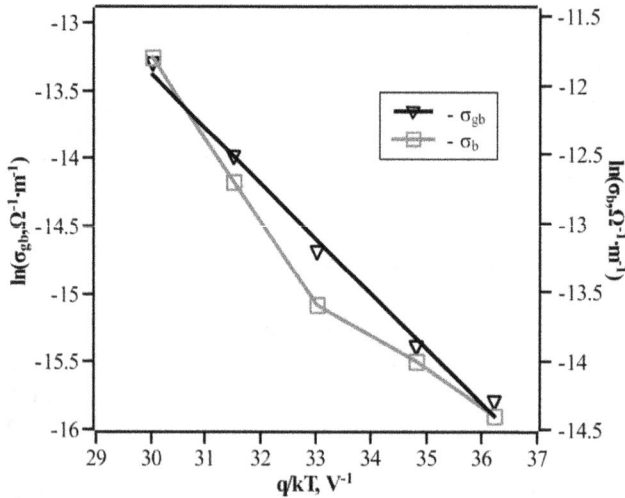

Figure 4.18. Temperature dependencies of the grain boundaries (σ_{gb}) and grain bulks (σ_b) of sample LN133 in Arrhenius coordinates.

Here, E_a is the activation energy, which can be obtained from the slope of a linear graph. The activation energy of $E_a = 0.4$ eV for sample LN133 is in good agreement with those determined for Si–LN heterostructures fabricated at lower working pressures in a reaction chamber [5, 25]. It was demonstrated that under low biases in those heterostructures, hopping conductivity dominates when carriers overcome an intergranular potential barrier of 0.4 eV. A similar value for the activation energy was obtained in [26], where conductivity was also limited by the hopping mechanism. Therefore, reactive gas pressure and the presence of the $LiNb_3O_8$ phase do not influence the intergranular barrier height. Apparently, defects segregated at the grain boundaries trap electrons from grains, causing Fermi level pinning at the interfaces.

Regarding the conductivity attributed to the bulk component of polycrystalline grains σ_b, it is affected by two activated processes with energies of 0.21 eV (at temperatures $T = 300$ K–350 K) and 0.77 eV (at temperatures $T = 350$ K–390 K) (see figure 4.18). From our perspective, conductivity σ_b is influenced by two mechanisms: correlated barrier-hopping (CBH) conductivity over CLCs with a barrier of 0.2 eV (at room temperature) and the thermal activation of carriers from deep centers with an energy of 0.7 eV in the bandgap of LN at temperatures $T = 350$ K–390 K. An activation energy of conductivity of $E_a = 0.7$ eV was reported by other authors [27, 28], who attributed it to trap centers associated with $Nb_{Li}^{4\bullet}$ antisite defects in LN.

As discussed in chapter 3, AC conductivity in disordered materials obeys the following 'universal' law:

$$\sigma_{ac}(\omega, T) = A(T)\omega^s. \tag{4.18}$$

Here, $A(T)$ is the frequency-independent parameter, ω is the angular frequency, and s is the exponent, which depends on the particular conductivity mechanism and can be derived from the slope of the $\ln\sigma - \ln\omega$ graph. This frequency dependence for sample LN133 is shown in figure 4.19.

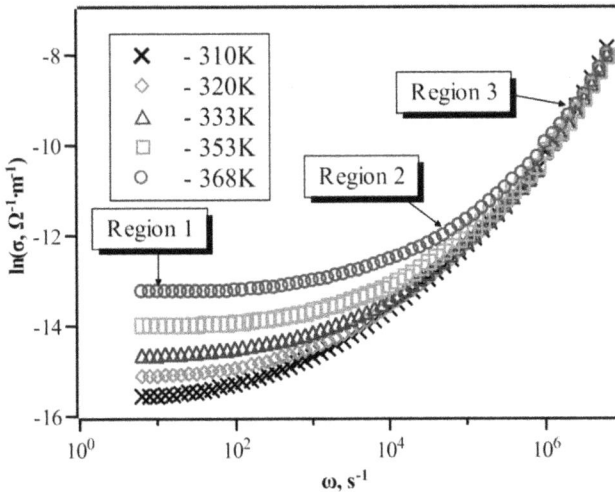

Figure 4.19. Frequency dependence of the conductivity of sample LN133 at various temperatures.

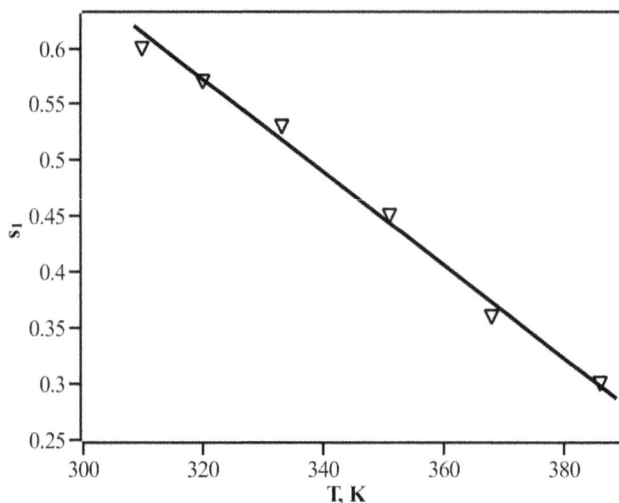

Figure 4.20. Temperature dependence of exponent s in equation (4.18), corresponding to region 2 in figure 4.19 for sample LN133.

Three regions, corresponding to different values of the exponent s in equation (4.18), are clearly visible in figure 4.19. The first region, in the low frequencies, corresponds to hopping DC conductivity with an activation energy corresponding to the barrier of $\varphi = 0.4\,\text{eV}$. Regions 2 and 3 in figure 4.19 correspond to two other charge transport mechanisms. Figure 4.20 illustrates the temperature dependence of the exponent s_1 in equation (4.18) for Region 2 in figure 4.19.

The diffusion-controlled relaxation (DCR) model, which describes charge transport [29], predicts an exponent of $s_1 \approx 0.5$, which is considerably lower than that for CBH conductivity ($s \approx 0.8$). According to the DCR model, charge transport occurs due to diffusive motion between two energetically stable states, and its activation energy has two components. The first component, i.e. the coulombic component, is attributed to the motion of a charge from its oppositely charged surroundings to a position between two adjacent sites. The second component is associated with the energy of deformation involved when a carrier penetrates a 'sluice' formed by bridge-like atoms separating two adjacent states. Within this model, the exponent s_1 is given by:

$$s_1 = 1 - \beta_{\text{CD}} \tag{4.20}$$

where β_{CD} is the Cole–Davidson parameter, which generally depends on the ratio τ/τ_d [29]. Here, τ and τ_d are relaxation times attributed to the diffusion-independent and -dependent processes, respectively. In the special case when $\omega \gg \tau_d/\tau^2$, we have $\beta_{\text{CD}} \to 0.5$, which is observed in our sample. For a more detailed description of charge transport mechanisms, let us analyze the temperature dependence of conductivity in the moderate frequency range ($\omega = 10^3 - 10^6\,\text{s}^{-1}$). Figure 4.21 shows the temperature dependence of conductivity for sample LN133 in Arrhenius coordinates, corresponding to the second range in figure 4.19.

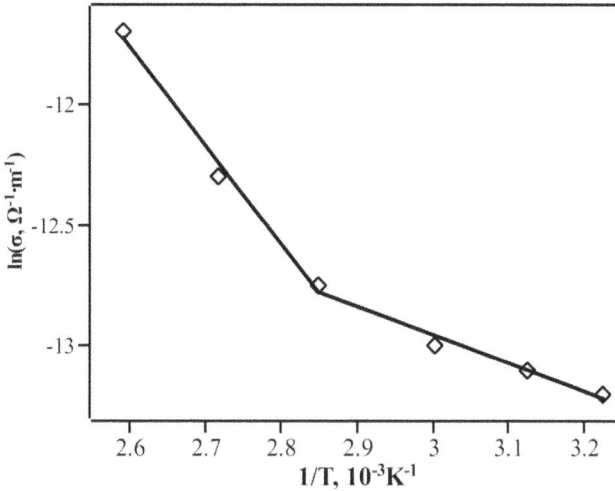

Figure 4.21. Temperature dependence of conductivity in Arrhenius coordinates for sample LN133, corresponding to region 2 in figure 4.19.

As shown in figure 4.21, in this frequency range, two activation processes occur with energies of $E_{a1} = 0.03$ eV and $E_{a2} = 0.2$ eV, corresponding to two linear sections at low and high temperatures, respectively. Some authors associate the process characterized by low activation energy E_{a1} with electronic hopping conductivity. The process with energy E_{a2} is also described within the framework of the DCR model, where the activation energy of AC conductivity E_a^{ac} is connected to that of DC conductivity E_a^{dc} as $E_a^{ac} \approx (1-s)E_a^{dc}$ [30], where s is the exponent in equation (4.18). In the studied frequency range, this relation holds because $E_a^{ac} = 0.2$ eV, $E_a^{dc} = 0.4$ eV, and $s = 0.5$.

Regarding the third region in figure 4.19, it was shown in our work [22] that the exponent s_2 in this range has a different temperature dependence, as shown in figure 4.22. This that indicates a new charge transport mechanism is involved.

According to the atomic double-well potential (ADWP) model discussed in [30], conductivity σ_{ac} occurs when carriers overcome exponentially distributed potential barriers. Within the framework of the ADWP model, the temperature dependence of the exponent s obeys the following law [30]:

$$s_2 = 1 - T/T_o \tag{4.21}$$

where T_0 is the characteristic temperature of the potential barrier distribution, which can be derived from the slope of a linear graph $s(T)$ (see figure 4.22). In our case, $T_0 = 312$ K, corresponding to an energy scattering of 0.03 eV.

Now let us analyze sample LN134. Unlike sample LN133, dielectric effects rather than conduction phenomena predominate here, as reflected in figure 4.23, which is similar to a Cole–Cole diagram.

The $C' - C''$ diagram is more useful than the traditional Cole–Cole diagram (which reflects the relationships between the imaginary and real parts of complex

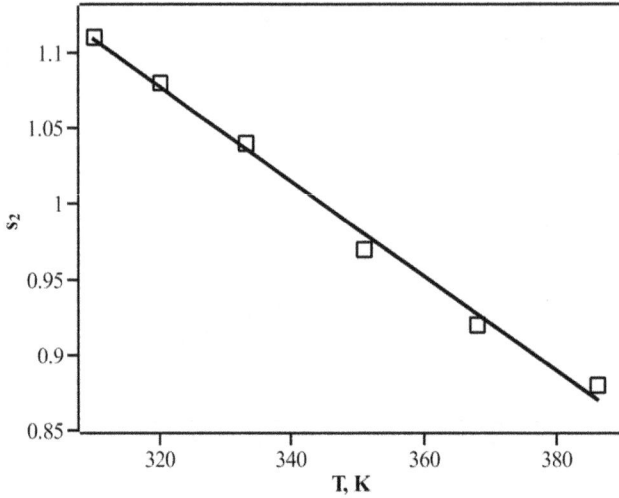

Figure 4.22. Temperature dependence of exponent s_2 in equation (4.18), corresponding to region 3 in figure 4.19 for sample LN133.

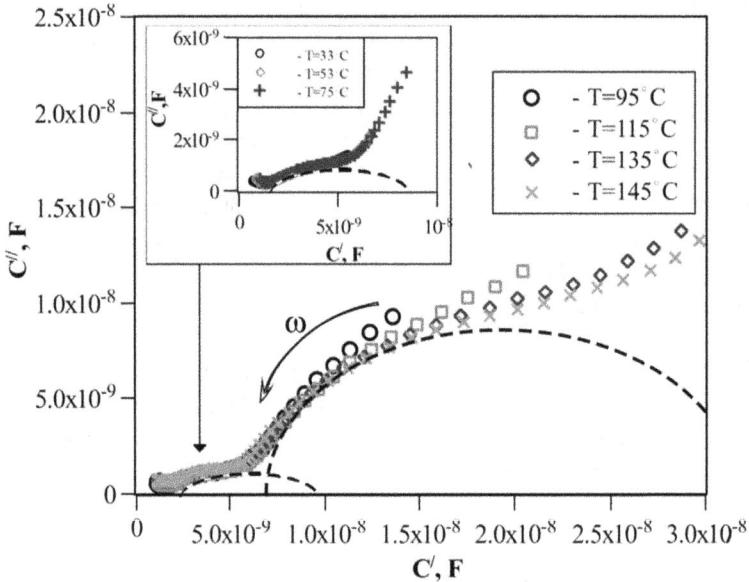

Figure 4.23. Frequency dependence of the real C' and imaginary C'' parts of complex capacitance $C^* = C' - jC''$ (a Cole–Cole-like diagram) for sample LN134 at different temperatures. The arrow indicates the direction of frequency increase.

dielectric permittivity) because it allows us to analyze how various components contribute to capacitance.

Two semicircle-like sections (marked by dashed lines), corresponding to low and high frequency ranges, are presented in figure 4.23. It is worth stressing that the large

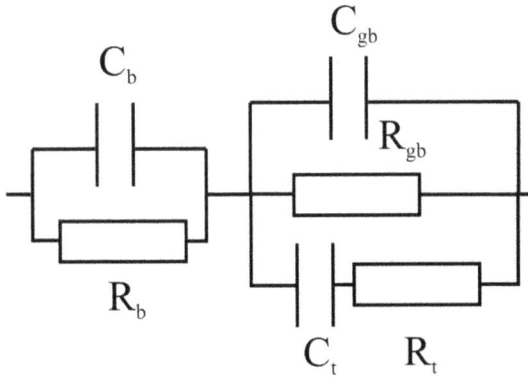

Figure 4.24. Modified equivalent circuit for LN films that considers the contribution of deep traps to the measured capacitance.

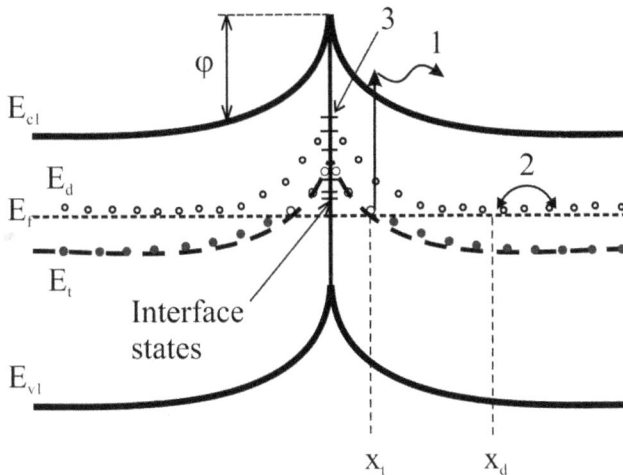

Figure 4.25. Band diagram of LN grains.

semicircle (a low-frequency response) is observed only in the temperature range of 95 °C–145 °C, and it disappears with a decrease in temperature when only one arc is presented. We now analyze the studied heterostructures using an equivalent circuit. However, in contrast to the circuit shown in the inset in figure 4.17, we use a modified equivalent circuit, displayed in figure 4.24.

In the equivalent circuit, R_b and C_b represent the bulk resistance and capacitance of grains, while R_{gb} and C_{gb} represent the resistance and capacitance of grain boundaries. R_t and C_t represent the resistance and capacitance resulting from a deep level in the bandgap of LN. Figure 4.25 shows the energy band diagram of two materials (two LN polycrystalline grains) containing deep traps and shallow donors with energies E_t and E_d, respectively, as well as interface states.

Among the numerous possible processes influencing relaxation phenomena in the studied films, the most frequently observed processes are marked by numbers in figure 4.25. At high frequencies, when only the charge at the edge of the depletion zone (corresponding to coordinate x_d in figure 4.25) follows the AC signal, the measured capacitance equals the capacitance of the depletion layer and corresponds to C_{gb} in figure 4.24. At lower frequencies and relatively high temperatures, when the thermal emission rate from deep traps exceeds the test frequency ω, the differential charge generated by the capture and re-emission of carriers from deep levels contributes to the capacitance. This change of charge occurs at the point where the Fermi level intersects the deep level E_t (the crossover point with coordinate x_t in figure 4.25). This carrier exchange, labeled '1' in figure 4.25, is accounted for by capacitor C_t and resistor R_t in the equivalent circuit shown in figure 4.24. Moreover, charge exchange with interface states (process '3' in figure 4.25) can also affect the measured capacitance. Thus, at low frequencies, the total measured capacitance is influenced by the 'slow' processes '1' and '3' in figure 4.25, occurring at the grain boundaries. At high frequencies, the main contribution is attributed to the bulk of the grains, where the 'fast' process of hopping conduction via shallow CLCs in the bandgap of LN dominates (see figure 4.25). In this model, each semicircle in figure 4.23 corresponds to one of the parallel elements in the equivalent circuit shown in figure 4.24. The complex capacitance corresponding to this equivalent circuit can be expressed as:

$$C_* = C_{gb} + \frac{C_t}{1 + j\omega\tau} - \frac{j}{\omega R_{gb}} \tag{4.22}$$

where $\tau = R_t C_t$ is the relaxation time corresponding to the response of a deep level with energy E_t in the bandgap of LN. The total capacitance of a depletion layer is determined by the deep level when the test signal frequency is [31]:

$$\omega_t = 2e'_n \left[2 - \frac{w - x_t}{w} \right] \tag{4.23}$$

where w is the width of the depletion layer and e_n' is the thermal excitation coefficient from deep levels, described by the following expression:

$$e'_n = C_n N_c \exp\left(-\frac{E_c - E_t}{kT} \right). \tag{4.24}$$

Here, C_n is the capture coefficient of a trap level. At the frequency $\omega = \omega_t$, a maximum should occur in both the frequency dependence of capacitance $C''(\omega)$ and the dielectric loss tangent $tg\delta(\omega)$. It can be shown that the real and imaginary parts of the complex capacitance in equation (4.22) are described by the following expressions:

$$C' = \frac{C_t}{1 + (\omega\tau)^2} + C_{gb}$$
$$C'' = \frac{C_t \omega\tau}{1 + (\omega\tau)^2}. \tag{4.25}$$

The capacitance associated with a deep-level response can be found from [32]:

$$C_t = \frac{q^2}{kT} N_t f_0 (1 - f_0) S \qquad (4.26)$$

where N_t is the density of CLCs, f_0 is the Fermi function, and S is the contact area. The imaginary part of the complex capacitance C'', described by equation (4.25), reaches a maximum when $\omega\tau = 1$, which directly gives the relaxation time $\tau = 1/\omega$. The second equation in (4.25) at the maximum is:

$$C''|_{\max} = \frac{C_t}{2}. \qquad (4.27)$$

Using equation (4.26) for C_t, we obtain the density of CLCs N_t through equation (4.27) in the following form:

$$N_t = \frac{8kT}{q^2 S} C''|_{\max}. \qquad (4.28)$$

The energy of deep traps E_t can be found from the slope of a linear graph $\ln(\omega_t)$–q/kT (see equations (4.23) and (4.24)). Figure 4.26 shows the temperature dependence of the dielectric loss tangent for sample LN134, in which two peaks are clearly observed: peak 1 at low frequencies and peak 2 at high frequencies.

The peaks in figure 4.26 are attributed to the 'slow' and 'fast' relaxation processes occurring at the grain boundaries and in the bulk of the grains, respectively. The temperature dependencies of the frequency ω_m (corresponding to the maximums of the spectra $tg\delta(\omega)$ for both peaks) are shown in figure 4.27.

As seen in figure 4.27, the temperature dependencies of ω_m for both peaks are linear functions $\ln(\omega_m)$ – q/kT with slopes giving activation energies of $E_{a1} = 0.9\,\mathrm{eV}$ and

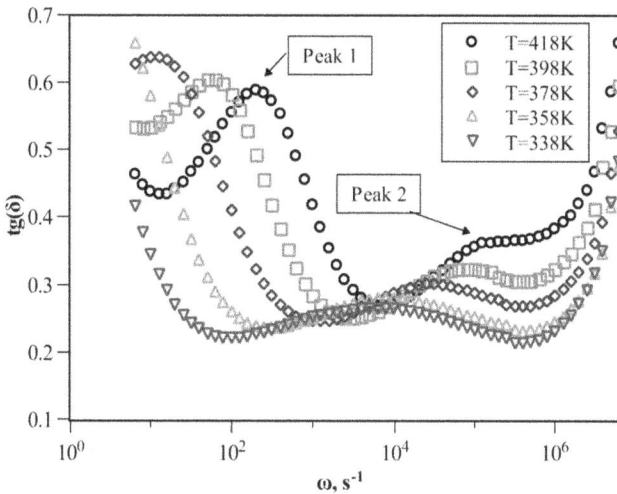

Figure 4.26. Frequency dependence of the dielectric loss tangent for sample LN134 at different temperatures.

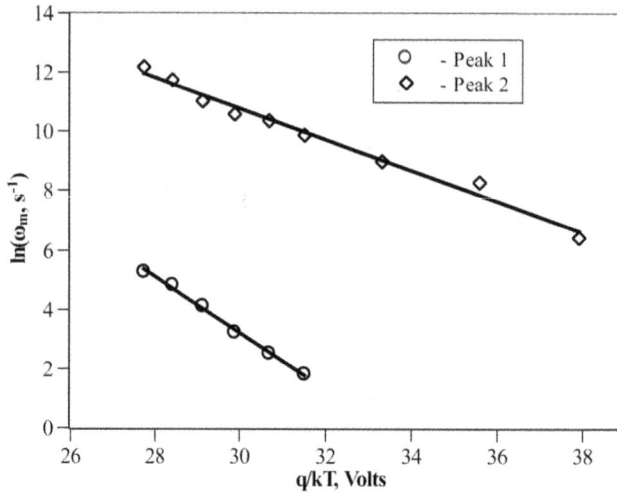

Figure 4.27. Temperature dependencies of the maximum frequency ω_m, corresponding to the two peaks in figure 4.26.

$E_{a2} = 0.5\,\text{eV}$ for peak 1 and peak 2, respectively, in full agreement with equation (4.24). Determining $C''|_{max}$ from the Nyquist diagrams and using equation (4.28), we estimated the density of states for traps with energy $E_{t1} = 0.9\,\text{eV}$ in sample LN134 to be $N_t = 7.8 \times 10^{17}\,\text{cm}^{-3}$[1].

Some researchers have demonstrated [33, 34] that an activation energy of 0.9 eV in LN corresponds to an energy level E_t associated with the presence of lithium vacancies. On the other hand, an activation energy of 0.5 eV is attributed to hopping polaron conductivity, which dominates in single crystals of LN according to many studies [35–38]. It was shown in earlier studies that in the case of oxygen deficit, charged oxygen vacancies and free electrons occur in nonstoichiometric LN single crystals [39, 40]. By contrast, some authors have stated that oxygen vacancies are not the major defects because in this case, oxygen and lithium diffuse, leaving antisite $Nb_{Li}^{4\bullet}$ and lithium vacancies, as reflected in the increased density of LN crystals [9]. However, a recent study demonstrated that TA of single crystals in pure oxygen leads to a decrease in DC conductivity, caused by the neutralization of oxygen vacancies [35]. The effect of oxygen deficit on DC conductivity can be described by the following equation [41]:

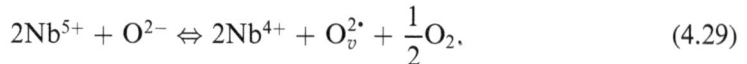

$$2Nb^{5+} + O^{2-} \Leftrightarrow 2Nb^{4+} + O_v^{2\bullet} + \frac{1}{2}O_2. \tag{4.29}$$

Here, $O_v^{2\bullet}$ is a doubly charged oxygen vacancy. Thus, free electrons are generated due to the ionization of Nb^{4+} ions according to the following reaction:

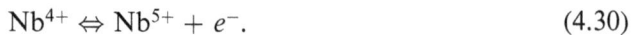

$$Nb^{4+} \Leftrightarrow Nb^{5+} + e^-. \tag{4.30}$$

---

[1] This value was recalculated from surface density into bulk concentration.

Consequently, electrons that escape from neutral oxygen become free or self-trapped at the positions of Nb ions, forming small polarons and influencing electrical conductivity. This viewpoint aligns with the n-type conductivity observed in single crystals of LN undergoing oxygen reduction [35]. Another study [36] clearly demonstrated that an activation energy of $W_h = 0.35$–0.4 eV corresponds to hopping polaronic conductivity, with polaron binding energies in the range of $W_p = 0.7$–0.8 eV. Thus, the thermally activated conductivity with an activation energy of 0.7 eV observed in our experiments can be attributed to the thermal activation of polarons associated with antisite defects $Nb_{Li}^{4:}$ [41]. Furthermore, some researchers have reported the hopping DC conductivity of small polarons with an energy of $E_a = 0.05$ eV (corresponding to the polaronic band in lithium niobate [36]), which is in good agreement with our results for samples similar to LN134, as discussed in chapter 3.

Nyquist diagrams for sample LN135 under three temperature regimes are shown in figure 4.28(a).

The Nyquist plot indicates that the response of the studied heterostructures to an AC signal is due to their conductivity rather than dielectric phenomena. As previously discussed, the studied sample can be reasonably represented by two Schottky barriers connected back-to-back in series (see figure 4.28(b)). For further analysis of the impedance spectra, we use the simplified equivalent circuit shown in figure 4.28(b). Here, C_1 and C_2 represent the capacitances of the depletion zones in LN and Si, respectively, while the resistors R_1 and R_2 correspond to their resistances. The complex impedance of such an equivalent circuit is described by an equation similar to (4.15). Both parallel RC elements in the equivalent circuit should be represented by two semicircles in the appropriate frequency ranges in the Nyquist diagrams. A large semicircle corresponds to the component with the largest resistance in the nonuniform system.

We have demonstrated that the conductivity of sample LN135 is described by the double depletion layer model and is affected by the Schottky barrier heights ϕ_1 and ϕ_2 (see figure 4.28(b)). From the I–V analysis, it follows that $\phi_1 > \phi_2$, so the highest resistance (and the largest semicircle in the Nyquist diagram) corresponds to the depletion zone in Si. At this temperature, carriers have sufficient energy to overcome only the lowest barrier. At room temperature, the larger semicircle masks the smaller one, corresponding to the response of the depletion zone in LN, which starts to dominate when the temperature increases (see figure 4.28(a)). To distinguish the contribution of each parallel RC element to the total impedance, we measured the frequency dependencies of the imaginary parts of the complex impedance Z^* and the dielectric modulus M^*, as shown in figure 4.29.

Both spectra clearly demonstrate peaks corresponding to maximal resistance (a peak in $Z''(\omega)$) and minimal capacitance (a peak in $M''(\omega)$). Based on the previously described approach, we determined the values of all elements in the equivalent circuit. Note that the coefficients α_1 and α_2, describing the distribution of relaxation times in equation (4.15), are close to zero, indicating nearly single characteristic times with values of $\tau_1 = 8 \times 10^{-3}$ s and $\tau_2 = 5 \times 10^{-2}$ s. The temperature dependencies of the resistances R_1 and R_2 in Arrhenius coordinates are shown in figure 4.30.

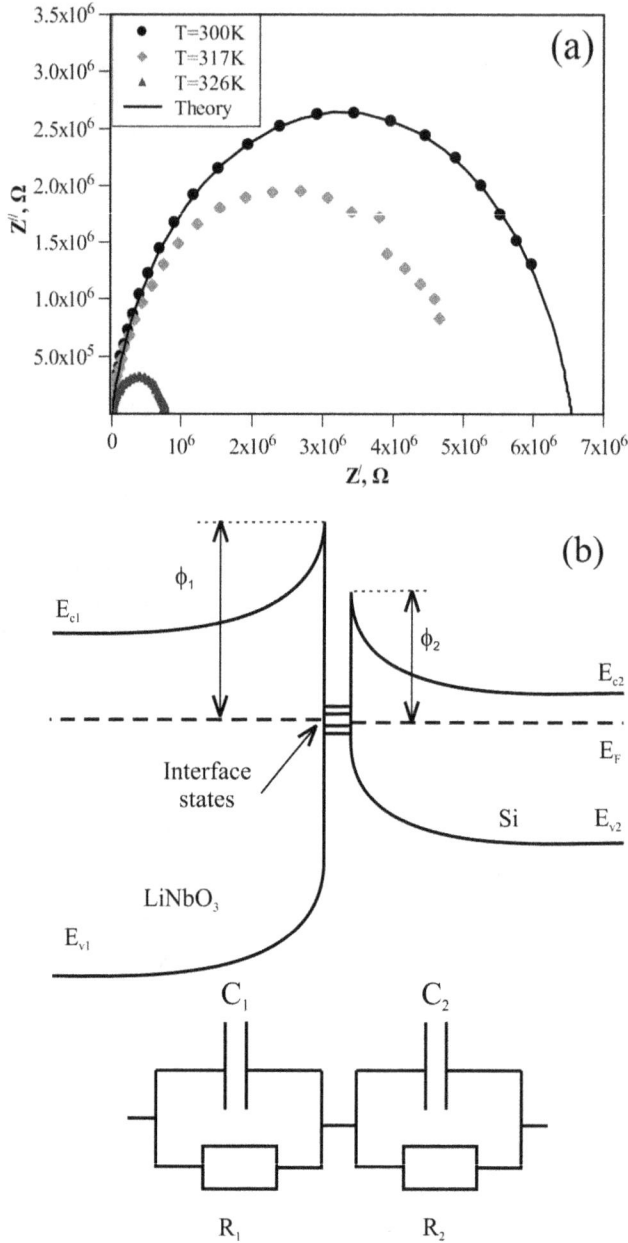

Figure 4.28. Nyquist plots at different temperatures (a), and a band diagram with an equivalent circuit (b) for sample LN135.

Activation energies determined from the slope of the corresponding straight lines in figure 4.30 have magnitudes of 0.4 and 0.26 eV, corresponding to the Schottky barriers ϕ_1 and ϕ_2 derived from I–V analysis. Table 4.4 summarizes the data associated with the influence of sputtering parameters on the electrical properties of the studied LN-based heterostructures.

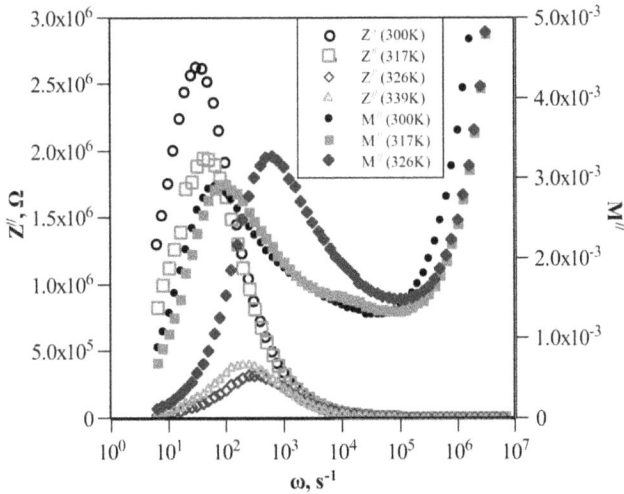

Figure 4.29. Frequency dependencies of imaginary parts of the complex impedance Z^* and the dielectric modulus M^* for sample LN135 at different temperatures.

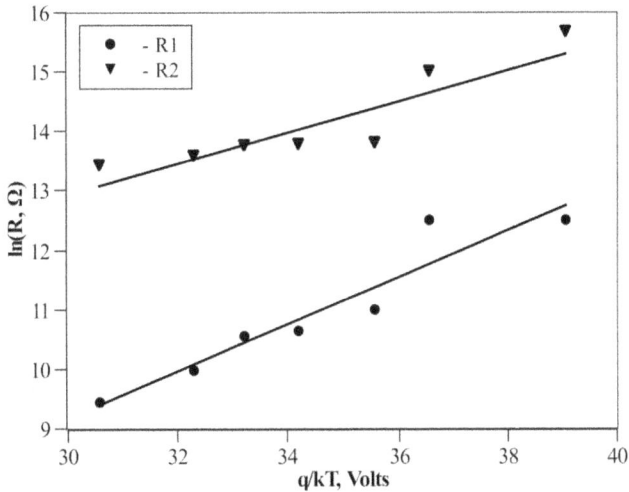

Figure 4.30. Temperature dependencies of resistances R_1 and R_2 in the equivalent circuit shown in figure 4.28(b).

4.2 Thermal annealing effect on electrical properties of Si–LiNbO$_3$ heterostructures

In chapter 2, it was demonstrated that TA positively influences the degree of crystallinity and surface roughness while reducing mechanical stress in synthesized films. Consequently, TA should also affect the electrical properties of Si–LN heterostructures. Similar to the study of structural properties, TA of the studied films was performed at temperatures of 600 °C–700 °C in a coaxial oxygen atmosphere oven for 1 h.

Table 4.4. Summarized data on the influence of reactive gas composition and pressure on the electrical properties of Si-LiNbO$_3$-Al heterostructures.

Properties	Sample #		
	Increase of pressure	The presence of oxygen in a chamber	
	LN133	LN134	LN135
Effective density of charge in a film, Q_{ef} (C/cm^2) (Possible sources: charged antisite defects $Nb_{Li}^{4\bullet}$ and oxygen vacancies)	$+9.0 \times 10^{-7}$	$+8.0 \times 10^{-7}$	$+5.1 \times 10^{-8}$
Effective density of CLC (defects), N_{ef}, cm^{-2}	5.7×10^{12}	5.0×10^{12}	3.2×10^{11}
Donor concentration in Si, N_d (cm^{-3}) (nominal $N_{do} = 1 \times 10^{15}$ cm^{-3}). Donor formation is cause by diffusion of oxygen and lithium	Declines exponentially from 2×10^{16} at the interface to N_{do} at the distance of 550 nm.	Declines exponentially from 2×10^{16} at the interface to N_{do} at the distance of 550 nm.	Declines exponentially from 2×10^{18} at the interface to N_{do}.
Donor concentration in LiNbO$_3$ film, N_d (cm^{-3})		7.0×10^{13} (I-V analysis)	7×10^{17} (C-V analysis)
Density of states at the Si/LiNbO$_3$ interface, D$_{ss}$ (eV^{-1}cm^{-2})	5.0×10^{12}	9.0×10^{12}	2.5×10^{13} (caused by formation of an intermediate layer at the Si/LiNbO$_3$ interface)
Band diagram	Figure 4.9	Figure 3.25(b) (see chapter 3)	Figure 4.28(b)
DC conductivity	Low electric fields		
	($E < 3.3$ kV/cm)	($E < 5$ kV/cm)	
	Hopping conductivity limited by the intergrannular potential barrier of 0.4eV in LiNbO$_3$ films. Hopping conductivity over CLC with the activation energy of 0.2eV. At high temperatures is caused by thermal ionization of carriers from deep levels ($E_t = 0.7$ eV)	Hopping polaronic conductivity with the activation energy of 0.05 eV over CLC	

Average electric fields		
($E = 3.3 - 10$ kV/cm)	($E = 5 - 30$ kV/cm)	($E = 3.3 - 10$ kV/cm)
At temperatures $T = 77 - 200$ K: the Fowler-Nordheim tunneling through the potential barrier of $\phi = 0.4$ eV at Si/LiNbO$_3$ interface. At $T \approx 300$ K: the Richardson-Schottky emission.	The Fowler-Norgheim tunneling through the potential barrier of $\varphi_b = 0.25$eV at Si/LiNbO$_3$ interface	The Richardson-Schottky emission over the Schottky barrier of $\phi_1 = 0.43$eV at Si/LiNbO$_3$ interface.

Strong electric fields		
($E > 10$ kV/cm)	($E = 30 - 90$ kV/cm)	($E > 10$ kV/cm)
At $T \approx 300$ K: non-activated hopping conductivity over CLC in LiNbO$_3$ films.	the Richardson-Schottky emission over the barrier of 0.02 eV.	The thermally-assisted tunneling through the intergrannular potential barrier of 0.7eV in LiNbO$_3$ films.

	Low frequency response		
AC conductivity and dielectric losses	($\omega = 10 - 10^3$ s^{-1})	($\omega = 10 - 10^3$ s^{-1})	
	From grain boundaries (apparently by LiNb$_3$O$_8$ phase) with the average relaxation time of $\tau_1 = 1 \times 10^{-3}$ s and with the correlated-barrier hopping conductivity.	From the grain boundaries is attributed to the trapping and re-emission of electrons from the deep traps with the relaxation time of $\tau_1 = 130$ s.	Hopping conductivity limited by the Schottky barriers at LiNbO$_3$/Si interface with the height of $\phi_1 = 0.43$ eV and $\phi_2 = 0.26$ eV (see the band diagram).

Average frequency response ($\omega = 10^3 - 10^5$ s^{-1})
Hopping conductivity over CLC (the diffusion-controlled relaxation mechanism) with the activation energy of 0.03 eV ($T = 300 - 350$ K) and 0.2 eV ($T = 350 - 390$ K).

(Continued)

Table 4.4. (*Continued*)

Properties	Sample #		
	Increase of pressure	The presence of oxygen in a chamber	
	LN133	LN134	LN135
	High frequency response		
	($\omega = 10^5 - 10^7$ s^{-1})	($\omega = 10^3 - 10^5$ s^{-1})	
	From bulk of grains (LiNbO$_3$ phase) with the average relaxation time of $\tau_2 = 4 \times 10^{-4}$ s. Hopping conductivity over the exponentially distributed potential barriers.	From bulk of grains (LiNbO$_3$ phase) with the relaxation time of $\tau_2 = 1.5 \times 10^{-3}$ s. Hopping conductivity of small polarons with the activation energy of 0.5 eV.	
Concentration of CLC (traps), N_t (cm^{-3})	6×10^{17} (I-V analysis)	7.8×10^{17}	4.0×10^{18} (C-V analysis)
Energy of traps in the bang gap of LiNbO$_3$ films	$E_{t1} = 0.2$eV, $E_{t2} = 0.7$eV (attributed to antisite defects $Nb_{Li}^{4\bullet}$)	$E_{t1} = 0.9$eV (attributed to lithium vacancies $V_{Li}^/$), $E_{t2} = 0.5$eV	$E_t = 0.7$eV (attributed to antisite defects $Nb_{Li}^{4\bullet}$)

4.2.1 Capacitance–voltage characteristics and ferroelectric properties of Si–LiNbO₃–Al heterostructures

Figure 4.31 shows the *P–E* loops of LN films after TA.

All parameters of the *P–E* loops shown in figure 4.31 are provided in table 4.5.

From table 4.5, it can be inferred that TA does not influence the remnant polarization despite the presence of a non-ferroelectric LiNb₃O₈ phase in all films after TA (see chapter 2). On the other hand, TA shifts all *P–E* loops along the vertical axis, regardless of the sputtering regime. This behavior can be explained by the preferential orientation of polarization in the films after TA.

In fact, the local domain structure of synthesized films changes significantly during TA. Figure 4.32 depicts the local domain structure of both as-grown and TA LN films, as revealed by the PFM method.

Three contrasts are observed in the vertical piezoresponse component (figure 4.32): light, dark, and intermediate. The light and dark extremes correspond to ferroelectric domains with a preferential vertical direction (up or down), whereas the intermediate contrast corresponds to polarization along the surface of the sample. As-grown LN films do not manifest a preferential domain orientation (figure 4.32(a)). The piezoresponse signal from the films after TA demonstrates a single vertical direction of polarization with few inclusions of the opposite direction, whereas a complex lateral domain structure is observed (figures 4.32(b) and (c)). This result explains the vertical shift of *P–E* loops in the films after TA, where preferentially oriented domains are observed, making them nonsymmetric in terms

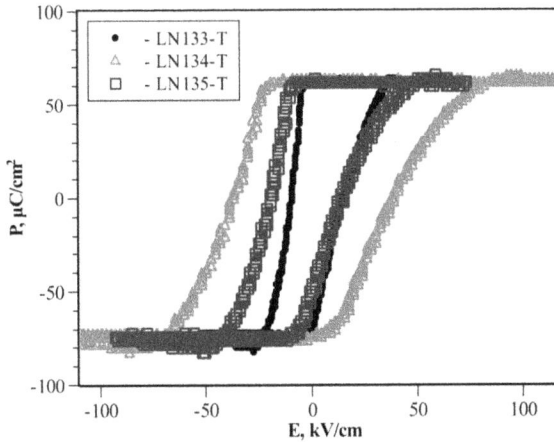

Figure 4.31. Ferroelectric hysteresis loops of LN films after TA.

Table 4.5. Parameters of experimental P–E loops of LN films after TA (the corresponding parameters of the as-grown films are given in brackets).

Sample #	Remnant polarization, P_r ($\mu C\,cm^{-2}$)	Coercive field, E_c ($kV\,cm^{-1}$)	Shift along the vertical axis, ΔP_r ($\mu C\,cm^{-2}$)	Built-in field, E_b ($kV\,cm^{-1}$)
LN133-T	68 (69)	12 (14.0)	−6 (−1)	2.4 (2.8)
LN134-T	69 (69)	37.8 (39.5)	−7 (−1)	1.2 (2.8)
LN135-T	69 (69)	17.4 (27.3)	−7 (−1)	−2.7 (2.0)

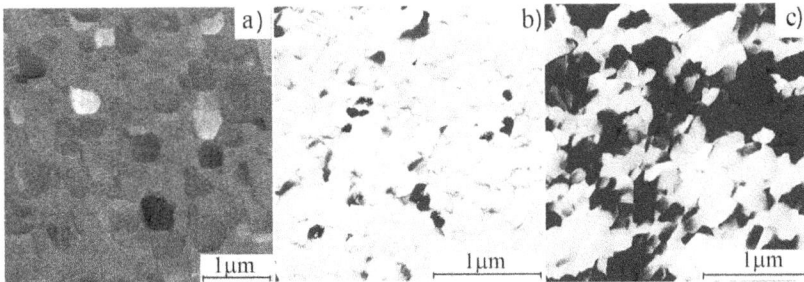

Figure 4.32. PFM images of as-grown LN film (a) and films after TA (b, c). Patterns (a) and (b) correspond to the vertical polarization component, while pattern (c) corresponds to the lateral polarization component. Reproduced from [42], with permission from Springer Nature.

of polarization reversal relative to the applied electric field. A similar behavior was observed in the local domain structure described in [43], which studied the influence of the annealing temperature on the ferroelectric properties of LN. An annealing temperature of 700 °C is optimal for preserving the local domain structure and the optimal LN/LiNb$_3$O$_8$ phase ratio.

Our results, shown in table 4.5, reveal other interesting properties of the studied films after TA. The coercive field of films fabricated by RFMS in an Ar(60%) + O_2(40%) gas mixture decreases dramatically after TA (sample LN135-T). Furthermore, TA of these films leads to the formation of a relatively high built-in field, which has the opposite direction to the fields of as-grown films. To investigate the possible nature of this phenomenon, we analyzed high-frequency $C-V$ characteristics.

It was shown earlier that films fabricated in an Ar atmosphere contain a positive oxide charge that does not depend on the reactive gas pressure in the chamber, the degree of orientation in the films, or the presence of the $LiNb_3O_8$ phase (samples LN133 and LN134). Therefore, we compared the $C-V$ characteristics of as-grown heterostructures, fabricated in a pure Ar environment and in an Ar + O_2 gas mixture (samples LN134 and LN135), with the same heterostructures after TA (samples LN134-T and LN135-T). The $C-V$ curves of the studied samples are shown in figure 4.33.

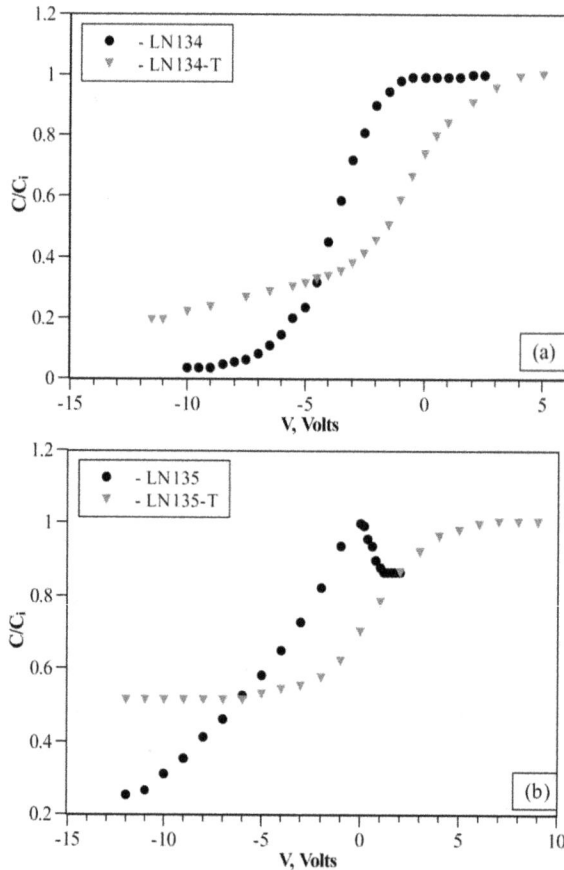

Figure 4.33. Typical high-frequency $C-V$ characteristics of as-grown heterostructures (samples LN134 and LN135) and heterostructures after TA (samples LN134-T and LN135-T), recorded at $T = 293$ K.

TA leads to a decrease in the positive oxide charge in the films for both samples, which is reflected in figure 4.33 as a decrease in the horizontal shift of the experimental $C\text{-}V$ curves compared to ideal $C\text{-}V$ curves. Furthermore, as shown in figure 4.33(b), TA leads to the disappearance of a branch attributed to the modulation of the depletion zone in LN, observed in sample LN135. Under conditions corresponding to sample LN135, an intermediate layer with variable composition forms between the Si substrate and the LN film. The absence of depletion zone modulation in the LN film in sample LN135-T results in the formation of an abrupt film–SiO_2 layer interface in the studied heterostructures after TA.

It is also important to note that the $C\text{-}V$ curves of all samples exhibit hysteresis after TA, as shown in figure 4.34.

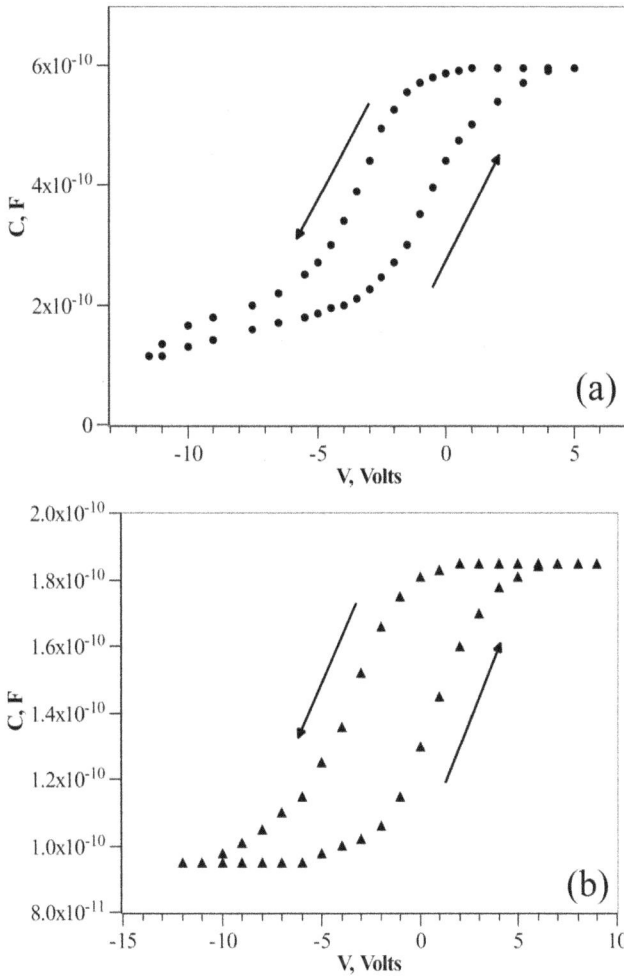

Figure 4.34. High-frequency $C\text{-}V$ hysteresis loops for samples LN134-T (a) and LN135-T (b) at a temperature of $T = 293$ K.

One possible explanation for this phenomenon is the migration of positively charged ions in the film, which leads to charge accumulation during measurements because the ions do not follow the change in voltage. However, this hypothesis is unrealistic in our case, because C–V hysteresis is observed at liquid nitrogen temperature where ions are 'frozen' and do not participate in charge transport. On the other hand, this counterclockwise hysteresis type can be caused by the capture of positive charge by traps. The magnitude of this charge can be estimated through the width of the 'window' ΔV on the hysteresis curves, using the following expression:

$$Q_{ot} = C_{FB} \cdot \Delta V_{FB}. \tag{4.31}$$

Here, C_{FB} is the flat band capacitance and ΔV_{FB} is the difference in the flat band voltages of a heterostructure during forward and reverse voltage sweeps. We will discuss the origin of this C–V hysteresis later. The results of C–V analysis for heterostructures after TA are provided in table 4.6 and figures 4.35 and 4.36.

As can be seen from figures 4.36 and 4.6(b) and (c), TA leads to a significant decrease in the energy distribution of surface states at the Si–film heterojunction. This decrease is likely due to a reduction in the defect concentration in the SiO_2 layer present at the heterojunction. Additionally, the donor distribution in the silicon substrate also changes (see figures 4.35 and 4.4). The concentration of donors in sample LN134-T rises with TA, whereas in sample LN135-T, TA leads to a decrease in N_d. This can be explained by the assumption that oxygen creates deep donors when diffusing into silicon. During the annealing process of Si–LN heterostructures fabricated in an Ar atmosphere (sample LN134-T), oxygen actively diffuses into Si under a concentration gradient, creating additional donor centers. In the case of heterostructures synthesized in an Ar(60%) + O_2(40%) gas mixture (sample LN135-T), the thick SiO_2 layer at the Si–film interface limits oxygen diffusion. Conversely, the reverse diffusion of O_2 from Si also reinforces this process (see chapter 2).

Oxygen diffusion is evidently responsible for the decrease in positive oxide charge formed in as-grown LN films (see table 4.6). Molecular oxygen, diffusing during TA, neutralizes positively charged oxygen vacancies in LN films, as proposed in [41]. The flat band regime is implemented in Si–LN heterostructures formed in an Ar environment after TA. This is important for applications of LN-based heterostructures where space charge effects in the substrate play a negative role. This result is particularly significant for practical applications of LN-based heterostructures in optoelectronics and nonvolatile memory units.

It is worth noting that TA changes the sign of the oxide charge from positive to negative in heterostructures fabricated in an Ar(60%) + O_2(40%) reactive gas mixture (see table 4.6, sample LN135-T). This phenomenon can be explained within the framework of complex formation. If molecular oxygen diffusing into the LN film settles near a vacancy, charge can be transferred from the vacancy to the oxygen according to the following reaction [44]:

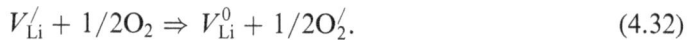

$$V_{Li}^{/} + 1/2O_2 \Rightarrow V_{Li}^{0} + 1/2O_2^{/}. \tag{4.32}$$

Table 4.6. Results of C–V analysis for samples LN134-T and LN135-T (the corresponding parameters for as-grown heterostructures are given in brackets).

Sample #	Dielectric constant of the film, ε	Flat band voltage, V_{FB} (volts)	Effective charge in the film, Q_{eff} (C cm^{-2})	Effective density of states, N_{eff} (cm^{-2})	Position of a charge centroid (relative to the film surface) in thickness units, d_c/d	Trapped charge, Q_{ot} (C cm^{-2})
LN134-T	25	-0.1 (-5.6)	$+5.5 \times 10^{-9}$ ($+8.0 \times 10^{-7}$)	3.3×10^{10} (5.0×10^{12})	0.97 (0.36)	1.7×10^{-7}
LN135-T	12	2.0 (-3.0)	-4.0×10^{-8} ($+5.1 \times 10^{-8}$)	2.4×10^{11} (3.2×10^{11})	0.71	8.5×10^{-8}

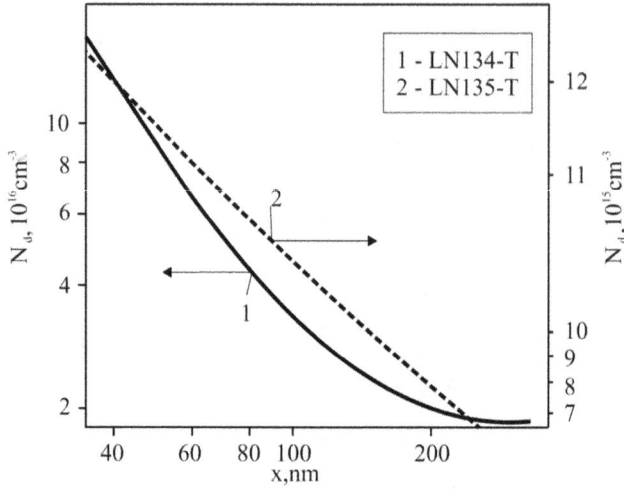

Figure 4.35. Doping profile of the Si–LN heterostructures after TA; x is the distance from the Si–LN interface toward the bulk.

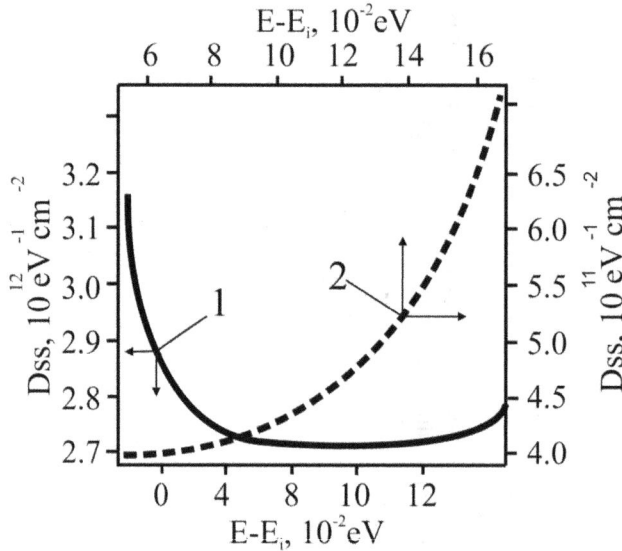

Figure 4.36. Energy distribution of surface states in the upper half of the Si bandgap for the studied Si–LN heterostructures after TA (1—sample LN134-T, 2—sample LN135-T).

Here, $V_{Li}^{/}$ and V_{Li}^{0} are negatively charged and neutral vacancies, respectively. The decrease of the built-in field in LN-based heterostructures after TA (see table 4.5) can also be explained in terms of the reduction of positive oxide charge in films during TA.

The C–V hysteresis shown in figure 3.34 can be attributed to charge trapping effects in LN-based heterostructures after TA (see table 4.6). The magnitude of the trapped charge does not significantly differ among heterostructures fabricated in different reactive gas environments. According to the classification of oxide charge given in chapter 3, capture and emission of carriers are possible only for the interface-trapped charge (Q_{it}) and the oxide-trapped charge (Q_{ot}). Q_{ot} can be captured by states present at the Si–oxide interface ('interface traps') or distributed near the interface ('border traps') at a distance of about 5 nm from the semiconductor surface [45]. Our structural study (see chapter 2) demonstrates that a 5 nm thick SiO_2 layer forms at the Si–LN interface in heterostructures fabricated in an Ar atmosphere by RFMS. In Si–LN heterostructures made in an Ar(60%) + O_2(40%) gas mixture, an extended intermediate layer forms with a composition close to that of SiO_2. Apparently, TA drives oxygen to diffuse toward the Si–LN interface, accompanied by the recrystallization of the deposited films. Oxygen, interacting with silicon, forms the extended intermediate layer, as confirmed by elemental composition studies (see chapter 2). The formation of this layer during TA can be a source of additional charge trapping centers, created not only due to specific technological regimes but also by applied electric fields during electrical measurements.

It is well known that the classification of defects depends on their location, electrical behavior, charge state, and physical structure. Primarily, the term 'trap' refers to the defect location: 'oxide traps' and 'interface traps' are defects located in the layer and at the substrate/layer interface, respectively. By contrast, the term 'state' is associated with electrical behavior, distinguishing between 'fixed' (defects whose charge state does not change) and 'switching' (defects which exchange charge with a substrate or an electrode). The 'fixed' states can be positive, negative, or neutral, whereas the 'switching' states change their charge state due to the capture and re-emission of an electron or a hole. They can be donor-like (acceptor-like), where the states are initially neutral and become positively charged due to the emission of an electron (capture of a hole), or where they become negatively charged due to the capture of an electron (emission of a hole). Finally, defects can be stable or unstable, depending on experimental conditions.

Defects in SiO_2 can be intrinsic, caused by the growth process (e.g. oxygen vacancies), or generated by irradiation or applied electric fields [46]. Numerous papers suggest that avalanche electron or hole injection from the Si substrate and Fowler–Nordheim electron injection from the substrate are major sources of defect generation in silicon dioxide in the Si–SiO_2 system. The kinetics of defect generation is quite complex due to the variety of their types: from positively and negatively charged fixed traps to 'border' and 'interface' states. Electrons injected from the Si substrate through Fowler–Nordheim tunneling gain energy due to high electric fields in the SiO_2 layer (electron heating) [47]. Hot electrons generate electron–hole pairs, provided the SiO_2 layer thickness exceeds 10–12 nm at fields >10 MV cm^{-1}. Generated holes tunnel into the SiO_2 layer due to their strong electric field. Some of these holes can be captured by traps in the bulk of the layer or at the interface, leading to an increase in the positive captured charge. Recombination of electrons with these captured holes can be the major source of neutral electron traps, border traps, and interface states at Si–SiO_2 interfaces [48]. The authors of [49, 50] revealed an important property: Fowler–Nordheim emission does not generate the minority carriers described above in thin SiO_2 films ($<20\,nm$) because this thickness is insufficient for the heating of tunneling electrons. Evidently, in our case, the thin ($\sim 5\,nm$) SiO_2 layer at the Si–LN interface in as-grown heterostructures (see chapter 2) is not thick enough for the generation of electron–hole pairs, even though Fowler–Nordheim emissions are frequently observed in the studied heterostructures [5, 51]. Therefore, C–V curves of as-grown Si–LN heterostructures do not exhibit positive charge capture at the border traps, and consequently, hysteresis does not appear. TA leads to the formation of an intermediate layer with a higher-quality Si–SiO_2 interface, enabling the described process of generation and capture of holes. To verify this assumption, we studied the I–V characteristics of Si–LN after TA.

4.2.2 Current–voltage characteristics of Si–LiNbO$_3$–Al heterostructures

Figure 4.37 presents the I–V characteristics of as-grown Si–LN–Al heterostructures formed by the RFMS method in an Ar reactive gas environment (sample LN134) and the same heterostructures after TA (sample LN134-T).

Two distinct sections can be seen in the I–V curves, separated by a dashed line. The first section, in the range of relatively low electric fields ($E < 2 \times 10^6\,V\,m^{-1}$), shows a fast-growing current associated with the barrier properties of the Si–LN heterostructure. The second, gentler slope at $E > 2 \times 10^6\,V\,m^{-1}$ is attributed to currents limited by the film's bulk properties. As shown in figure 4.37, TA decreases the conductivity of as-grown films fabricated in an Ar environment. Furthermore, the linear sections in the I–V characteristics, plotted in Fowler–Nordheim

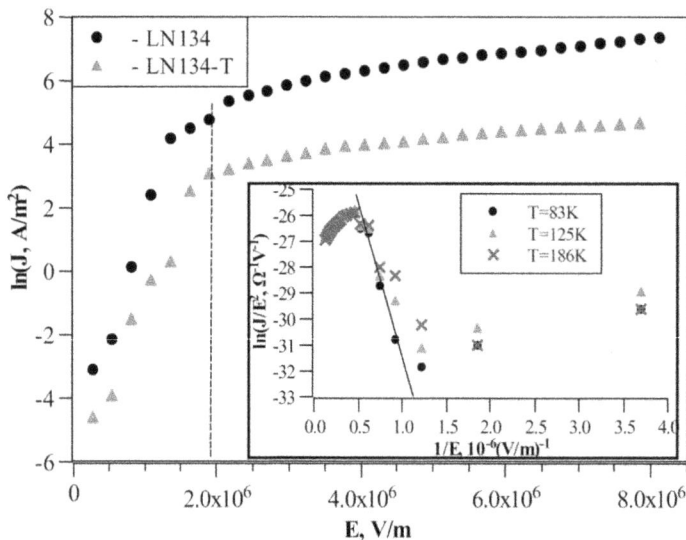

Figure 4.37. I–V characteristics of as-grown Si–LN–Al heterostructures fabricated by RFMS method in an Ar environment (sample LN134) and the same heterostructures after TA (sample LN134-T). Measurements were recorded at room temperature. The insert shows the I–V characteristics of sample LN134-T in Fowler–Nordheim coordinates at different temperatures.

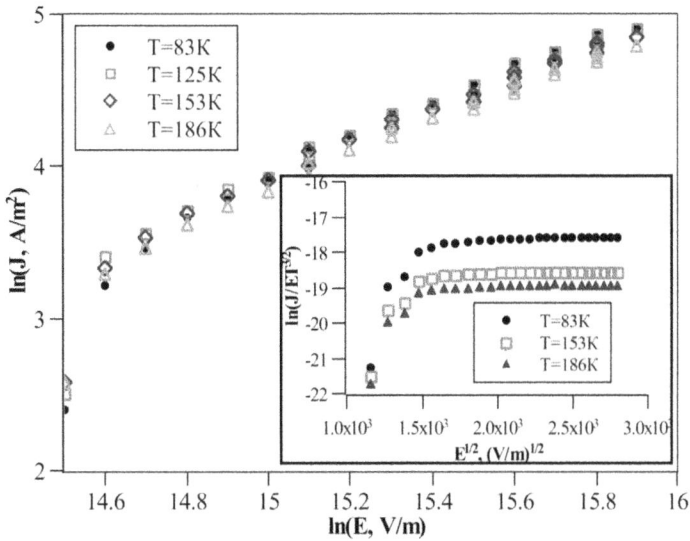

Figure 4.38. I–V characteristics of sample LN134-T in $\ln(J)$–$\ln(E)$ coordinates at different temperatures. The insert shows I–V curves in Simmons coordinates (explanations are given in the text).

coordinates (see inset in figure 4.37), indicate that tunneling dominates in this range of applied electric fields. Using equation (4.4), we determine the height of the potential barrier, corresponding to traps with an energy of 1.7 eV in the bandgap of LN. This value is almost exactly equal to the energy determined in chapter 3 and our previous work [52] for as-grown Si–LN heterostructures. The observed Fowler–Nordheim emission supports the model discussed above, which associates the generation of holes in the SiO_2 layer with their subsequent capture by border traps. This trapping and re-emission is responsible for the C–V hysteresis shown in figure 4.34.

Under strong electric fields ($E > 2 \times 10^6 \, \mathrm{V \, m^{-1}}$), ohmic conductivity is observed over a wide temperature range, as indicated by the linear I–V dependence in $\ln(J)$–$\ln(E)$ coordinates with a slope of one (see figure 4.38).

The most frequently observed Richardson–Schottky emission is described by equation (1.8) in chapter 1, which indicates that I–V characteristics should be linear in Simmons coordinates. However, if conductivity is limited by the bulk of a material, the pre-exponential term dominates in equation (1.8):

$$J_0 = 2q\left(\frac{2\pi mkT}{h^2}\right)^{3/2}\mu E_0. \tag{4.33}$$

Here, q is the elementary charge, m is the carrier effective mass, k is Boltzmann's constant, T is temperature, μ is carrier mobility, and E_0 is the applied electric field. Thus, if the pre-exponential factor dominates, the I–V characteristic should be a horizontal line in Simmons coordinates, which is observed for sample LN134-T (see the inset in figure 4.38). Furthermore, as follows from equation (4.33), the intercept

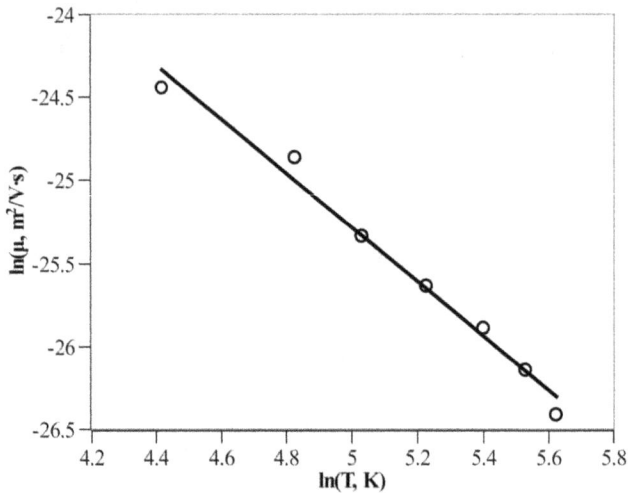

Figure 4.39. Temperature dependence of drift mobility for carriers in the film of sample LN134-T.

of the horizontal branch of the I–V characteristic with the vertical axis is proportional to the carrier mobility in the film. The temperature dependence of this parameter is determined only by the temperature dependence of mobility, which depends on scattering mechanisms. Figure 4.39 shows the temperature dependence of drift mobility in $\ln(\mu)$–$\ln(T)$ coordinates, calculated using equation (4.33).

The mobility at room temperature is $\mu = 1.3 \times 10^{-12}\,\mathrm{m}^2\,\mathrm{V}^{-1}\,\mathrm{s}^{-1}$, which is close to the drift mobility in single-crystal lithium niobate ($\mu = 8 \times 10^{-13}\,\mathrm{m}^2\,\mathrm{V}^{-1}\,\mathrm{s}^{-1}$) [53]. It is important to note that, as follows from the slope of the graph shown in figure 4.39, the drift mobility of carriers in sample LN134-T declines with temperature according to the $T^{-3/2}$ law, which is attributed to phonon scattering [54].

Thus, considering the results of structural analysis, we can conclude that TA of LN films leads to their recrystallization, increasing the size of grains and minimizing the influence of intergranular barriers on charge transport. As a result, the electrical conductivity of LN films results in hopping conductivity with phonon scattering of electrons, similar to the conductivity observed in single-crystal lithium niobate.

As demonstrated earlier by C–V analysis, TA of Si–LN heterostructures fabricated in an Ar + O_2 gas mixture leads to a lower depletion zone at the substrate/film interface. Thus, in these heterostructures, we can expect the absence of a blocking contact at the Si-films interface, which improves its injection properties, a required condition for space-charge-limited current (SCLC). Indeed, the I–V characteristics of sample LN135-T and other LN-based heterostructures [55, 56] are linear in double logarithmic coordinates and have several sections attributed to SCLC (see figure 4.40).

The power dependence of the I–V characteristics is indicative of SCLC and can be represented by $J \propto V^{\alpha}$, where the coefficient α (the differential slope of the I–V curves) depends on the concentration and distribution of traps. In many cases, when monoenergetic traps are present in the bandgap of a material, a linear section with

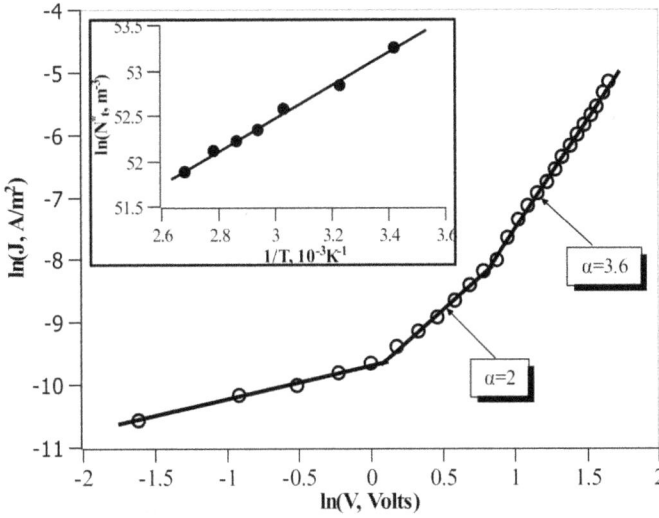

Figure 4.40. Typical *I–V* characteristic of sample LN135-T in double logarithmic coordinates at a temperature of 300 K. The inset depicts the temperature dependence of trap concentration in the bandgap of LN.

$\alpha = 2$ is observed in the $\ln(J)$–$\ln(V)$ graph, serving as a 'fingerprint' of SCLC. The differential slope of *I–V* characteristics is defined as:

$$\alpha = \frac{d(\ln J)}{d(\ln V)}. \tag{4.34}$$

In polycrystalline and amorphous dielectrics, electron traps are not monoenergetic but are distributed in the bandgap of the material. Let us assume there is an exponential energy distribution of traps:

$$N_t(E) = \frac{N_t}{E_o} \exp\left(-\frac{E - E_t}{E_o}\right). \tag{4.35}$$

Here, N_t is the bulk concentration of traps in the dielectric, E_t is the energy level from which the distribution is exponential (relative to the bottom of the conduction band), and E_o is the characteristic energy of distribution, given by the following equation:

$$E_o = lkT. \tag{4.36}$$

In this case, the *I–V* characteristics are described as follows [57]:

$$J = q^{(1-l)}\mu N_c\left(\frac{2l + 1}{l + 1}\right)^{(l+1)}\left(\frac{l\varepsilon\varepsilon_0}{(1 + l)N_t^*}\right)^l \frac{V^{l+1}}{d^{2l+1}}. \tag{4.37}$$

Here, q is the elementary charge, μ is the carrier mobility, N_c is the effective density of states at the lower edge of the conduction band, ε_0 is the electric constant, and ε and d are the dielectric constant and thickness of the dielectric layer, respectively. The concentration N_t^* is given by:

$$N_t^* = N_t \exp\left(\frac{E_t}{lkT}\right). \tag{4.38}$$

The parameter l in equations (4.36) and (4.37) can be found from $l = \alpha_m - 1$, where α_m is the maximum magnitude of this parameter, given by equation (4.34). The critical voltage of the transition from an ohmic to a 'quadratic' section of the I–V curve is given by the following expression [57]:

$$V_x = \frac{qd^2 N_t^*}{\varepsilon \varepsilon_0} \left(\frac{n_0}{N_c}\right)^{1/l} \frac{l+1}{l} \left(\frac{l+1}{2l+1}\right)^{(l+1)/l}. \tag{4.39}$$

Here, n_0 is the concentration of free carriers. The critical voltage, corresponding to the transition from the 'quadratic' law to the 'trap-filled-limit' law, is defined as follows [57]:

$$V_{\text{TFL}} = \frac{qd^2}{\varepsilon \varepsilon_0} \left(\frac{9(N_t^*)^l ((l+1)/l)^l ((l+1)/(2l+1))^{l+1}}{8N_c}\right)^{1/(l-1)}. \tag{4.40}$$

Using the magnitudes of V_x and V_{TFL}, determined from experimental I–V curves (figure 4.40), and the power exponent l, calculated using the experimental parameter α, we can solve equations (4.39) and (4.40) for N_t^* and n_0. The temperature dependence of N_t^* (T) is a straight line in $\ln(N_t^*) - 1/T$ coordinates (see the inset in figure 4.40), indicating the correctness of our assumption about the exponential distribution of traps. The magnitude of E_t is determined by the slope of the graph, i.e. $\ln(N_t^*) - 1/T$, and N_t is obtained by extrapolating this curve to $1/T \to 0$. We have determined the following magnitudes: $n_0 = 7 \times 10^{13} \, \text{cm}^{-3}$, $N_t = 3.0 \times 10^{14} \, \text{cm}^{-3}$, and $E_t = 0.4 \, \text{eV}$. The distribution of traps in the bandgap of the film is shown in figure 4.41.

Thus, TA of Si–LN heterostructures fabricated in an $Ar + O_2$ gas mixture results in a decrease in the bulk concentration of traps, which is in good agreement with other studies [55, 56]. This decrease explains the twofold decline in the coercive field

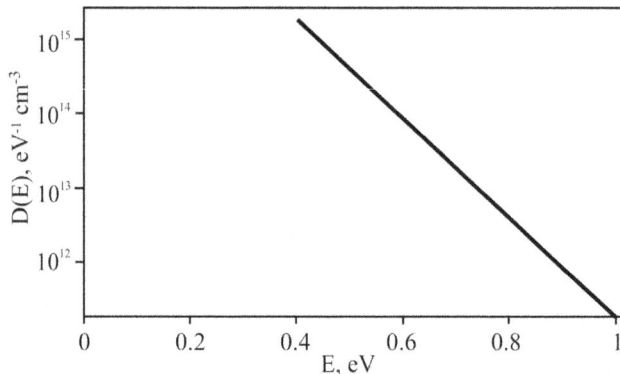

Figure 4.41. Distribution of traps in the bandgap of LN film after TA.

of sample LN135-T compared to as-grown heterostructures (see table 4.5). The results of [44] demonstrated a similar effect in the TA process of LN films in an oxygen atmosphere, explained by a decrease in the concentration of oxygen vacancies segregated at grain boundaries. The influence of the 'parasitic' $LiNb_3O_8$ phase, formed during TA, on electrical phenomena in the studied heterostructures can be revealed by the IS technique.

4.3 Impedance spectroscopy of Si–LiNbO₃–Al heterostructures after thermal annealing

To utilize IS methods, we selected heterostructures in films containing both the LN and $LiNb_3O_8$ phases before TA (sample LN133). Nyquist diagrams and the frequency dependencies of the imaginary parts of the complex impedance and dielectric modulus of sample LN133-T are shown in figure 4.42. Our analysis of the

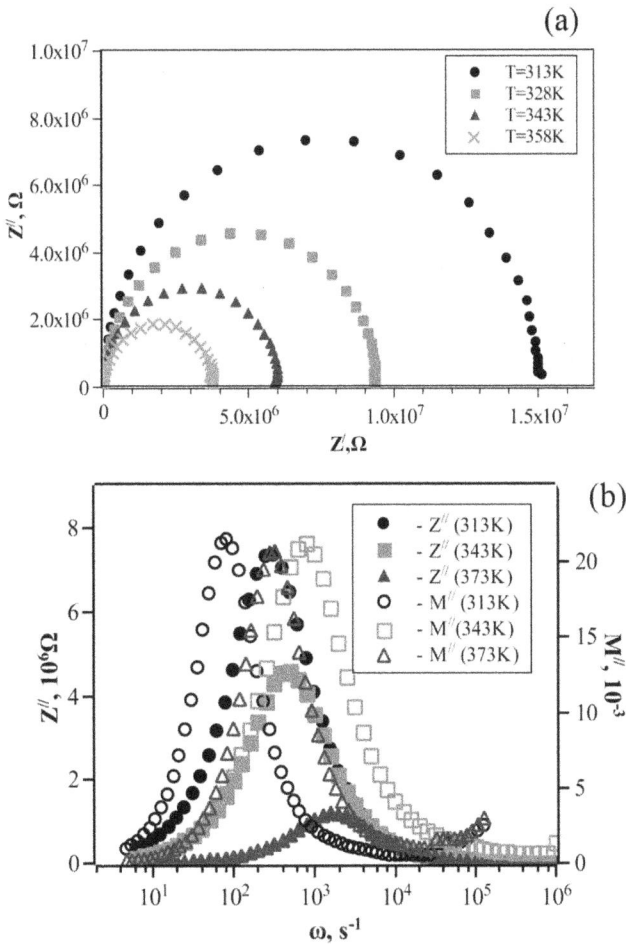

Figure 4.42. Nyquist diagrams (a) and frequency dependencies of imaginary parts of the complex impedance and dielectric modulus (b) of sample LN133-T at different temperatures.

IS spectra was conducted using an equivalent circuit similar to that used for sample LN133, as shown in the inset of figure 4.17.

The Nyquist diagrams in figure 4.42(a) are nearly ideal semicircles, indicating single relaxation times rather than the distribution observed for sample LN133. This suggests that TA improves the homogeneity of synthesized films in the studied heterostructures. We analyzed $Z'(\omega)$ and $M'(\omega)$ spectra using the same method as for sample LN133. Considering that TA increases the average grain size to $d_g = 80$ nm (see chapter 2 and our work [2]), we obtained equal average magnitudes for both the thickness of the bulk zone of a grain and its grain boundary, $d_b \approx d_{gb} = 40$ nm ($d_b/d_{gb} = 1$) [22]. This finding aligns with our previous work [2], which showed an increase in the amount of the $LiNb_3O_8$ phase in deposited films after TA.

According to the model proposed in [58], atmospheric oxygen penetrates the film during TA and neutralizes oxygen vacancies, reducing the migration ability of atoms. Evaporation of Li_2O occurs during the incubation period, leading to the formation of niobium vacancies, Nb_v. This process causes atoms to move in their neighborhood, resulting in the nucleation of $LiNb_3O_8$–LN pairs and phase separation. Therefore, the desorption of Li_2O is the limiting factor for the crystallization rate. Once $LiNb_3O_8$ and LN nuclei are formed, crystallization and phase transitions continue at a rate dependent on the annealing parameters. The oxidation of Li, forming the light compound Li_2O, increases the loss of Li in the film, which plays an important role in TA.

Figure 4.43 shows the temperature dependencies of conductivity (in Arrhenius coordinates) associated with grain boundaries ($\sigma_{gb}(T)$) and the grain bulk ($\sigma_b(T)$).

From figure 4.43, it is evident that both components of conductivity are activated processes with activation energies of $E_{a1} = 0.3$ eV and $E_{a2} = 0.05$ eV for $\sigma_{gb}(T)$ and $\sigma_b(T)$, respectively. Apparently, during TA, nonstoichiometric oxygen accumulates at

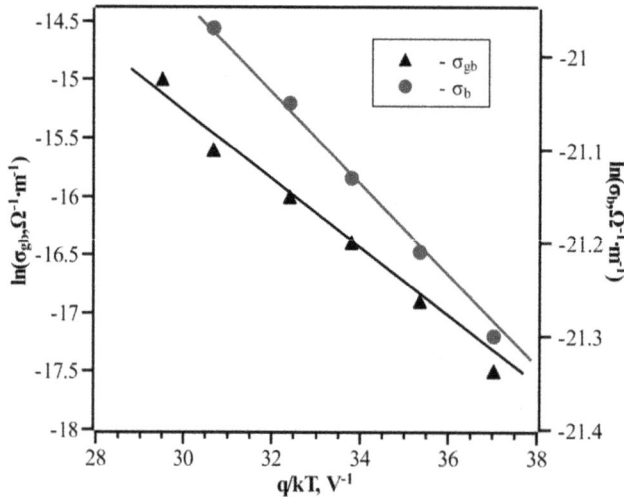

Figure 4.43. Temperature dependencies of conductivity (in Arrhenius coordinates), associated with grain boundaries $\sigma_{gb}(T)$ and the grain bulk $\sigma_b(T)$ for sample LN133-T.

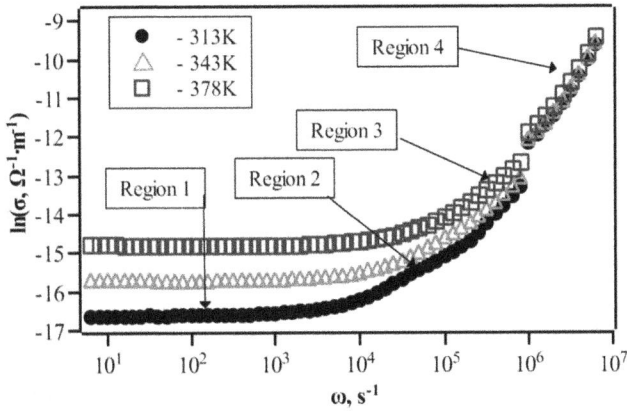

Figure 4.44. Frequency dependence of conductivity for sample LN133-T at three temperatures.

the grain boundaries, serving as the oxygen source for the LN → LiNb$_3$O$_8$ phase transition described above, which slightly decreases the height of the intergranular barrier. Consequently, the intergranular region of grains (the LiNb$_3$O$_8$ phase) increases to the point where grain boundaries and bulk regions play equal roles in conductivity, making the film properties more uniform. Figure 4.44 shows the frequency dependence of conductivity for sample LN133-T at different temperatures.

Four segments corresponding to different power exponents in equation (4.18) and different conductivity mechanisms are observed in the $\sigma(\omega)$ dependencies (see sample LN133-T). The first segment at low frequency is influenced by DC conductivity, and as shown earlier, its activation energy is 0.3 eV. This magnitude matches the activation energy determined above for $\sigma_{gb}(T)$, representing the intergranular barrier height. In region 2 in figure 4.44, conductivity is a thermally activated process with an activation energy of 0.2 eV, determined from the slope of the Arrhenius graph shown in figure 4.45 [22]. This value agrees with those obtained earlier for sample LN133 within the DCR model framework.

Figure 4.46 demonstrates the temperature dependence of the power exponent s in equation (4.18), corresponding to the third and fourth regions in figure 4.44.

The temperature dependence of $s_2(T)$ is typical for conductivity caused by the resonant absorption of quanta with energy $\hbar\omega$, corresponding to the energy of the applied electric field $E(\omega)$ [59]. Since this mechanism is not observed in as-grown heterostructures (sample LN133), we can infer that it is due to the increased amount of the LiNb$_3$O$_8$ phase in the films after TA. The fourth region in the frequency dependence $\sigma(\omega)$ (see figure 4.44) can be described within the framework of the ADWP model. However, the parameter T_0 (see equation (4.21)) in this case is $T_0 = 1436$ K. Since this parameter describes the dispersion of the bulk potential, it is apparent that TA leads to a redistribution of defects in the bulk of the grains.

Thus, TA results in a slight decrease in the intergranular barrier, enhancing the role of bulk electrical conductivity, which decreases due to the extension of the LiNb$_3$O$_8$ into the bulk areas of grains.

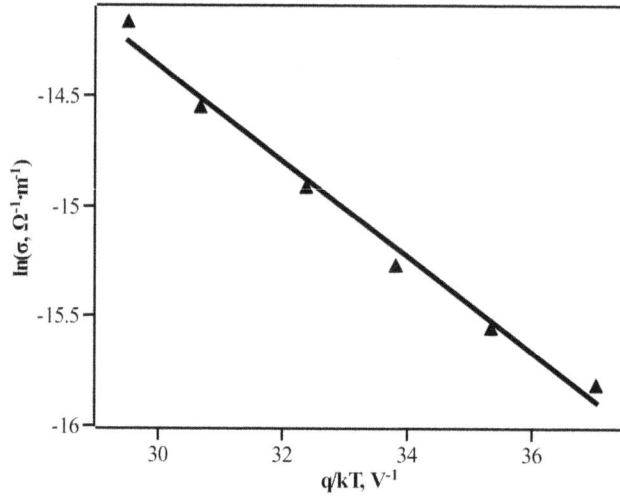

Figure 4.45.

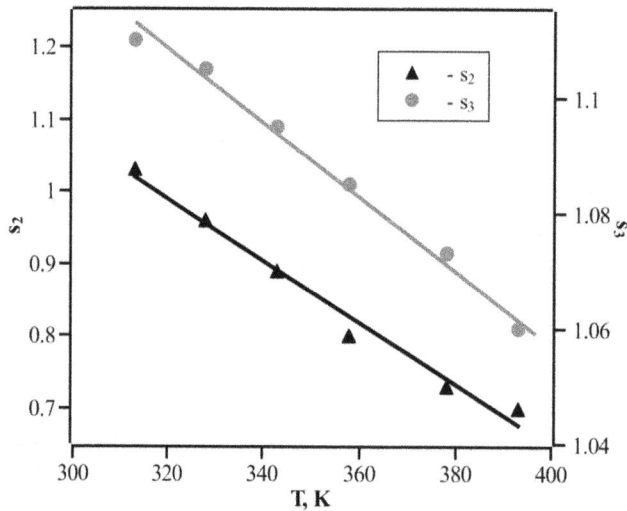

Figure 4.46. Temperature dependence of the power exponent s in equation (4.18) corresponding to regions 3 and 4 in figure 4.44.

4.4 Optical bandgap of thin LiNbO$_3$ films produced by different fabrication regimes

4.4.1 Dependence of the optical bandgap shift in thin LiNbO$_3$ films on RFMS conditions and subsequent thermal annealing

The bandgap of the materials in a semiconductor heterostructure is a crucial parameter that affects its electrical and optical properties. Despite the well-documented optical

Table 4.7. Deposition regimes used to produce samples for our optical investigation.

Sample #	LN-1	LN-2	LN-3	LN-4	LN-4-T
Deposition technique	IBS	RFMS	RFMS	RFMS	RFMS
Magnetron power/ supply power (W)	2000	100	100	100	100
Substrate–target distance (cm)	6	6	6	6	6
Substrate temperature (°C)	Unheated	Unheated	550	550	550
Substrate position		Offset from the target erosion zone	Over the target erosion zone	Offset from the target erosion zone	Offset from the target erosion zone
Annealing	—	—	—	—	+

properties of lithium niobate, recent studies have revealed variations in these properties. Specifically, the theoretically and experimentally obtained magnitudes of the optical bandgap vary from 3.57 [60] to 4.7 eV [61]. The bandgap of thin oxide films is influenced by several factors, such as the size of polycrystalline grains [62], mechanical stress caused by the mismatch of crystal lattice parameters between the substrate and the film [63], and the defect (vacancy) concentration [64]. Furthermore, as shown in chapter 2 and our other studies [2, 56], TA, one of the most effective post-deposition treatments, significantly affects the structure and composition of LN films. Given the broad range of practical applications of LN-based heterostructures in optoelectronics, studying the effect of sputtering conditions on their optical properties is extremely important.

Thin LN films were deposited using the RFMS and IBS methods under the conditions specified in table 4.7 (where shaded cells indicate the influencing parameters used for comparison). Cleaved fluorphlogopite wafers were used as transparent substrates for optical measurements, and subsequent TA was performed at 650 °C for 60 min.

Single-phase polycrystalline LN films fabricated using the sputtering regimes given in table 4.7 have grain sizes, textures, and surface morphologies that are influenced by the plasma properties [1, 2]. Thus, it is expected that their optical properties should strongly depend on the synthesis regime. Figure 4.47 shows the dependence of the absorption coefficient on incident photon energy for the studied LN films [65].

As seen in figure 4.47, the absorption coefficient α initially rises steadily with incident photon energy, and a sharp increase is observed in the energy range from 4 to 4.5 eV. The fundamental interband transition in single-crystal lithium niobate is attributed to the valence band maximum at the Γ point and the conduction band minimum at the 0.4Γ–K point of the Brillouin zone [66]. It is known that both direct and indirect transitions are observed in single-crystal LN. The absorption coefficient

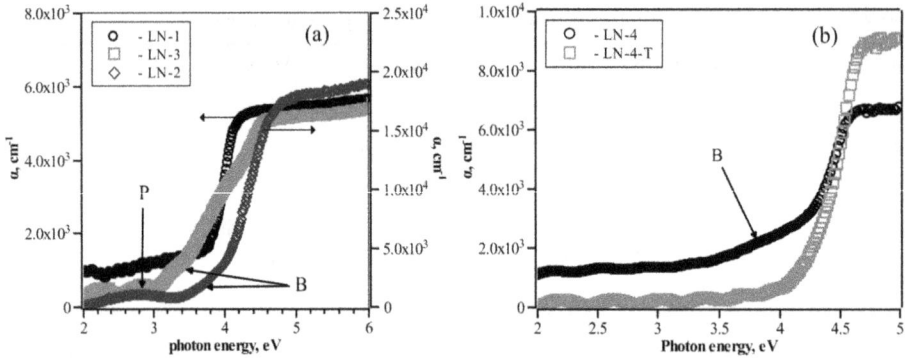

Figure 4.47. Dependence of the absorption coefficient α on the incident photon energy for samples LN-1, LN-2, LN-3 (a) and LN-4, LN-4-T (b). Reprinted from [65], Copyright (2017), with permission from Elsevier.

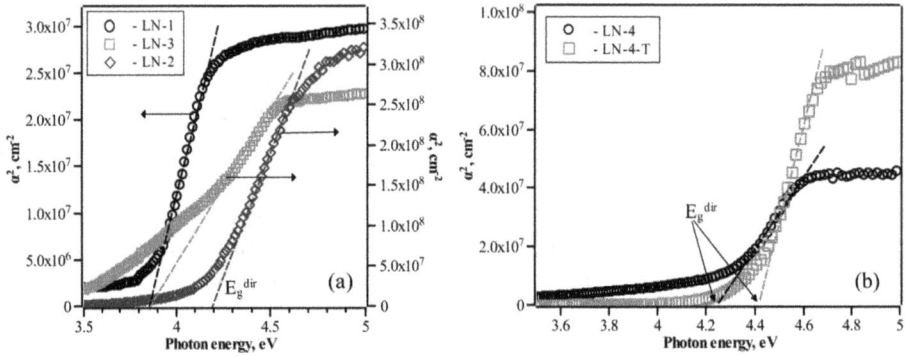

Figure 4.48. Dependence of the absorption coefficient α^2 on the incident photon energy for samples LN-1, LN-2, LN-3 (a) and LN-4, LN-4-T (b). Reprinted from [65], Copyright (2017), with permission from Elsevier.

α depends on the band structure of a semiconductor or dielectric and is described by the following equation [67]:

$$\alpha(\nu) \propto B\left(h\nu - E_g\right)^r. \tag{4.41}$$

Here, B is a frequency-independent factor, h is the Planck constant, ν is the frequency, E_g is the bandgap, and r is a parameter ($r = 1/2$ for allowed direct transitions and $r = 2$ for allowed indirect transitions). The segments of rapid increase between 4.0 and 4.5 eV in figure 4.47 can be attributed to the allowed direct transitions.

According to equation (4.41), these segments should be straight lines in $\alpha^2-h\nu$ coordinates. Figure 4.48 demonstrates a linear dependence of α^2 vs. $h\nu$ graphs in the range of 4.0–4.5 eV, indicating that allowed direct transitions occur in this energy span [65].

The energies of the allowed direct transitions E_g^{dir}, obtained from the intercepts of the linear segments of the $\alpha^2 - h\nu$ graphs with the horizontal axis, are provided in table 4.8.

Table 4.8. Direct and indirect bandgap energies and CLC concentrations for the studied samples.

Sample #	Direct energy gap, E_g^{dir} (eV)	Indirect energy gap, E_g^{indir} (eV)	Trap concentration, N_t (cm^{-3})
LN-1	3.8	—	8×10^{19} [55]
LN-2	4.2	2.8	—
LN-3	3.8	2.5	7×10^{18} [25]
LN-4	4.2	2.0	2×10^{18} [55]
LN-4-T	4.4	—	3×10^{14} [55]

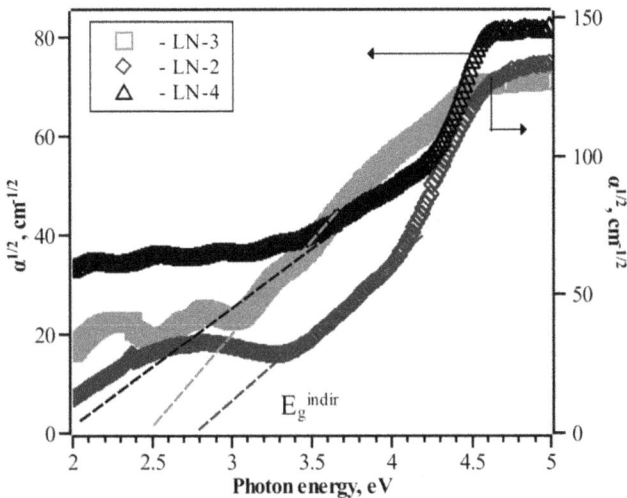

Figure 4.49. Dependence of $\alpha^{1/2}$ on incident photon energy for samples LN-2, LN-3, and LN-4. Reprinted from [65], Copyright (2017), with permission from Elsevier.

The second segment, marked 'B' in figure 4.47 for samples LN-2, LN-3, and LN-4, can be attributed to allowed indirect transitions in the studied films. According to equation (4.41), absorption spectra should be linear in $\alpha^{1/2} - h\nu$ coordinates, as shown in figure 4.49.

Experimental data indicate that the direct bandgap energies are 3.60 and 3.68 eV for congruent and stoichiometric lithium niobate, respectively [68], which are close to the theoretically predicted magnitude of 3.57 eV [60]. However, by considering electron–hole interactions and corrections applied to existing models, the authors of [66] obtained a magnitude of 4.7 eV. It has been reported that interband optical transitions are limited by the presence of deep-level centers (DLCs) in the bandgap of the material [69]. Therefore, the variation in E_g^{dir} energies reflected in table 4.8 indicates the strong influence of sputtering parameters on the bandgap of LN films, likely due to the transition of carriers from these states to the bands.

The structural and electrical properties of thin LN films fabricated under the conditions given in table 4.7 are influenced by the RFMS and IBS parameters [1, 2, 5]. As revealed in chapter 2 and our work [1], films fabricated by the IBS method exhibit a lower O:Nb ratio than those deposited by RFMS. This fact correlates with the results shown in table 4.8, where sample LN-1, fabricated by the IBS technique, demonstrates a narrower bandgap (but a higher concentration of DLCs) compared to sample LN-2, deposited by RFMS. The substrate temperature does not significantly influence the direct bandgap E_g^{dir} in the studied films (see samples LN-2 and LN-4 in table 4.8).

In contrast, table 4.8 (samples LN-3 and LN-4) shows that the position of the substrate relative to the target, a key parameter of RFMS, has a profound effect on direct optical transitions in LN thin films. As demonstrated in chapter 2 and our work [1], all samples located above the target erosion zone are influenced by plasma (the plasma effect), causing the formation of c-oriented LN films, in contrast to films fabricated outside the target erosion zone, which contain arbitrarily oriented polycrystalline grains. Furthermore, films deposited under the same conditions as sample LN-3 (see table 4.7) have a relatively high concentration of traps (defects) ($N_t = 7 \times 10^{18}\,\text{cm}^{-3}$) [25] compared to films fabricated without the ion assist effect (sample LN-4), which have a trap concentration of ($N_t = 2 \times 10^{18}\,\text{cm}^{-3}$) [55]. Intensive bombardment under the ion assist effect forms a surface layer with a high defect concentration, which participates in optical absorption and affects the direct bandgap E_g^{dir}.

The effect of TA on the bandgap, reflected in table 4.8, can originate from a decrease in defect concentration and a reduction in mechanical strain during TA. Specifically, it has been reported that the bandgap of single-crystal LN increases as the defect concentration decreases [70]. Moreover, TA reduces mechanical strain, causing a blueshift in the optical bandgap of annealed LN films [63]. Our results suggest (see table 4.8) that the direct bandgap E_g^{dir} of LN films increases from 4.2 to 4.4 eV after TA (samples LN-4 and LN-4-T). This finding aligns with our earlier study [55], which showed that TA decreases the concentration of DLCs from 2×10^{18} to $3 \times 10^{14}\,\text{cm}^{-3}$ in films grown and annealed under regimes similar to those used for samples LN-4 and LN-4-T. The post-TA disappearance of texture in films, accompanied by increased average grain size (see chapter 2 and our studies [2, 56]), indicates a decline in mechanical strain in deposited LN films. Furthermore, the electrical conductivity and dielectric properties of annealed LN films are influenced by the bulk properties of grains rather than grain boundaries [22], making their optical properties similar to those of single-crystal lithium niobate.

It was shown in [71] that the absorption edge below 3.8 eV has an indirect nature in single-crystal LN. Indirect transitions occur with photon participation when the bottom of the conduction band and the top of the valence band are not at the same wave vector K, as observed in bulk lithium niobate. LN films exhibit high defect concentrations, creating band tails in the bandgap and the possibility of indirect band–band tail transitions. The results of [71] suggest that band tails are caused by electric fields produced by oxygen vacancies, leading to indirect optical transitions with an energy of 3.5 eV. We have demonstrated that TA of as-grown LN films decreases the defect concentration (oxygen vacancies or antisite defects $\text{Nb}_{\text{Li}}^{+4}$) from

1×10^{18} to $5 \times 10^{16} \, \text{cm}^{-3}$ [2]. However, the indirect bandgap energies in our experiments were lower than 3.5 eV (see table 4.8). There is evidence that such low-energy transitions can be associated with polaron hopping conductivity, affecting the absorption coefficient [72]. The absorption band at energy E_{opt} can be expressed as

$$E_{\text{opt}} = 4W. \tag{4.42}$$

Here, W is the activation energy of conductivity.

It is important to note that the peak with an energy of $E_p \approx 2.8 \, \text{eV}$, labeled '$P$' in figure 4.47(a), has been reported by other researchers. They emphasize that the amplitude of a broad peak observed at an energy of 2.5 eV in their absorption spectra $\alpha(\omega)$ depends on the degree of reduction in LN single crystals [64] and attribute it to polaron hopping conductivity. Based on a detailed analysis of various polaronic models, another group of authors concluded that a broad peak in the absorption coefficient $\alpha(\omega)$ shifts its position within the low-energy range depending on the specific type of polaron involved in optical absorption via hopping conductivity. The following relationships exist between the observed peaks and different types of polarons: a peak at 0.94 eV is associated with free small polarons, 1.64 eV with bound polarons, and 2.50 eV with bound bipolarons [38]. Since the energy of the 'P' peak in the $\alpha(\omega)$ spectra from samples LN-2 and LN-3 (see figure 4.47) is close to 2.5 eV, we can conclude that bound bipolarons are responsible for optical absorption at this energy in our samples.

4.4.2 Effect of pulsed photon treatment on the optical bandgap of LiNbO$_3$ films

A major technological challenge limiting the practical applications of LN films is the synthesis of heterostructures with LN films that exhibit properties closely resembling those of bulk LN. One potential solution to this issue involves creating amorphous films under optimal conditions, followed by subsequent crystallization [73]. However, the properties of the synthesized LN films are not only influenced by sputtering conditions but also by post-deposition treatments [2, 44, 58].

In addition to traditional TA, an innovative method that utilizes photon treatment of as-grown films has been successfully employed to enhance the crystallization process [74, 75]. Specifically, PPT results in an accelerated crystallization process, a higher degree of dispersion in the formed structures, and a reduction in temperature thresholds for phase formation. In our recent study, we demonstrated that amorphous Li–Nb–O films, ultimately comprising nanostructured clustered nuclei of niobium oxides, lithium oxides, and LN, crystallize rapidly under PPT, leading to the formation of a high concentration of vacancies [76]. Although several studies have reported the structural evolution of thin films on various substrates under PPT, to the best of our knowledge, only a few papers have analyzed how the optical properties of materials evolve in this process [77].

As an efficient post-growth treatment, PPT significantly impacts the structural properties of LN films. For instance, the grain size observed in LN films subjected to

PPT was notably larger compared to those treated with TA, and they underwent recrystallization [73].

The objective of the study presented in this section is to investigate how PPT affects the optical bandgap of LN films grown using the RFMS method in various reactive gas environments. This research is particularly significant due to the growing demand for high-quality LN-based heterostructures in integrated optics and multifunctional devices. These devices leverage a combination of the piezo-electric, ferroelectric, acousto-optic, and nonlinear optical properties inherent in LN.

The novelty of this work lies in addressing the knowledge gap related to the impact of PPT, an innovative post-growth treatment, on the optical properties of LN films.

Thin LN films were synthesized through RFMS of LN targets. The magnetron power and vacuum conditions were set to 50 W and 0.1 Pa, respectively. This process was conducted in different environments, namely Ar, Ar(90%) + O_2(10%), and Ar(60%) + O_2(40%). Transparent substrates for optical measurements were obtained by employing cleaved plates of fluorphlogopite (Continental Trade, Poland).

In accordance with our recent publication [76], the formation of amorphous Li–Nb–O films involved using non-heated substrates strategically positioned to offset the target erosion zone. Conversely, for the growth of polycrystalline films, substrates were heated to 550 °C. After the film synthesis, the sample surfaces underwent PPT using the UOL.P-1M experimental setup, with the chamber depicted schematically in figure 4.50.

The PPT was carried out by irradiating a sample with powerful radiation (spectral range: $\lambda = 0.2$–$1.2\,\mu$m, figure 4.1(b)) in air using the following modes: single and double irradiation with a packet of 10^{-2} s pulses for 1.0 s (which corresponds to a

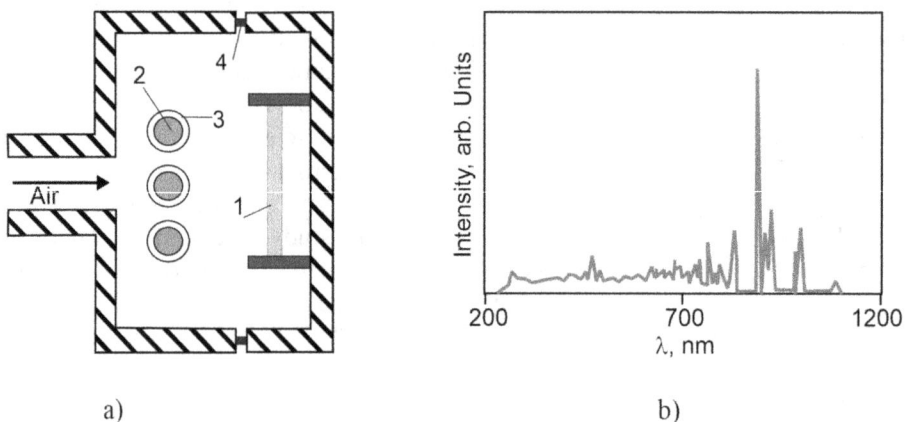

a)   b)

Figure 4.50. The PPT chamber (a) and the emission spectrum of the xenon lamps (b) used in this study: 1—a sample, 2—xenon lamps, 3—quartz tubes, 4—sealing.

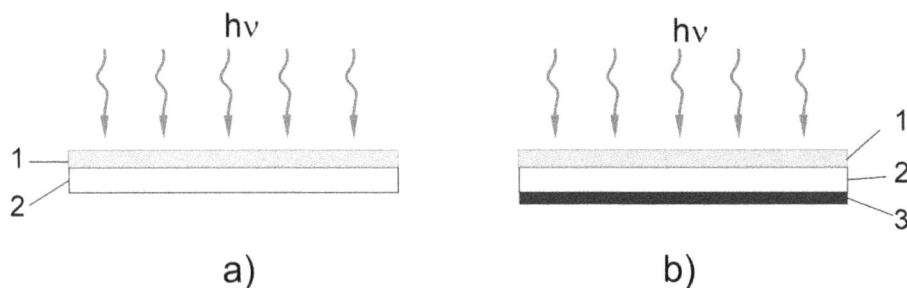

Figure 4.51. The normal (a) and enhanced (b) PPT treatment. 1—LN film, 2—substrate, 3—mirror.

supplied radiation energy of $E_I \sim 140 \, \text{J cm}^{-2}$). Additionally, a metal layer, serving as a mirror and reflecting the transmitted radiation, was deposited onto the back side of some studied samples to enhance the PPT effect (see figure 4.51). The structure of the synthesized films was investigated by high-resolution transmission electron microscopy (HRTEM, Titan 80–300) and scanning electron microscopy (SEM, FEI HELIOS Nanolab) using the focused ion-beam (FIB, Ga + with an energy of 30 keV) technique for heterostructure preparation—the 'cross-section' method. A cross-section of Li–Nb–O was obtained using the following process: a protective layer containing C and Pt ($1.5 \times 3 \times 16 \, \mu\text{m}^3$) was deposited onto the studied area by a gas injection system (GIS).

The sample material was then sputtered onto both sides of the protective layer by the FIB at a depth of $4 \, \mu\text{m}$. The lamella fabricated in this manner was cut along a perimeter by the FIB to detach the cross-section from the bulk of the sample. We then extracted the lamella using a micromanipulator. After that, the lamella was fixed to a copper holder by the GIS and thinned by the FIB. The composition of the studied films was investigated by x-ray diffraction (XRD, ARL X'TRA Thermo Techno) and energy dispersive x-ray microanalysis (EDXMA). The structure, surface morphology, and phase composition were studied by SEM (TESCAN Mira 3 lMH). The scanning TEM (STEM) images were obtained in two ways. The first method was STEM with ring detectors that detected electrons inelastically scattered at low angles ($\sim 3°$), which registers bright-field images (bright-field STEM, BF-STEM). The second method is the registration of electrons elastically scattered by atomic nuclei at high angles ($\sim 12°$), yielding dark-field images (high-angle annular dark-field STEM, HAADF-STEM with Z-contrast). Comparative analysis of images obtained by BF-STEM and HAADF-STEM techniques allows the visualization of columns of light and heavy atoms.

The transmission spectra were studied using a single-beam spectrophotometer (USB2000) equipped with a lamp (USB-DT) operating in the spectral range of 200–800 nm, both manufactured by Ocean Optics, USA. To exclude the influence of the substrate (fluorphlogopite), its spectrum was subtracted from the original spectrum measured from the studied sample. The possible interference pattern that could have been caused by the substrate was not observed because of its relatively high thickness

Figure 4.52. In-plane (a) and cross-sectional (b), (c) SEM patterns of the as-grown Li–Nb–O films on Si. Reproduced from [78], with permission from Springer Nature.

($\sim$0.1 mm) compared to the spectral width of the slit (1 nm) and due to the presence of a diffusive component of light reflected from the film.

We have previously examined the structure and composition of LN films fabricated through the RFMS technique in an Ar(60%) + O_2(40%) environment, followed by TA, as discussed in earlier chapters and our published articles [2, 4]. In this section, our focus is directed towards the structural properties of films cultivated in a gas mixture of Ar(90%) + O_2(10%). Figure 4.52 shows SEM patterns of the as-grown films deposited onto non-heated Si substrates. As depicted in figure 4.52, the film thickness measures approximately 300 nm, and inhomogeneities lead to a height variation not exceeding 10 nm.

An uneven contrast in the cross-sectional pattern acquired by detecting reflected electrons using the HAADF-STEM technique (figure 4.52(c)) indicates nonuniformity in the chemical composition perpendicular to the growth direction of the as-grown films. The Z-contrast displayed in figure 4.52(c) is observed concurrently with the absorption contrast. This is attributed to the thickness change resulting from the FIB preparation of the studied film, which is characterized by a columnar growth texture. However, we employed a precise approach to sample preparation—sliding FIB, wherein the intensity of FIB gradually decreases, leading to a decline in the absorption contrast intensity. Thus, the image shown in figure 4.52(c) represents the Z-contrast, supported by the bright band at the substrate–film interface where a layer of silicon dioxide is situated.

Figure 4.53 presents high-resolution TEM and microdiffraction patterns (see inset) of an ultrathin cross-sectional segment of the as-grown heterostructures. The absence of periodic contrast in the TEM image, along with the blurred reflections in the diffraction pattern, indicates the amorphous structure of the as-grown Li–Nb–O films.

Analyzing the XRD pattern of Li–Nb–O films grown on fluorphlogopite after TA, we demonstrated that TA induces the crystallization of initially amorphous Li–Nb–O

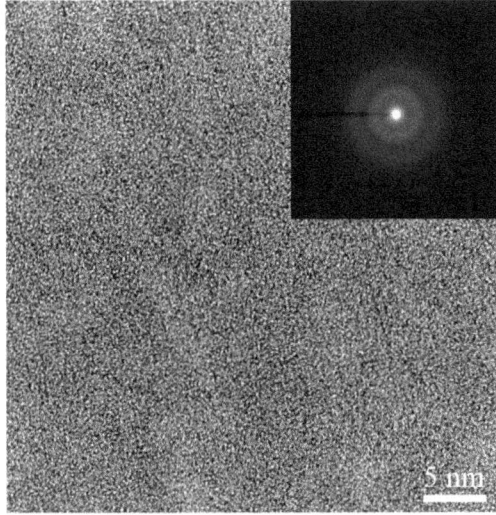

Figure 4.53. TEM image and diffraction pattern (see inset) of the as-grown Li–Nb–O films on Si. Reproduced from [78], with permission from Springer Nature.

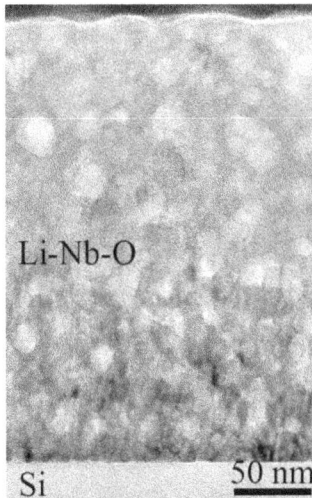

Figure 4.54. Cross-sectional TEM pattern of ultrathin Li–Nb–O films after TA. Reproduced from [78], with permission from Springer Nature.

films, transforming them into the LN phase [78]. Moving on to figure 4.54, the cross-sectional TEM pattern of an ultrathin Li–Nb–O film post-TA is presented. The amplitude contrast, indicative of various crystallite orientations, suggests a grain size ranging between 10 and 30 nm. The intricate contrast depicted in figure 4.54 results from a combination of amplitude and diffraction contrasts. The crystallites, which are coherent scattering areas formed during TA, exhibit an isotropic morphology with sizes ranging from 10 to 20 nm. These LN crystallites, positioned sequentially (the

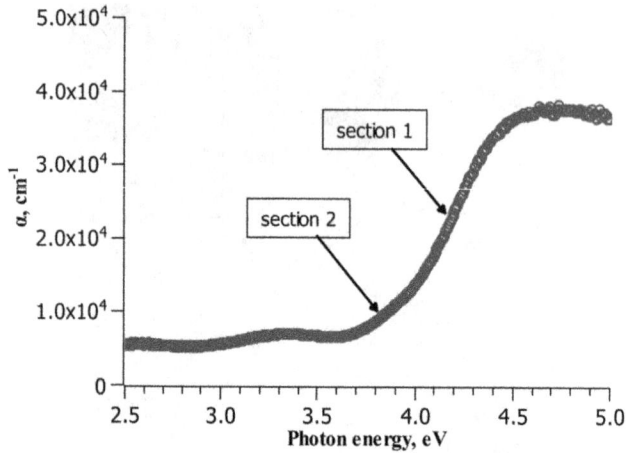

Figure 4.55. Dependence of the absorption coefficient α on the incident photon energy for the as-grown amorphous Li–Nb–O films fabricated in an Ar(90%) + O_2(10%) gas mixture.

thickness of the foil prepared for the study is approximately 60 nm), give rise to a complex TEM pattern characteristic of polycrystalline LN.

Examining the XRD pattern of the amorphous Li–Nb–O film on fluorphlogopite following PPT treatment reveals peaks corresponding not only to fluorphlogopite but also to the hexagonal crystal phase [78]. The lattice parameters ($a = 5.22$ Å, $c = 13.87$ Å) closely resemble those of LN ($a = 5.24$ Å, $c = 13.86$ Å). The average size of crystallites, estimated using the Scherrer formula, is approximately 40 nm, which agrees with the size observed for films after TA. Consequently, PPT induces structural transformations in amorphous Li–Nb–O films, facilitating their crystallization into LN.

Figure 4.55 illustrates the absorption coefficient (α) as a function of the incident photon energy for the as-grown samples under study. As depicted in figure 4.55, the absorption coefficient exhibits a steady increase with the photon energy, and it experiences a notable sharp rise within the energy range of 3.7–4.5 eV.

The essential interband transition in bulk LN crystals is correlated with the valence band maximum at the C point and the conduction band minimum situated at 0.4C–K in reciprocal space [66]. Both direct and indirect optical transitions are integral components of the fundamental optical transitions in LN, as emphasized in [66].

A semiconductor's absorption coefficient can be determined based on its band structure using formula (4.41). Sections 1 and 2 in figure 4.55 can be ascribed to direct and indirect transitions, respectively. In accordance with equation (4.41), these sections should be depicted by straight lines in graphs of α^2 versus $h\nu$ and $\alpha^{1/2}$ versus $h\nu$, yielding corresponding bandgaps as intercepts on the energy axis.

Parts (a) and (b) of figure 4.56 illustrate the dependence of the absorption coefficients α^2 and $\alpha^{1/2}$ on the incident photon energy for as-grown amorphous LN films synthesized in an Ar(90%) + O_2(10%) gas mixture. As previously mentioned, we determined the direct and indirect bandgaps (E_g^{dir} and E_g^{indir}) of the studied films

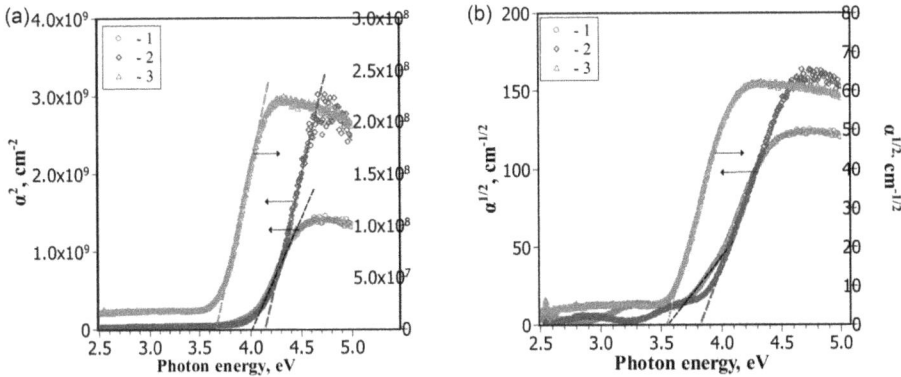

Figure 4.56. Dependence of the absorption coefficients α^2 (a) and $\alpha^{1/2}$ (b) on the incident photon energy for amorphous LN films fabricated in an Ar(90%) + O$_2$(10%) gas mixture. 1—as-grown films, 2—films after PPT, 3—films after PPT (with mirrored rear surface).

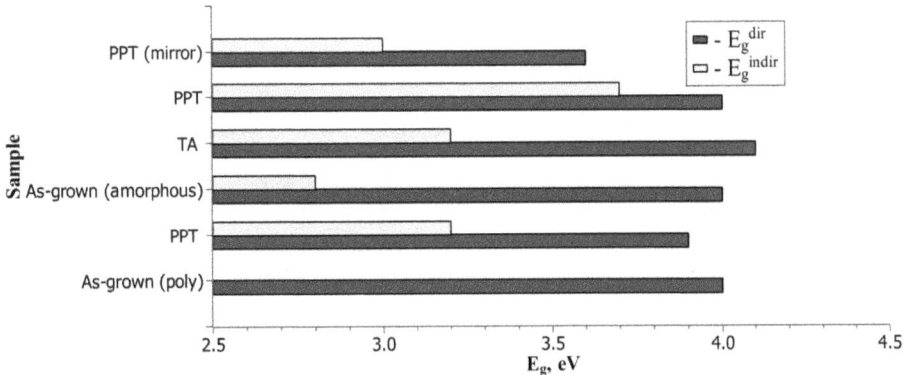

Figure 4.57. Direct E_g^{dir} and indirect E_g^{indir} bandgaps of the studied LN films grown in an Ar(90%) + O$_2$(10%) gas mixture after various treatments. PPT—films after PPT treatment, TA—films after TA, PPT(mirror)—films after more intense PPT with a mirrored rear surface (see figure 4.51(b)).

from linear sections similar to those shown in figure 4.55. The obtained results for the as-grown films, fabricated in both Ar(90%) + O$_2$(10%) and Ar(60%) + O$_2$(40%) gas environments and subjected to various treatments, are presented in figures 4.57 and 4.58.

In figure 4.57, it is evident that the as-grown amorphous and polycrystalline (poly-) LN films produced in an Ar(90%) + O$_2$(10%) gas mixture exhibit a similar magnitude of $E_g^{dir} = 4\,\text{eV}$, closely matching that of single-crystal LN ($E_g = 4.12\,\text{eV}$) [68]. However, unlike poly-LN, amorphous LN films demonstrate indirect transitions with an energy of $E_g^{indir} = 2.8\,\text{eV}$. Both PPT and TA have a marginal effect on E_g^{dir}, while the E_g^{indir} of amorphous LN films increases after these treatments, with PPT exhibiting a more pronounced influence. Specifically, PPT increases E_g^{indir} by 32%, in contrast to the increase of 14% associated with TA. Additionally, more intense PPT

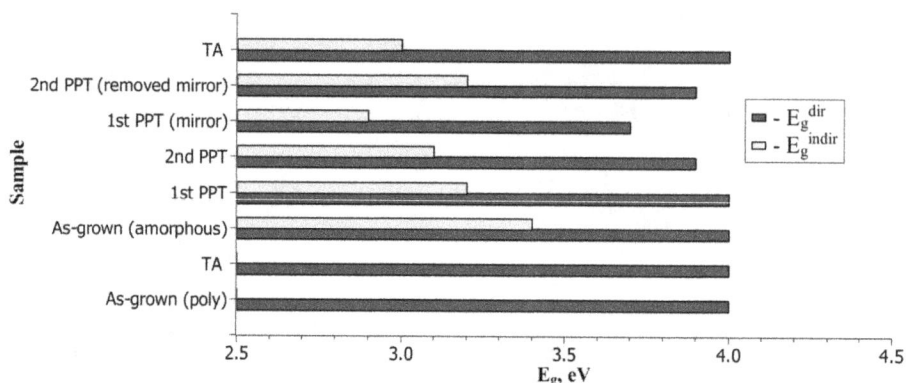

Figure 4.58. Direct E_g^{dir} and indirect E_g^{indir} bandgaps of the studied LN films grown in an Ar(60%) + O$_2$(40%) gas mixture after various treatments. 1st PPT—films after the first PPT treatment, 2nd PPT—films after the second PPT treatment, TA—films after TA, 1st PPT(mirror)—films after the first more intense PPT with a mirrored rear surface (see figure 4.51(b)), 2nd PPT (removed mirror)—films after the second PPT without a mirror.

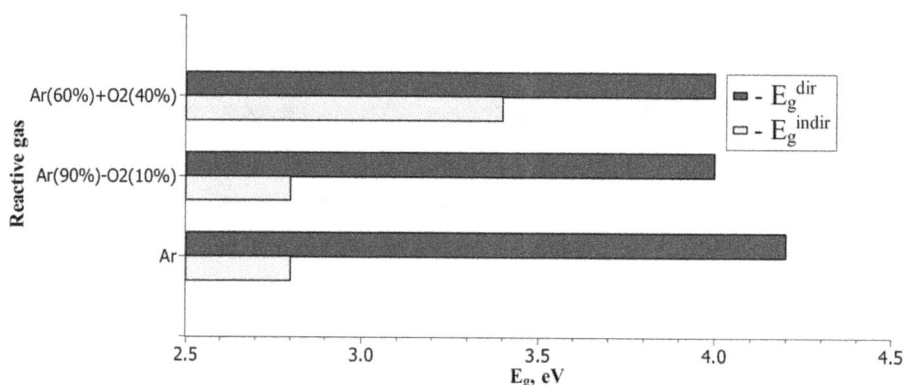

Figure 4.59. Direct E_g^{dir} and indirect E_g^{indir} bandgaps of the studied as-grown amorphous Li–Nb–O films fabricated in various reactive gas environments.

conducted on the 'mirrored' amorphous samples (refer to figure 4.51(b)) increases E_g^{indir} by 7% and decreases E_g^{dir} by 10% (see PPT(mirror) in figure 4.59).

Figure 4.58 illustrates that both the as-grown amorphous and poly-LN films, produced in an Ar(60%) + O$_2$(40%) gas mixture, exhibit an equivalent E_g^{dir} of 4 eV. This finding is consistent with the findings for films fabricated in an Ar(90%) + O$_2$(10%) reactive environment. Moreover, an indirect band with an energy of $E_g^{indir} = 3.4$ eV is observed in amorphous LN films, which is lower than the energies of films produced in an Ar(90%) + O$_2$(10%) gas mixture (see figures 4.57 and 4.58).

Consecutive PPT treatments (1st PPT and 2nd PPT in figure 4.58) applied to the as-grown films result in a 6% and 3% reduction in E_g^{indir}, respectively, while E_g^{dir} remains largely unchanged. A more vigorous PPT that includes the use of a mirror (1st PPT(mirror) in figure 4.58) has a more pronounced effect, reducing both E_g^{indir}

and E_g^{dir} by 15% and 7%, respectively. Subsequent PPT after removing the mirror (see 2nd PPT(removed mirror) in figure 4.58) increases the magnitude of E_g^{indir} by 10%. Regarding TA, this treatment causes a more significant decrease in E_g^{indir} compared to the second PPT (compare TA and 2nd PPT in figure 4.58). Figure 4.59 illustrates the impact of a reactive gas environment on the E_g^{dir} and E_g^{indir} of as-grown amorphous Li–Nb–O films. As depicted in figure 4.59, the presence of oxygen leads to a slight reduction in E_g^{dir}, and this effect remains consistent regardless of the oxygen concentration. In terms of the indirect bandgap, E_g^{indir} exhibits no variation for LN films grown in either a pure Ar environment or an Ar(90%) + O_2(10%) gas mixture.

Upon increasing the oxygen content by up to 60%, E_g^{indir} experiences a notable 21% increase. It is noteworthy that the influence of PPT on E_g^{indir} of LN films is opposite, depending on the oxygen content in the reactive plasma. Building on the findings presented in figures 4.57 and 4.58, PPT enhances E_g^{indir} in LN films produced in an Ar(90%) + O_2(10%) gas mixture, while it diminishes the bandgap when an Ar(60%) + O_2(40%) atmosphere is employed. As documented in various scholarly publications, the absorption edge in LN is influenced by the electric fields generated by charged defects [79]. To be more specific, certain researchers have illustrated that the absorption edge corresponding to low incident photon energy can be ascribed to the existence of lithium vacancies [70]. The absorption coefficient of LN matches the shape of the well-established Urbach law, as widely observed in the literature [79]:

$$\alpha = \alpha_0 \exp\left(\frac{\sigma(hv - E_g)}{kT}\right) \tag{4.43}$$

with an absorption slope given by [79]:

$$s = \frac{\sigma}{kT}. \tag{4.44}$$

Here, s is the slope of the $\ln(\alpha)$–hv graph, k is the Boltzmann constant, T is room temperature, and σ is the fitting parameter of the Urbach law, which corresponds to impurities or defects. We determined the value of σ by analyzing the slope (s) derived from $\ln(\alpha)$–hv graphs of the studied samples. The results obtained, showcasing the relationship between E_g^{indir} and σ, are depicted in figure 4.60. As illustrated in figure 4.60, the indirect bandgap increases with σ, indicating a corresponding rise in the defect concentration within the analyzed films.

Previous studies have demonstrated that the oxygen concentration in the reaction chamber significantly influences the properties of synthesized LN films. Specifically, research has shown that utilizing an Ar(90%) + O_2(10%) gas mixture results in the formation of oxygen-deficient LN films. To achieve the correct LN composition, it has been reported that the partial oxygen pressure must be increased [80]. The elevated oxygen concentration enhances the bombardment of the deposited film, promoting the insertion of oxygen into the interstices of the crystal lattice [7].

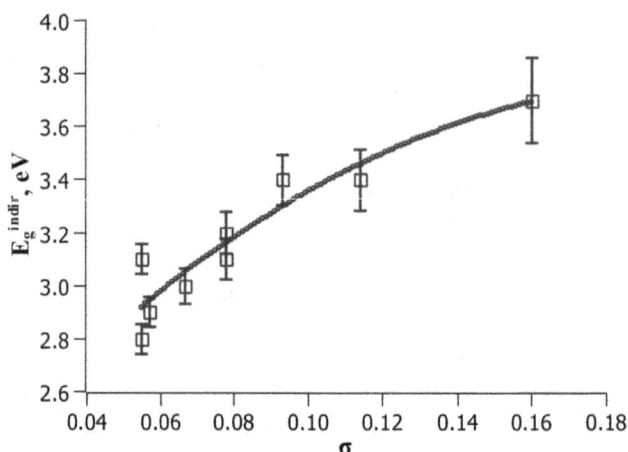

Figure 4.60. Indirect bandgap E_g^{indir} as a function of the fitting parameter σ for the studied films.

The defect states in LN are predominantly associated with antisite defects such as Nb_{Li} and oxygen vacancies. The indirect transitions responsible for photon absorption at an energy level of approximately 3 eV, observed in LN, are attributed to the presence of defect states (V_{Li} or Nb_{Li} [70]) within the bandgap, along with disorders present in the lattice [81]. Specifically, electronic defects such as bound polarons (Nb_{Li}^{4+}), bipolarons (Nb_{Li}^{4+}–Nb_{Nb}^{4+}), and Q-polarons (resulting from interactions between two bipolarons) are accountable for indirect transitions with energy levels of 1.6, 2.5, and 3.6 eV, respectively [38]. The observed energy value of E_g^{indir} = 2.8 eV in the as-grown amorphous Li–Nb–O films, fabricated in an Ar atmosphere and an Ar(90%) + O_2(10%) gas mixture, can be attributed to bipolarons. Conversely, the as-grown amorphous Li–Nb–O films deposited in an Ar(60%) + O_2(40%) reactive atmosphere exhibit an E_g^{indir} value of 3.4 eV, primarily associated with the existence of Q-polarons. In chapter 2, we demonstrated that these amorphous films comprise nanostructured precursors (clustered nuclei), ultimately consisting of crystalline phases such as niobium oxides, lithium oxides, and LN as the synthesis products [73]. Consequently, the oxygen content within the reaction chamber significantly influences the composition of these clustered nuclei.

As illustrated in figures 4.59 and 4.60, the as-grown LN films fabricated under a low partial oxygen pressure (Ar(90%) + O_2(10%) gas mixture) exhibit a lower defect concentration, resulting in a reduced magnitude of E_g^{indir} when compared to films fabricated in an Ar(60%) + O_2(40%) reactive atmosphere. At elevated reactive gas pressures, such as in an Ar(60%) + O_2(40%) reactive gas mixture, the heightened frequency of collisions impedes the direct access of Li atoms to the substrate surface. Consequently, this hinders the formation of a Li-deficient phase, ultimately promoting the generation of defects and distortions due to the insertion of oxygen atoms.

During the PPT process, the defect concentration in lithium niobate LN films deposited in a Ar(60%) + O_2(40%) gas mixture decreases concurrently with the decrease in E_g^{indir}, as illustrated in figure 4.58. In contrast, PPT results in an elevated concentration of Nb_{Li} defects in LN films deposited in an environment consisting of 10% Ar and 90% O_2. This increase is attributed to the lower O/Nb ratio in these films, as discussed earlier, and aligns with findings from other researchers [80]. It is noteworthy that the indirect bandgap is closely linked to the presence of defects and distortions in the films.

This section has been reproduced with permission from [78].

4.5 Temperature-induced transition from p-type to n-type conduction in LiNbO₃/Nb₂O₅ polycrystalline films

One interesting issue to discuss at the end of this chapter is the temperature-induced transition from p-type to n-type conduction in LN films. While the study of this phenomenon is extensive and complex, we have made an attempt to explain it. Thin films (300 nm) were fabricated using the RFMS method without the ion assist effect. Sputtering was performed at a magnetron power of 100 W in an Ar atmosphere and in an Ar + O_2 gas mixture at Ar/O_2 ratios of 60/40 and 80/20 ($P = 1.5 \times 10^{-1}$ Pa). The substrates were positioned 5 cm away from the target erosion zone. N-type silicon wafers (001)Si with a resistivity of 7.2 $\Omega \cdot$cm were used as substrates, which were heated to 550 °C during the film synthesis process.

Figure 4.61 shows the XRD patterns of the studied films, which had a thickness of 300 nm. All XRD patterns contain peaks attributed to the rhombohedral cell of LN.

The peaks corresponding to $2\Theta = 24.3°$ ($\bar{3}11$) and $2\Theta = 30.3°$ ($\bar{1}11$) in spectra 2 and 3 (see figure 4.61) correspond to the monoclinic lattice of Nb_2O_5 ($a = 12.74$ Å,

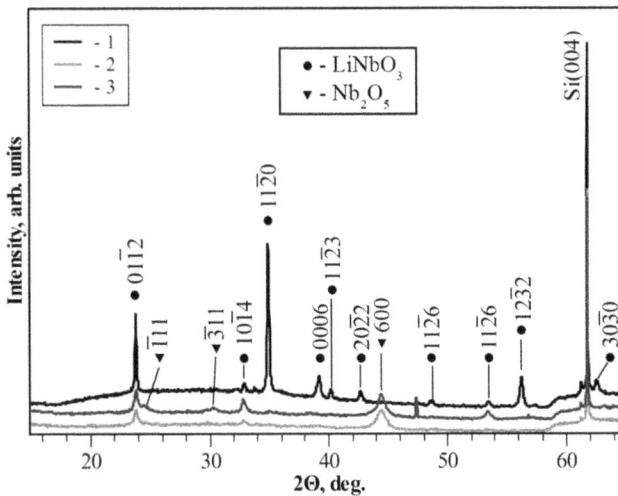

Figure 4.61. XRD patterns of the studied films fabricated by the RFMS method in different reactive gas environments: 1—pure Ar, 2—Ar + O_2 gas mixture at an Ar/O_2 ratio of 60/40, 3—Ar + O_2 gas mixture at an Ar/O_2 ratio of 80/20.

$b = 4.88\,\text{Å}$, $c = 5.56\,\text{Å}$). The relative intensity of the peaks attributed to the LN phase in figure 4.61 corresponds to the intensity of peaks for polycrystalline lithium niobate listed in the crystallographic database (see appendix), indicating arbitrary grain orientation in all studied films. Some Nb_2O_5 grains are oriented randomly, while others exhibit a single-axis texture (i.e. the crystallographic plane (600) is parallel to the substrate plane). The quantitative ratio of the LN and Nb_2O_5 phases was estimated using the reference intensity ratio (RIR) for non-textured phases. In the film deposited using an Ar + O_2 gas environment (80/20), the mass fractions of the LN phase and non-textured Nb_2O_5 are about 65% and 35%, respectively. In films fabricated in an Ar + O_2 gas mixture with an Ar/O_2 ratio of 60/40, these fractions are around 55% and 45%, respectively. Thus, the amount of non-textured Nb_2O_5 phase in the film declines by 20%, while the amount of the textured Nb_2O_5 phase (based on the intensity of (600) peaks) decreases fourfold when the oxygen content is reduced from 40% to 20% in a reactive plasma.

The grain sizes determined from XRD patterns (coherent scattering areas) are $l = 21\,\text{nm}$ and $l = 10\,\text{nm}$ for LN and Nb_2O_5 grains, respectively [82]. These results coincide with the TEM and microdiffraction patterns shown in figures 4.62 and 4.63.

Figure 4.62 presents a TEM microdiffraction pattern (a) and a bright-field TEM image (b) of a 100 nm thick film deposited by RFMS in an Ar atmosphere. All reflections in the microdiffraction pattern are attributed to the LN phase with arbitrary grain orientation. The bright-field TEM pattern (figure 4.62(b)) indicates that the film contains grains at sizes from 50 to 100 nm. Figure 4.63 shows a TEM microdiffraction pattern (a), bright-field (b), and dark-field (c, d) images of a thin (100 nm) film deposited by RFMS in an Ar + O_2 environment at an Ar/O_2 ratio of 60/40. All reflections in this pattern correspond to the LN and Nb_2O_5 phases with randomly oriented grains. Analysis of the TEM patterns indicates that the films

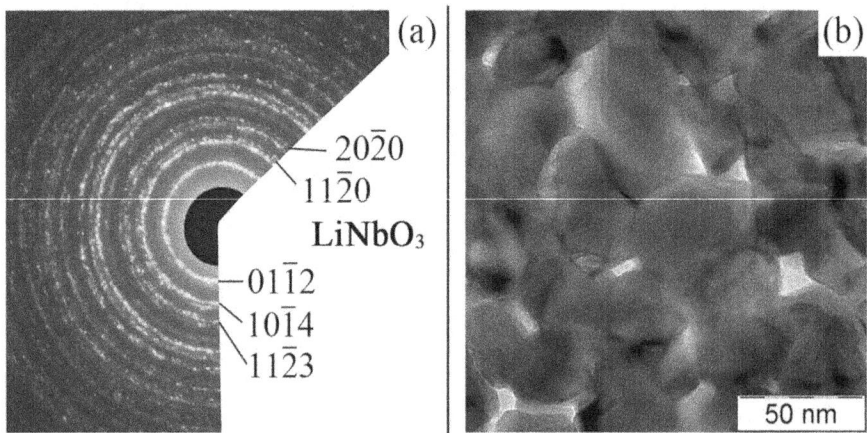

Figure 4.62. TEM diffraction pattern (a) and a bright-field (b) TEM image of the thin (100 nm) film deposited by the RFMS method in an Ar atmosphere. Reprinted from [82], Copyright (2017), with permission from Elsevier.

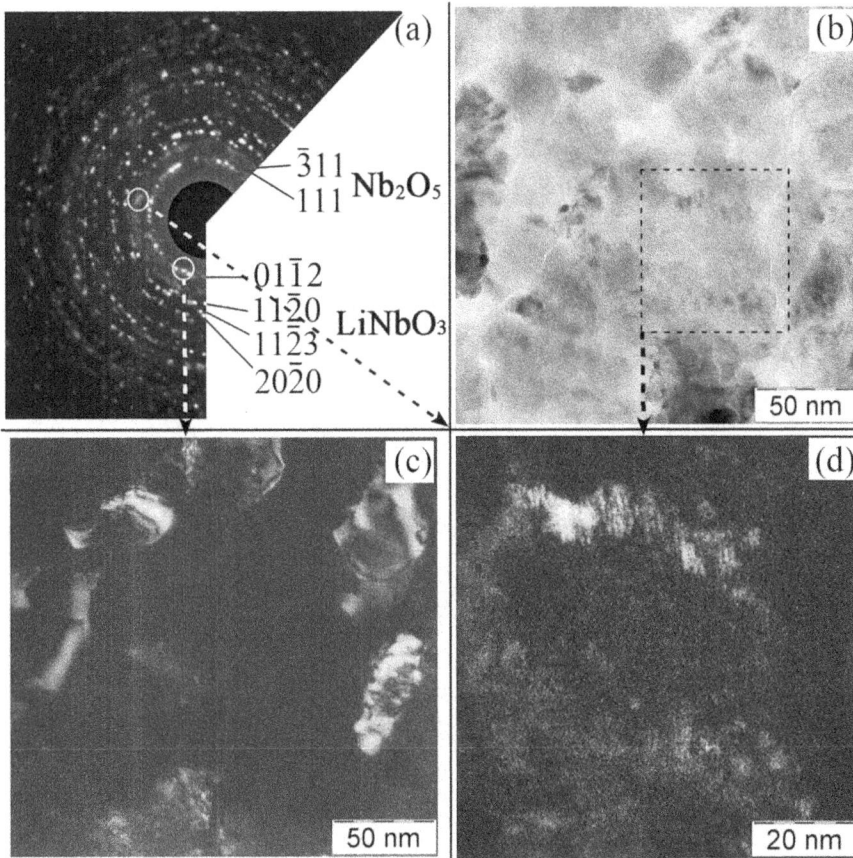

Figure 4.63. TEM microdiffraction pattern (a), bright-field (b) and dark-field (c, d) images of the thin (100 nm) film deposited by the RFMS method in an Ar + O$_2$ environment at an Ar/O$_2$ ratio of 60/40. Reprinted from [82], Copyright (2017), with permission from Elsevier.

fabricated under these conditions contain single-phase crystalline LN blocks at sizes of 50–100 nm and blocks composed of LN subgrains and Nb$_2$O$_5$ nanograins of the same size (figure 4.63(b)). The circled area in figure 4.63(b) illustrates Nb$_2$O$_5$ nanograins at sizes of 5–10 nm within a single LN block. Their sizes were estimated based on the total sum of intensities ($\bar{3}$11) and ($\bar{1}$11) (figure 4.63(d)). Further detailed analysis of the TEM images reveals that some blocks are composed solely of Nb$_2$O$_5$ nanograins (figure 4.64).

According to the mechanism proposed in [83], Nb$_2$O$_5$ oxide is formed as a product of the following dissociation: $2\text{LiNb}_3\text{O}_8 \rightarrow 3\text{Nb}_2\text{O}_5 + \text{Li}_2\text{O}$. Since LN-based heterostructures fabricated in an Ar + O$_2$ reactive gas environment can be represented as two back-to-back Schottky barriers connected in series, charge transport from the film to a Si substrate at low applied voltage is energetically unfavorable. Thus, during Hall measurements, a conductive substrate does not contribute.

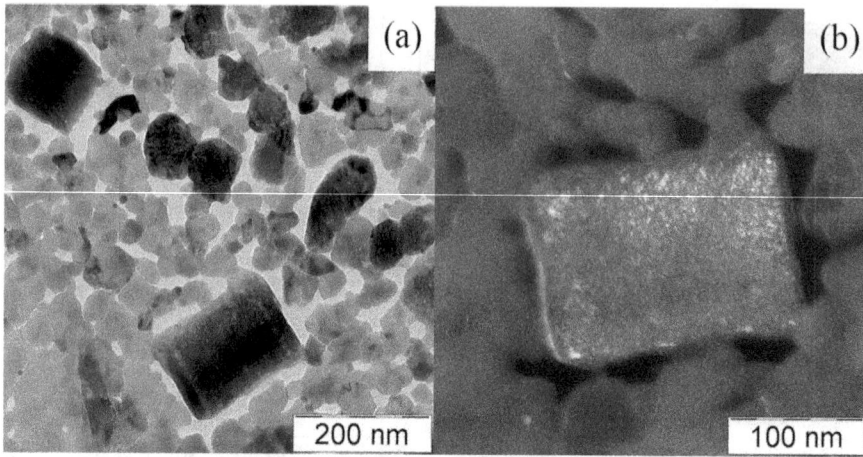

Figure 4.64. TEM patterns of the LN/Nb$_2$O$_5$ thin (100 nm) film fabricated by the RFMS method in an Ar + O$_2$ environment at an Ar/O$_2$ ratio of 60/40. Patterns (a) and (b) represent single crystalline blocks composed entirely of Nb$_2$O$_5$ nanograins. Reprinted from [82], Copyright (2017), with permission from Elsevier

If only one type of carrier is present in a material, the Hall coefficient (RH) is inversely proportional to the free-carrier concentration (n for electrons or p for holes) according to the following expressions [84]:

$$R_H = \frac{1}{q \cdot p} r - \text{for a p–type semiconductor}$$

$$R_H = \frac{1}{q \cdot n} r - \text{for a n–type semiconductor} \tag{4.45}$$

where p and n are the concentrations of holes and electrons, respectively, and r is the scattering factor defined as $r = \mu_H / \mu_d$ (μ_H and μ_d are the Hall and drift mobility, respectively). The sign of R_H corresponds to the sign of the majority carriers in a semiconductor: $R_H > 0$ for a p-type semiconductor and $R_H < 0$ for an n-type semiconductor. Equations (4.45) are used to determine the free-carrier concentration based on the experimental Hall coefficient R_H. The scattering coefficient typically ranges from one for metals to two for semiconductors.

The Hall coefficient in the films fabricated by RFMS in an Ar + O$_2$ environment with an Ar/O$_2$ ratio of 60/40 exhibits a complex temperature dependence, as shown in Arrhenius coordinates in figure 4.65.

As seen in figure 4.65, R_H is almost temperature independent at low temperatures and decreases sharply when $T > 100$ K, becoming negative at 150 K. After that, the Hall coefficient rises sharply, changing its sign from negative to positive and monotonically increasing at temperatures close to 300 K.

Interpreting Hall measurements in polycrystalline semiconductors is challenging due to the presence of trapped charge at the interfaces. Grain boundaries contain high concentrations of CLCs (border traps), which capture or scatter carriers from

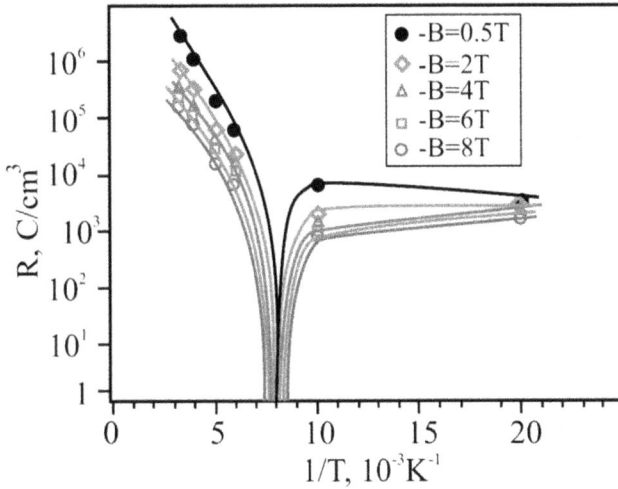

Figure 4.65. Temperature dependence of the Hall coefficient at different applied magnetic fields B for the films fabricated by RFMS in an Ar + O$_2$ atmosphere at an Ar/O$_2$ ratio of 60/40.

the bulk. The maximum charge that can be captured depends on both the surface density (N_t) and energy position (E_t) of the traps, influencing the activation energy. The depletion zone can be smaller than the average grain size or extend through the entire volume, so the carrier free path can be shorter or longer than the grain size. Based on the model proposed in [84], when the depletion zone extends only partially into a grain, the free-carrier concentration derived from Hall measurements reflects the bulk concentration. The Hall mobility is given by [84]:

$$\mu_b = \mu_0 \exp\left(-\frac{\varphi_b}{kT}\right) \tag{4.46}$$

where φ_b is the band bending at the grain boundaries, and μ_0 is a parameter dependent on grain size l as $\mu_0 \approx 10^6 l$ at room temperatures (l is measured in m, μ_0— in m^2 V s^{-1}). The band bending associated with the doping level N reaches a maximum when the trap density $N_t = N \cdot l$ and the mobility is minimal.

Since the Hall coefficient is almost temperature independent in the range of $T = 50$ K–100 K, the barrier height φ_b is low, and the depletion area width is small compared to the grain size. Thus, carrier mobility in this temperature range is limited by the bulk properties of grains, and the concentration of free carriers determined from the Hall measurements reflects the concentration in the bulk of the grains. The concentration of major carriers (holes) in this temperature range, estimated using equation (4.45), is $N_p = 2.2 \times 10^{15}$ cm^{-3}.

Some investigators note that oxygen vacancies, which are responsible for the formation of hole polarons, can also account for p-type conductivity in LN [85]. In contrast, LN films deposited under the same conditions but at lower partial oxygen pressure exhibit n-type conductivity and high donor concentrations [86]. The change in the sign of the Hall coefficient observed in our experiments can occur when both types of carriers (electrons and holes) exist in a film. At low temperatures, when the concentration of holes exceeds that of electrons ($p \gg n$), p-type conductivity occurs, and the Hall coefficient is positive. The observed decrease in temperature dependence shown in figure 4.65 at $T > 100\,K$ can result from the emission of electrons from interface states, contributing to the total conductivity at these temperatures. The transition from p-type to n-type conductivity occurs due to the higher mobility of electrons compared to holes. In this situation, the Hall coefficient is given by the following equation [87]:

$$R_H = \frac{1 - xb^2}{qp(1 + xb)^2} r.$$ (4.47)

Here, b is a parameter equal to the ratio of electron and hole mobilities μ_n/μ_p, $x = n/p$, and r is a scattering factor. The Hall coefficient becomes zero when the electron and hole contributions compensate each other ($1 = xb^2$) and goes through a minimum when $x = (b + 2)/b^2$. The minimal magnitude of R_H is given by:

$$R_{min} = -\frac{1}{4} \frac{b^2 \cdot r}{(b + 1)pq}.$$ (4.48)

Thus, when only holes participate in conductivity,

$$\left| \frac{R_{max}}{R_{min}} \right| = \frac{4(b + 1)}{b^2}$$ (4.49)

where R_{max} is the maximum Hall coefficient at low temperatures, given by equation (4.45). Solving equation (4.49) for b and using experimental values of R_{max} and R_{min}, we obtain $b = 49$ for the studied films.

Furthermore, the Nb_2O_5 phase can contribute to n-type conductivity at these temperatures. It has been demonstrated that n-type conductivity in Nb_2O_5 is highly sensitive to the partial pressure of oxygen used during the synthesis process [88]. Oxygen vacancies $V_O^{2\bullet}$ and cation interstitials can capture electrons, creating deep traps in the bandgap of Nb_2O_5, which cause its n-type conductivity. The conductivity of this oxide is affected by the partial oxygen pressure P_{O_2} according to the relationship:

$$\sigma = const \cdot P_{O_2}^{-0.25} \text{ [88]}.$$

Thus, changing the Ar/O_2 ratio in the reaction chamber is an effective tool for fabricating LN/Nb_2O_5 films with the desired electrical properties.

In polycrystalline materials, grain boundaries play a significant role in charge transport phenomena. The increase in the Hall coefficient with temperature observed in the range of $150\,K$–$300\,K$ can be attributed to intergranular barriers at the interfaces. It has been shown that mobility increases with temperature in bulk materials and decreases with temperature in thin films [89]. Mobility is a rising function of temperature according to equation (4.46), provided that charge transport

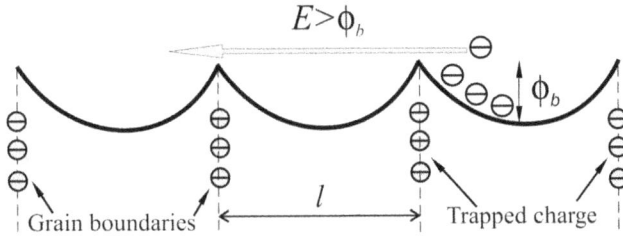

Figure 4.66. Schematic representation of a polycrystalline n-type semiconductor where the mean free path exceeds the average grain size l.

in polycrystalline films is limited by intergranular barriers φ_b. In the temperature range of 200 K–300 K, the Hall coefficient is influenced by intergranular barriers and scattering at interfaces. There are three main scattering mechanisms in thin films: surface scattering, dislocation scattering, and grain boundary scattering. Since the thickness of the studied films (300 nm) exceeds the mean free path of carriers, surface scattering does not influence charge transport in our case. Evidently, grain boundary scattering at barriers with height φ_b is the dominant scattering mechanism. In [84], the following equation for the maximum potential barrier height was proposed when $N_l = N_t$:

$$\varphi_b^{\max} = \frac{q^2 N_t l}{8\varepsilon\varepsilon_0} \tag{4.50}$$

where N_t is the trap density and l is the average grain size. Based on the model developed in [84], only electrons with energies higher than the potential barrier height φ_b can overcome multiple grains without considerable scattering, as shown schematically in figure 4.66.

Only electrons with energy $E > \varphi_b$ are deflected by an applied magnetic field, so the concentration determined through Hall measurements differs from the bulk concentration and can be expressed as:

$$n = n_1 \cdot \exp\left(-\frac{\varphi_b}{kT}\right). \tag{4.51}$$

Thus, when charge transport is limited by intergranular barriers, the temperature dependence of concentration derived from the Hall effect should be a linear function in Arrhenius coordinates, with a slope of φ_b. As seen in figure 4.65, the Hall coefficient is a thermally activated process with an activation energy of $\varphi_b = 0.13$ eV in the temperature range of 200 K–300 K. Since the average grain size in the studied films is $l = 20$ nm, and considering $\varepsilon = 28$, we estimated the density of border traps using equation (4.50) and obtained $N_t = 8.0 \times 10^{12}$ cm^{-2}. Another contribution to this activated process may be associated with the thermal activation of electrons from traps in Nb_2O_5 with an activation energy of 0.15 eV [88].

The scattering factor (or the Hall factor) r describes the scattering of carriers. The magnetic field dependence of the Hall coefficient is determined by the respective

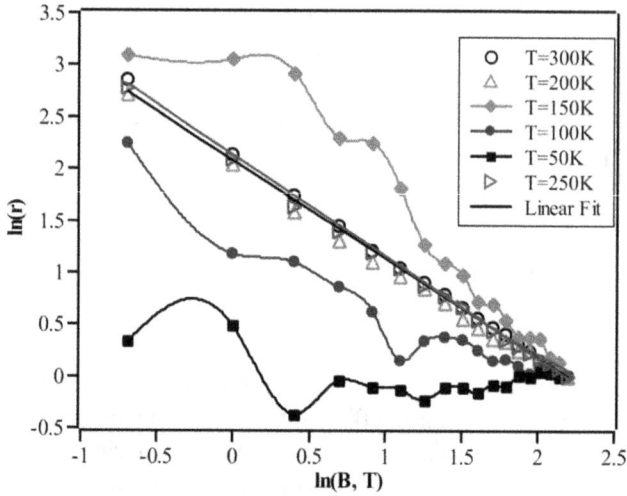

Figure 4.67. Dependence of the Hall factor of the studied LN/Nb_2O_5 films on applied magnetic fields at different temperatures.

function for the Hall factor $r(B)$. The following expression has been proposed to determine the Hall factor from experimental measurements:

$$r = \frac{R(0)}{R(\infty)} \tag{4.52}$$

where $R(0)$ and $R(\infty)$ are the Hall coefficients recorded under low and high magnetic fields, respectively. Figure 4.67 demonstrates the field dependence of the Hall factor $r(B)$ for the studied films in a double logarithmic scale $\ln(r)$–$\ln(B)$.

As can be seen in figure 4.67, the Hall factor approaches one under strong applied magnetic fields and does not depend on temperature, in full agreement with theory [54, 89]. Furthermore, figure 4.67 shows that the dependence of the Hall factor on the applied field varies with temperature. The Hall factor r can have various magnitudes depending on the scattering mechanisms occurring in the material. Specifically, in bulk semiconductors, r can be 1.18 for thermal phonon scattering or 1.93 for ionized impurity scattering [54]. In polycrystalline semiconductors, the Hall factor can differ from that of the bulk material when other scattering mechanisms dominate. This is because the Hall factor, as a function of energy, is influenced by these scattering mechanisms. The field dependence of the Hall factor in the temperature range of 200 K–300 K is given by:

$$r = A(T) \cdot B^s. \tag{4.53}$$

Here, B is the magnetic field, and the parameters $A(T)$ and s can be found from the intercept and slope, respectively, of the experimental $\ln(r)$–$\ln(B)$ graph. The temperature dependence of the parameter $A(T)$ in equation (4.53) forms a straight line in Arrhenius coordinates, as shown in figure 4.68. The activation energy $E_a = 0.03$ eV,

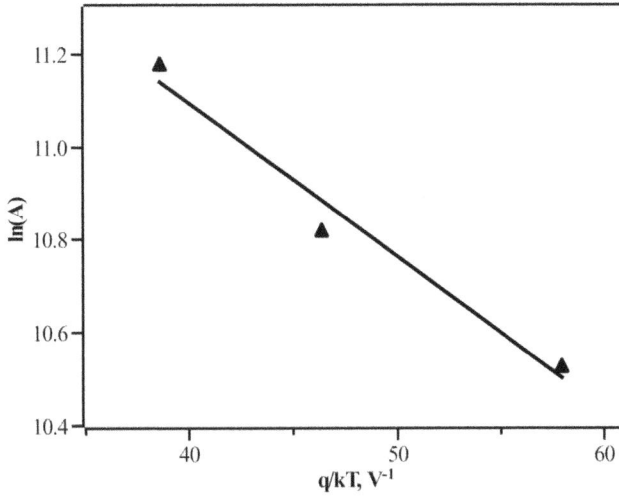

Figure 4.68. Temperature dependence of the pre-exponential factor $A(T)$ in equation (4.53) for the LN/Nb$_2$O$_5$ films in the temperature range of 200 K–300 K in Arrhenius coordinates.

obtained from the slope of this graph, can be interpreted as the difference in activation energies for the Hall and drift mobilities [82].

Regarding the power exponent s, it is temperature independent and equals $s = -1$.

As can be seen from figure 4.67, the field dependence of the Hall factor exhibits a complex behavior in the range of 50 K–200 K. Previously, an analysis of the Hall coefficient $R_H(T)$ demonstrated that carrier mobility is limited by hole mobility in the grain bulk at a temperature of 50 K. In the range of low magnetic fields, the Hall coefficient follows the law [54, 89]:

$$R_H(B) = R_H(0)[1 - \alpha \cdot \mu^2 B^2]. \tag{4.54}$$

Here, $R(0)$ is the Hall coefficient at zero field, and α is the nonlinearity factor that depends on the scattering mechanism. Specifically, $\alpha = 0.99$ for acoustic phonon scattering, $\alpha = 0.49$ for ionized impurity scattering, and $\alpha = 0$ for neutral impurity scattering. Conversely, under strong magnetic fields, the Hall coefficient is given by [54, 89]:

$$R_H(B) = R_H(\infty)\left[1 + \frac{\alpha^/}{\mu^2 B^2}\right]. \tag{4.55}$$

Here, $R_H(\infty)$ is the Hall coefficient under strong magnetic fields (the high field limit), and $\alpha^/$ is the nonlinearity factor in the high field limit. Note that under low magnetic fields, carriers with higher mobility dominate.

Figure 4.69 shows the field dependence of the Hall coefficient for the studied LN/Nb$_2$O$_5$ films at $T = 150$ K.

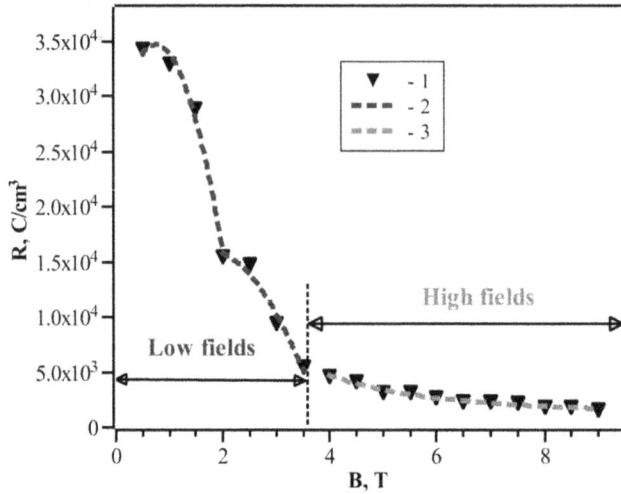

Figure 4.69. Field dependence of the Hall coefficient for the studied LN/Nb$_2$O$_5$ films at a temperature of 150K. 1—experimental results, 2—theoretically calculated using equations (4.54), 3—theoretically calculated using equation (4.55).

As demonstrated in [54], if several types of carriers coexist in a semiconductor and their mobilities differ significantly, the field dependence $R_H(B)$ manifests as many points of inflection (quadratic segments) as there are types of carriers contributing to the Hall measurements. The two quadratic sections $R_{H1}(0)$ and $R_{H2}(0)$, observed under low fields in figure 4.69, correspond to two types of carriers, fully in agreement with equation (4.52). These may be electrons with high mobility (contributing R_H at lower fields) and holes with lower mobility (contributing R_H at higher fields). By assuming the same scattering mechanism occurs, we obtained the following magnitudes using equation (4.54): $N_n = 1.8 \times 10^{14}\,\mathrm{cm}^{-3}$, $N_p = 3.9 \times 10^{14}\,\mathrm{cm}^{-3}$, $\mu_{Hn} = 0.6\,\mathrm{m^2V^{-1}s^{-1}}$, $\mu_{Hp} = 0.45\,\mathrm{m^2V^{-1}s^{-1}}$, $\mu_{Hn} = 0.6\,\mathrm{m^2V^{-1}s^{-1}}$, and $\mu_{Hp} = 0.45\,\mathrm{m^2V^{-1}s^{-1}}$ (the carrier concentrations were estimated from $R_{H1}(0)$ and $R_{H2}(0)$ through equation (4.47)) [82]. Under high magnetic fields (see figure 4.58), the Hall coefficient decreases with the field according to equation (4.55). Assuming $\alpha' = 1$, we calculated the Hall coefficient ($R_H(\infty) = 966\,\mathrm{C\,cm}^{-3}$), the Hall mobility ($\mu_{Hp} = 0.13\,\mathrm{m^2V^{-1}s^{-1}}$), and the concentration of holes ($N_p = 6.5 \times 10^{15}\,\mathrm{cm}^{-3}$), which are in good agreement with the findings in [90].

This section has been reproduced with permission from [82].

4.6 Summary and discussion

1. **Positive oxide charge in LN films:**

    LN films fabricated by RFMS in an Ar reactive gas environment exhibit positive oxide charge that is independent of working pressure, the presence of the LiNb$_3$O$_8$ phase, and defects created by bombardment. The most likely sources of this charge are antisite defects (Nb$_{Li}^{4\cdot}$) and oxygen vacancies.

2. **Ferroelectric properties and plasma effect:**

   Ferroelectric $\langle 0001 \rangle$-textured LN films deposited by RFMS with the plasma effect demonstrate remnant polarization similar to that of bulk lithium niobate, irrespective of the reactive plasma composition. Increased reactive gas pressure decreases the coercive field because films fabricated under these conditions exhibit random grain orientation.

3. **Non-ferroelectric LiNb$_3$O$_8$ phase:**

   In films fabricated at higher Ar pressure, the non-ferroelectric LiNb$_3$O$_8$ phase forms at the grain boundaries, while the bulk grains consist of the LN phase. The ratio between the grain boundaries and the bulk core widths is $d_{gb}/d_b = 0.4$. Both phases contribute to electrical conductivity.

4. **Intergranular barriers in polycrystalline LN films:**

   Polycrystalline LN films with intergranular barriers of 0.4 eV are formed during RFMS, regardless of the reactive gas pressure. The barrier height does not depend on the presence of the LiNb$_3$O$_8$ phase.

5. **Reactive gas mixtures and built-in field:**

   Thin LN films deposited by RFMS in an Ar + O$_2$ (60/40) reactive gas mixture exhibit the lowest built-in field, attributed to decreased oxygen vacancy concentration. Oxygen in the reaction chamber reduces lithium vacancies, reducing the formation of positively charged complexes. Lithium and oxygen atoms diffuse into the silicon substrate, forming deep donors in Si–LN heterostructures.

6. **Electrical conductivity mechanisms:**

   Both DC and AC conductivity mechanisms, along with dielectric relaxation mechanisms, were studied for LN-based heterostructures fabricated under different RFMS conditions. The electrical properties of Si–LN heterostructures synthesized in an Ar atmosphere are influenced by traps in the LN bandgap with activation energies of $E_a = 0.7$ eV (antisite defects) and $E_a = 0.9$ eV (lithium vacancies). The presence of oxygen in the reaction chamber (Ar/O$_2$ = 60/40) eliminates the 0.9 eV center but results in a high density of traps at the Si–LN interface.

7. **Domain orientation and TA:**

   As-grown LN films do not exhibit a preferred domain orientation. Post-TA, the films show preferentially oriented domains, making them non-symmetric in terms of polarization reversal relative to the applied electric field. This effect is independent of the sputtering regime.

8. **Impact of TA on positive effective charge:**

   TA reduces the positive effective charge by neutralizing oxygen vacancies. It also decreases interface states at the film–substrate interface, making Si–LN heterostructures behave similarly to MIS structures in the flat band regime.

9. **Improvement in ferroelectric properties due to TA:**

   TA positively affects the ferroelectric properties of LN films, reducing built-in fields responsible for the asymmetry of P–E loops. The post-TA reduction in trap concentration reduces the coercive field, especially for films fabricated in an Ar + O_2 gas mixture.

10. **Hysteresis of $C-V$ characteristics post-TA:**

The hysteresis of the $C-V$ characteristics of Si–LN heterostructures after TA is due to hole generation and subsequent re-emission to border traps at the Si–LN interface. Electron–hole pairs in the SiO_2 intermediate layer are generated by Fowler–Nordheim emission during electrical measurements.

11. **Trap energy and recrystallization:**

TA does not change the energy position of traps with $E_t = 1.7\,\text{eV}$ in the LN bandgap. However, TA leads to film recrystallization, increasing polycrystalline grains and minimizing intergranular barrier effects on charge transport. Consequently, conductivity in a wide range of electric fields is limited by thermal phonon scattering of electrons, which mirrors the behavior of LN single crystals.

12. **Optical properties:**

The optical absorption band edge of thin LN films fabricated by IBS and RFMS is highly sensitive to the sputtering regime. These films show both direct optical transitions with energies $E_g^{\text{dir}} = (3.8\text{–}4.2)\,\text{eV}$ and indirect transitions depending on the technological sputtering regime. Films deposited under ion assist show the smallest direct band E_g^{dir} due to high concentrations of traps and mechanical strain. Indirect transitions are caused by band 'tails' generated by defects. TA of as-grown films increases the bandgap to $E_g = 4.4\,\text{eV}$, close to that of single-crystal LN.

13. **Amorphous lithium niobate films:**

Amorphous lithium niobate films deposited by RFMS under various gas environments (pure Ar, Ar(90%) + O_2(10%), Ar(60%) + O_2(40%)) exhibit both direct and indirect optical transitions. The direct bandgap of as-grown films closely approximates that of single-crystal LN, reaching a maximum of 4.2 eV in films fabricated in an Ar atmosphere with the maximum Li/Nb ratio.

14. **Indirect transitions and polarons:**

Indirect transitions in amorphous LN films are attributed to bipolarons in films grown in Ar and Ar(90%) + O_2(10%) environments and Q-polarons in films grown in Ar(60%) + O_2(40%). The Ar(90%) + O_2(10%) gas composition is superior, ensuring a lower indirect bandgap E_g^{indir}.

15. **PPT:**

PPT is more efficient than conventional TA in crystallizing amorphous LN films and especially in tuning the indirect bandgap. While PPT does not affect the direct bandgap, it decreases E_g^{indir} in films grown in Ar(90%) + O_2(10%) and increases E_g^{indir} for films fabricated in Ar (60%) + O_2(40%).

16. **RFMS-deposited films and phase composition:**

Films deposited by RFMS in an Ar + O_2 atmosphere without plasma effects contain LN and Nb_2O_5 phases. The phase composition depends on the Ar/O_2 ratio, and a maximum phase ratio of LN (55%)/Nb_2O_5 (45%) is observed for Ar/O_2 at a ratio of 60/40. These films consist of crystalline blocks with LN subgrains (20 nm) and Nb_2O_5 nanograins (10 nm). For the first time, a p- to n-type conduction transition in LN/Nb_2O_5 films has been observed. The electron-to-hole drift mobility ratio is $\mu_n/\mu_p = 49$. At high magnetic fields, holes with a concentration of $N_p = 6.5 \times 10^{15}\,cm^{-3}$ play a crucial role in Hall measurements. As the temperature increases, electrons trapped in LN that contribute to the n-type conductivity of the Nb_2O_5 phase affect conductivity, limited by grain boundary scattering associated with an intergranular barrier of 0.13 eV. The effect of the conductivity type change depends on the Ar/O_2 ratio in the reaction chamber and requires more detailed study.

References

[1] Sumets M, Kostyuchenko A, Ievlev V, Kannykin S and Dybov V 2015 Sputtering condition effect on structure and properties of $LiNbO_3$ films *J. Mater. Sci., Mater. Electron.* **26** 4250–6

[2] Sumets M, Kostyuchenko A, Ievlev V, Kannykin S and Dybov V 2015 Influence of thermal annealing on structural properties and oxide charge of $LiNbO_3$ films *J. Mater. Sci., Mater. Electron.* **26** 7853–9

[3] Soergel E 2011 Piezoresponse force microscopy (PFM) *J. Phys. D: Appl. Phys.* **44** 464003

[4] Sumets M, Ievlev V, Kostyuchenko A, Vakhtel V, Kannykin S and Kobzev A 2014 Electrical properties of $Si-LiNbO_3$ heterostructures grown by radio-frequency magnetron sputtering in an $Ar + O_2$ environment *Thin Solid Films* **552** 32–8

[5] Sumets M, Ievlev V, Kostyuchenko A and Kuz'mina V 2014 Influence sputtering conditions on electrical characteristics of $Si-LiNbO_3$ heterostructures formed by radio-frequency magnetron sputtering *Mol. Cryst. Liq. Cryst.* **603** 202–15

[6] Hao L Z, Zhu J, Luo W B, Zeng H Z, Li Y R and Zhang Y 2010 Electron trap memory characteristics of $LiNbO_3$ film/AlGaN/GaN heterostructure *Appl. Phys. Lett.* **96** 032103

[7] Shandilya S, Tomar M and Gupta V 2012 Deposition of stress free c-axis oriented $LiNbO_3$ thin film grown on (002) ZnO coated Si substrate *J. Appl. Phys.* **111** 10–6

[8] Gordillo-Vázquez F J and Afonso C N 2002 Influence of Ar and O_2 atmospheres on the Li atom concentration in the plasma produced by laser ablation of $LiNbO_3$ *J. Appl. Phys.* **92** 7651

[9] Smyth D M 1983 Defects and transport in $LiNbO_3$ *Ferroelectrics* **50** 93–102

[10] van Opdorp C and Kanerva H K J 1967 Current–voltage characteristics and capacitance of isotype heterojunctions *Solid-State Electron.* **10** 401–21

[11] Sze S M and Kwok K N 2006 *Physics of Semiconductor Devices* (New York: Wiley)

[12] Gosele U and Tan T Y 1982 Oxygen diffusion and thermal donor formation in silicon *Appl. Phys.* A **28** 79–92

[13] Gosele U M 1988 Fast diffusion in semiconductors *Annu. Rev. Mater. Sci.* **18** 257–82

[14] Iyevlev V, Kostyuchenko A, Sumets M and Vakhtel V 2011 Electrical and structural properties of $LiNbO_3$ films, grown by RF magnetron sputtering *J. Mater. Sci., Mater. Electron.* **22** 1258–63

[15] Choi S-W, Choi Y-S, Lim D-G, Moon S-I, Kim S-H, Jang B-S and Yi J 2000 Effect of RTA treatment on $LiNbO_3$ MFS memory capacitors *Korean J. Ceram* **6** 138–42

[16] Wemple S H, DiDomenico M and Camlibel I 1968 Relationship between linear and quadratic electro-optic coefficiens in $LiNbO_3$, $LiTaO_3$, and other oxygen-octahedra ferroelectrics based on direct measurement of spontaneous polarization *Appl. Phys. Lett.* **12** 209–11

[17] Kim S, Gopalan V and Gruverman A 2002 Coercive fields in ferroelectrics: a case study in lithium niobate and lithium tantalate *Appl. Phys. Lett.* **80** 2740–2

[18] Zubko P, Jung D J and Scott J F 2006 Space charge effects in ferroelectric thin films *J. Appl. Phys.* **100** 114112

[19] Maissel L I and Glang R 1970 *Handbook of Thin Film Technology* (New York: McGraw-Hill)

[20] Saxena A N 1969 Forward current–voltage characteristics of Schottky barriers on n-type silicon *Surf. Sci.* **13** 151–71

[21] Kashirina N I and Lakhno V D 2010 Large-radius bipolaron and the polaron–polaron interaction *Phys. Usp.* **53** 431–53

[22] Sumets M, Kostyuchenko A, Ievlev V and Dybov V 2016 Electrical properties of phase formation in $LiNbO_3$ films grown by radio-frequency magnetron sputtering method *J. Mater. Sci., Mater. Electron.* **27** 7979–86

[23] Akazawa H and Shimada M 2006 Precipitation kinetics of $LiNbO_3$ and $LiNb_3O_8$ crystalline phases in thermally annealed amorphous LiNbO3 thin films *Phys. Status Solidi* **203** 2823–7

[24] Rost T A, Lin H, Rabson T A, Baumann R C and Callahan D L 1992 Deposition and analysis of lithium niobate and other lithium niobium oxides by rf magnetron sputtering *J. Appl. Phys.* **72** 4336–43

[25] Ievlev V, Sumets M, Kostyuchenko A and Bezryadin N 2013 Dielectric losses and ac conductivity of $Si-LiNbO_3$ heterostructures grown by the RF magnetron sputtering method *J. Mater. Sci., Mater. Electron.* **24** 1651–7

[26] Shandilya S, Tomar M, Sreenivas K and Gupta V 2009 Purely hopping conduction in c-axis oriented $LiNbO_3$ thin films *J. Appl. Phys.* **105** 094105

[27] Graça M P F, Prezas P R, Costa M M and Valente M A 2012 Structural and dielectric characterization of $LiNbO_3$ nano-size powders obtained by Pechini method *J. Sol–Gel Sci. Technol.* **64** 78–85

[28] Shandilya S, Tomar M, Sreenivas K and Gupta V 2009 Structural and interfacial defects in c-axis oriented $LiNbO_3$ thin films grown by pulsed laser deposition on Si using Al: ZnO conducting layer *J. Phys. D: Appl. Phys.* **42** 095303

[29] Elliott S R and Owens A P 1989 The diffusion-controlled relaxation model for ionic transport in glasses *Philos. Mag.* B **60** 777–92

[30] Elliott S R 1994 Frequency-dependent conductivity in ionically and electronically conducting amorphous solids *Solid State Ionics* **70–71** 27–40

[31] Casey H C, Cho A Y, Lang D V, Nicollian E H and Foy P W 1979 Investigation of heterojunctions for MIS devices with oxygen-doped $Al_xGa_{1-x}As$ on n-type GaAs *J. Appl. Phys.* **50** 3484–91

[32] Nicollian E H and Goetzberger A 1967 The $Si-SiO_2$ interface–electrical properties as determined by the metal-insulator-silicon conductance technique *Bell Syst. Tech. J.* **46** 1055–133

[33] Donnerberg H, Tomlinson S M, Catlow C R A and Schirmer O F 1989 Computer-simulation studies of intrinsic defects in $LiNbO_3$ crystals *Phys. Rev.* B **40** 11909–16

[34] Chen R H, Chen L-F and Chia C-T 2007 Impedance spectroscopic studies on congruent LiNbO$_3$ single crystal *J. Phys. Condens. Matter* **19** 086225

[35] Dhar A, Singh N, Singh R K and Singh R 2013 Low temperature dc electrical conduction in reduced lithium niobate single crystals *J. Phys. Chem. Solids* **74** 146–51

[36] Akhmadullin I S, Golenishchev-Kutuzov V A, Migachev S A and Mironov S P 1998 Low-temperature electrical conductivity of congruent lithium niobate crystals *Phys. Solid State* **40** 1190–2

[37] Dhar A and Mansingh A 1990 Polaronic hopping conduction in reduced lithium niobate single crystals *Philos. Mag. B* **61** 1033–42

[38] Schirmer O F, Imlau M, Merschjann C, Schoke B and D 2009 Electron small polarons and bipolarons in LiNbO$_3$ *J. Phys. Condens. Matter* **21** 123201

[39] Bergmann G 1968 The electrical conductivity of LiNbO$_3$ *Solid State Commun.* **6** 77–9

[40] Jorgensen P J and Bartlett R W 1969 High temperature transport processes in lithium niobate *J. Phys. Chem. Solids* **30** 2639–48

[41] Bollmann W 1977 The origin of photoelectrons and the concentration of point defects in LiNbO$_3$ Crystals *Phys. Status Solidi* **40** 83–91

[42] Ievlev V, Shur V, Sumets M and Kostyuchenko A 2013 Electrical properties and local domain structure of LiNbO$_3$ thin film grown by ion beam sputtering method *Acta Metall. Sin.* **26** 630–4

[43] Kiselev D A, Zhukov R N, Bykov A S, Voronova M I, Shcherbachev K D, Malinkovich M D and Parkhomenko Y N 2014 Effect of annealing on the structure and phase composition of thin electro-optical lithium niobate films *Inorg. Mater.* **50** 419–22

[44] Simões A Z, Zaghete M A, Stojanovic B D, Gonzalez A H, Riccardi C S, Cantoni M and Varela J A 2004 Influence of oxygen atmosphere on crystallization and properties of LiNbO$_3$ thin films *J. Eur. Ceram. Soc.* **24** 1607–13

[45] Fleetwood D M, Winokur P S, Reber R A, Meisenheimer T L, Schwank J R, Shaneyfelt M R and Riewe L C 1993 Effects of oxide traps, interface traps, and 'border traps' on metal-oxide-semiconductor devices *J. Appl. Phys.* **73** 5058–74

[46] Lai S K and Young D R 1981 Effects of avalanche injection of electrons into silicon dioxide —generation of fast and slow interface states *J. Appl. Phys.* **52** 6231–40

[47] Arnold D, Cartier E and DiMaria D J 1994 Theory of high-field electron transport and impact ionization in silicon dioxide *Phys. Rev. B* **49** 10278–97

[48] Cartier E 1998 Characterization of the hot-electron-induced degradation in thin SiO$_2$ gate oxides *Microelectron. Reliab.* **38** 201–11

[49] Warren W L and Lenahan P M 1987 Fundamental differences between thick and thin oxides subjected to high electric fields *J. Appl. Phys.* **62** 4305–8

[50] Warren W L and Lenahan P M 1986 Electron spin resonance study of high field stressing in metal-oxide-silicon device oxides *Appl. Phys. Lett.* **49** 1296–8

[51] Ievlev V, Sumets M and Kostyuchenko A 2013 Conduction mechanisms in Si-LiNbO$_3$ heterostructures grown by ion-beam sputtering method *J. Mater. Sci.* **48** 1562–70

[52] Ievlev V, Sumets M, Kostyuchenko A, Ovchinnikov O, Vakhtel V and Kannykin S 2013 Band diagram of the Si-LiNbO$_3$ heterostructures grown by radio-frequency magnetron sputtering *Thin Solid Films* **542** 289–94

[53] Buné A and Pashkov V A 1986 Electron-drift mobility in lithium niobate crystals *Fiz. Tverd. Tela* **28** 3024–7

[54] Kireev P 1978 *Semiconductor Physics* (Moscow: Mir Publishers)

[55] Iyevlev V, Sumets M and Kostyuchenko A 2012 Current–voltage characteristics and impedance spectroscopy of LiNbO$_3$ films grown by RF magnetron sputtering *J. Mater. Sci., Mater. Electron.* **23** 913–20

[56] Ievlev V M, Sumets M P and Kostyuchenko A V 2012 Effect of thermal annealing on electrical properties of Si-LiNbO$_3$ *Mater. Sci. Forum.* **700** 53–7

[57] Lampert M A and Mark P 1970 *Current Injection in Solids* (New York: Academic)

[58] Akazawa H and Shimada M 2007 Mechanism for LiNb$_3$O$_8$ phase formation during thermal annealing of crystalline and amorphous LiNbO$_3$ thin films *J. Mater. Res.* **22** 1726–36

[59] Elliott S R 1987 A.c. conduction in amorphous chalcogenide and pnictide semiconductors *Adv. Phys.* **36** 135–217

[60] Ching W Y, Gu Z-Q and Xu Y-N 1994 First-principles calculation of the electronic and optical properties of LiNbO$_3$ *Phys. Rev.* B **50** 1992–5

[61] Schmidt W G, Albrecht M, Wippermann S, Blankenburg S, Rauls E, Fuchs F, Rödl C, Furthmüller J and Hermann A 2008 LiNbO$_3$ ground- and excited-state properties from first-principles calculations *Phys. Rev.* B **77** 035106

[62] Yang J Y, Li W S, Li H, Sun Y, Dou R F, Xiong C M, He L and Nie J C 2009 Grain size dependence of electrical and optical properties in Nb-doped anatase TiO$_2$ *Appl. Phys. Lett.* **95** 213105

[63] Satapathy S, Mukherjee C, Shaktawat T, Gupta P K and Sathe V G 2012 Blue shift of optical band-gap in LiNbO$_3$ thin films deposited by sol–gel technique *Thin Solid Films* **520** 6510–4

[64] Dhar A and Mansingh A 1990 Optical properties of reduced lithium niobate single crystals *J. Appl. Phys.* **68** 5804–9

[65] Sumets M, Ovchinnikov O, Ievlev V and Kostyuchenko A 2017 Optical band gap shift in thin LiNbO$_3$ films grown by radio-frequency magnetron sputtering *Ceram. Int.* **43** 13565–8

[66] Thierfelder C, Sanna S, Schindlmayr A and Schmidt W G 2010 Do we know the band gap of lithium niobate? *Phys. Status Solidi* **7** 362–5

[67] Fox M 2010 *Optical Properties of Solids* (New York: Oxford University Press)

[68] Bhatt R, Bhaumik I, Ganesamoorthy S, Karnal A K, Swami M K, Patel H S and Gupta P K 2012 Urbach tail and bandgap analysis in near stoichiometric LiNbO$_3$ crystals *Phys. Status Solidi* **209** 176–80

[69] Wemple S H 1965 Some transport properties of oxygen-deficient single-crystal potassium tantalate (KTaO$_3$) *Phys. Rev.* **137** A1575–82

[70] Li X, Kong Y, Liu H, Sun L, Xu J, Chen S, Zhang L, Huang Z, Liu S and Zhang G 2007 Origin of the generally defined absorption edge of non-stoichiometric lithium niobate crystals *Solid State Commun.* **141** 113–6

[71] Jiangou Z *et al* 1992 Optical absorption properties of doped lithium niobate crystals *J. Phys. Condens. Matter* **4** 2977–83

[72] Kitaeva G K, Kuznetsov K A, Penin A N and Shepelev A V 2002 Influence of small polarons on the optical properties of Mg:LiNbO$_3$ crystals *Phys. Rev.* B **65** 054304

[73] Ievlev V M, Belonogov E K, Dybov V A, Kannykin S V, Serikov D V, Sitnikov A V and Sumets M P 2019 Synthesis of lithium niobate during crystallization of amorphous Li–Nb–O film *Inorg. Mater.* **55** 1237–41

[74] Salim E T, Fakhri M A, Ismail R A, Abdulwahhab A W, Salim Z T, Munshid M A and Hashim U 2019 Effect of light induced heat treatment on the structural and morphological properties of LiNbO$_3$ thin films *Superlattices Microstruct.* **128** 67–75

[75] Luo J, He X, Zhou J, Wang W, Xuan W, Chen J, Jin H, Xu Y and Dong S 2015 Flexible and transparent surface acoustic wave microsensors and microfluidics *Procedia Eng* **120** 717–20

[76] Sumets M, Ievlev V, Dybov V, Kostyuchenko A, Serikov D, Kannykin S, Kotov G and Belonogov E 2019 Electrical properties of amorphous films and crystallization of Li–Nb–O system on silicon *J. Mater. Sci., Mater. Electron.* **30** 15662–9

[77] Dharmadasa R, Lavery B, Dharmadasa I M and Druffel T 2014 Intense pulsed light treatment of cadmium telluride nanoparticle-based thin films *ACS Appl. Mater. Interfaces* **6** 5034–40

[78] Sumets M, Belonogov E, Dybov V, Serikov D, Kannykin S, Kostyuchenko A and Ievlev V 2021 Pulsed photon treatment effect on the optical bandgap of $LiNbO_3$ films grown by radio-frequency magnetron sputtering method *J. Mater. Sci., Mater. Electron.* **32** 4290–9

[79] Redfield D and Burke W J 1974 Optical absorption edge of LiNbO3 *J. Appl. Phys.* **45** 4566–71

[80] Griffel G, Ruschin S, Hardy A, Itzkovitz M and Croitoru N 1985 Characterization of sputtered $LiNbO_3$ films for integrated optics applications *Thin Solid Films* **126** 185–9

[81] Tumuluri A, Bharati M S S, Rao S V and James Raju K C 2017 Structural, optical and femtosecond third-order nonlinear optical properties of $LiNbO_3$ thin films *Mater. Res. Bull.* **94** 342–51

[82] Sumets M, Dannangoda G C C, Kostyuchenko A, Ievlev V, Dybov V and Martirosyan K S S 2017 Temperature transition of p- to n-type conduction in the $LiNbO_3/Nb_2O_5$ polycrystalline films *Mater. Chem. Phys.* **191** 35–44

[83] Tan S, Gilbert T, Hung C-Y, Schlesinger T E and Migliuolo M 1996 Sputter deposited c-oriented $LiNbO_3$ thin films on SiO_2 *J. Appl. Phys.* **79** 3548

[84] Orton J W and Powell M J 1980 The Hall effect in polycrystalline and powdered semiconductors *Rep. Prog. Phys.* **43** 1263

[85] Wilkinson A P, Cheetham A K and Jarman R H 1993 The defect structure of congruently melting lithium niobate *J. Appl. Phys.* **74** 3080

[86] Lim D, Jang B, Moon S, Won C and Yi J 2001 Characteristics of $LiNbO_3$ memory capacitors fabricated using a low thermal budget process *Solid-State Electron.* **45** 1159–63

[87] Ling C H, Fisher J H and Anderson J C 1972 Carrier mobility and field effect in thin indium antimode films *Thin Solid Films* **14** 267–88

[88] Greener E H, Whitmore D H and Fine M E 1961 Electrical conductivity of near-stoichiometric α-Nb_2O_5 *J. Chem. Phys.* **34** 1017–23

[89] Popovic R S 2003 *Hall Effect Devices* 2nd edn (Boca Raton, FL: CRC Press)

[90] Wang X, Liu X, Bo F, Chen S, Chen J, Kong Y, Xu J and Zhang G 2015 Photo-Hall effect in highly Mg-doped lithium niobate crystals *Appl. Phys. Lett.* **107** 191102

IOP Publishing

Lithium Niobate-Based Heterostructures (Second Edition)
Synthesis, properties, and electron phenomena
Maxim Sumets

Chapter 5

Oxide charge: localization, evolution, and related phenomena at heterojunctions

We have demonstrated that the plasma composition, its spatial inhomogeneity, and the relative target–substrate position significantly influence the electrical properties of Si–LiNbO$_3$ heterostructures through the charge centers in the bandgap of LiNbO$_3$ (LN) films and at the Si–LN interface. Using an Ar + O$_2$ reactive gas mixture results in a reduction in positive oxide charge and coercive field. The barrier properties of the Si–LN heterojunction are affected by the plasma composition. Thermal annealing (TA) of the fabricated films leads to an increase in the amount of the LiNb$_3$O$_8$ phase due to the diffusion of oxygen atoms toward the grain boundaries. Post-growth treatments, such as TA and pulsed photon treatment (PPT), have a profound effect on the optical properties of LN films.

LN is characterized as a ferroelectric material with distinctive properties, including a high Curie temperature ($T_c = 1210\,°C$), a wide bandgap ($E_g = 3.7\,eV$), and relatively elevated electro-optic coefficients. Thin films of LN are attractive due to their potential integration with existing silicon technology. Si–LN heterostructures form the basis of ferroelectric memory units, optoelectronic devices, and other integrated electronics. Diverse fabrication methods have proven successful in producing thin LN films, and the radio-frequency magnetron sputtering (RFMS) process stands out as one of the most effective deposition techniques, as it is particularly suitable for complex oxides such as LN. The integration of thin LN films with existing silicon technology presents opportunities for creating heterostructures with a broad range of practical applications.

However, the practical application of Si–LN heterostructures faces limitations due to certain issues. Notably, the presence of positive oxide charge adversely affects ferroelectric domain switching and memory effects. It is widely acknowledged that the properties of LN films are exceedingly sensitive to the deposition method employed. For instance, studies have demonstrated that the RFMS regime used significantly impacts these properties [1–3]. Additionally, TA, a method effective in enhancing the properties of as-grown polycrystalline films, induces significant changes in their characteristics [3–5].

Our research has revealed that the effective positive charge in LN films is independent of both the substrate orientation and the reactive gas pressure utilized in the RFMS process [1, 6]. Furthermore, TA at a temperature of 650 °C results in a reduction of the effective positive charge in Si–LN heterostructures [7]. The presence of oxygen in the reaction chamber also contributes to this reduction [8]. The existence of this charge in LN-based heterostructures proves detrimental, constraining their practical applications. Specifically, it has been emphasized that this charge impedes the proper motion of charged domain walls in domain wall memories [9]. Moreover, the origin and location of this charge remain unclear. Undoubtedly, the effective positive charge forms during the film's crystallization process, and TA influences it by altering the composition and structure of LN films.

The primary objective of this chapter is to reveal the physical nature, location, and evolution of the effective charge in Si–LN heterostructures, as well as the electronic phenomena associated with it.

5.1 Electrical properties and crystallization of amorphous Li–Nb–O films on silicon

Si–LN and Si–SiO$_2$–LN heterostructures are of particular interest, as they hold promise for various elements in integrated silicon electronics and optoelectronics. The functional characteristics of these devices depend on two key factors: electronic processes in LN films and, more significantly, the film–substrate interface. A study by Gudkov *et al* [10] emphasised the impact of the 0.5 eV potential barrier in the LN–Si heterojunction on charge transport under low electric fields. They found that under strong applied fields, the dominant mechanism is Poole–Frenkel emission, which is governed by electronic processes within the bulk of the film. TA in vacuum or in an oxygen-rich atmosphere efficiently alters the film's structure and its electrical characteristics and thus the structure and electrical characteristics of thin film devices. It remains crucial to develop optimal conditions for synthesizing ferroelectric LN films with low optical losses and efficient electrical conductivity. Moreover, fabricating an LN–substrate heterojunction with the ability to modulate space charge areas and control charge injection is an actual task in the development of practical electro-optic device applications, such as optical resonators and filters.

Studies [1, 11] have demonstrated that the structure of polycrystalline LN films fabricated by the RFMS technique significantly influences the electrical characteristics of LN–Si heterostructures. This includes the magnitude of the positive fixed oxide charge and the barrier properties of the LN–Si heterojunction. Therefore, it is essential to investigate the changes in electrical properties and the structural transformations initiated by TA. It has been found that the crystallization temperature of amorphous films depends on various factors. For instance, amorphous films fabricated by RFMS on non-heated quartz, Si(100), and SiO$_2$ substrates begin crystallization at temperatures of 500 °C, 550 °C, and 650 °C, respectively [5]. The crystallization processes of amorphous films deposited by the sol–gel method [12] and the spray pyrolysis technique [13] commence at 450 °C and 400 °C, respectively, resulting in the formation of single-phase polycrystalline LN films An increase in TA temperature leads to the formation of the lithium-deficient $LiNb_3O_8$ phase, as highlighted in some studies [5, 14].

Limited research has focused on how structural transformations that take place during TA affect the electrical properties of LN films and interfaces. The deposition of layers onto an unheated substrate with subsequent crystallization of an amorphous film is an approach that allows the quantitative estimation of the electrical properties of a heterostructure. It includes the effects of crystal structure dispersion and reduces the loss of components due to high vapor pressure, such as Li at a saturated steam pressure of 13 Pa under TA at 870 °C. This section aims to reveal the nature and characteristics of the electric properties of heterostructures formed by the crystallization of amorphous films in the Li–Nb–O system on the surfaces of Si wafers.

The examined heterostructures were manufactured by depositing a 300 nm thick layer of the Li–Nb–O system onto silicon wafers (Si(001)) with n-type conductivity, $\rho = 4.5\,\Omega \cdot cm$) through RFMS of LN targets.[1] The key parameters of the RFMS process were as follows: a vacuum level of around $\sim 10^{-1}\,Pa$, a reactive gas environment of Ar (0.75 Pa), and an incident power at the target of $15\,W \cdot cm^{-2}$. TA of the as-grown heterostructures was conducted in air at temperatures of $T = 300$ °C, 400 °C, 450 °C, 500 °C, 550 °C, and 600 °C, with each temperature maintained for 1 h.

The changing compositions of the amorphous Li–Nb–O films during TA were investigated *in situ* in air at temperatures of 350 °C, 400 °C, 425 °C, 450 °C, 475 °C, 500 °C, and 525 °C, each for 30 min, by recording x-ray diffraction (XRD) spectra. The heating rates within the temperature ranges of 25 °C–350 °C and 350 °C–525 °C were set to 50 and 10 °C min^{-1}, respectively. The fraction of the crystalline phase was determined using the following expression [15]:

$$V_{cr} = \frac{I_{cr}}{I_{cr} + \alpha I_a}. \tag{5.1}$$

Here, I_{cr} and I_a are the integral intensities of reflections from the crystalline and amorphous phases, respectively, and α is Huang's parameter ($\alpha = 1$ in our case). The electrical properties of the investigated heterostructures were analyzed using high-frequency ($f = 10^5\,Hz$) capacitance–voltage (C–V) characteristics and current–voltage (I–V) characteristics within the temperature range of $T = 100$–300 K. The top contacts ($S = 8 \times 10^{-7}\,m^2$) used for the electrical measurements were deposited by thermal evaporation and condensation of Al in a vacuum. The bottom ohmic contacts were fabricated by applying an In–Ga eutectic alloy to the Si substrate.

As indicated by the XRD patterns presented in figure 5.1, the as-grown films maintain their amorphous nature during TA at temperatures of up to 400 °C.

---

[1] The targets were fabricated at the Institute of Chemistry and Technology of Rare Elements and Mineral Raw Materials of the Russian Academy of Science (Apatity).

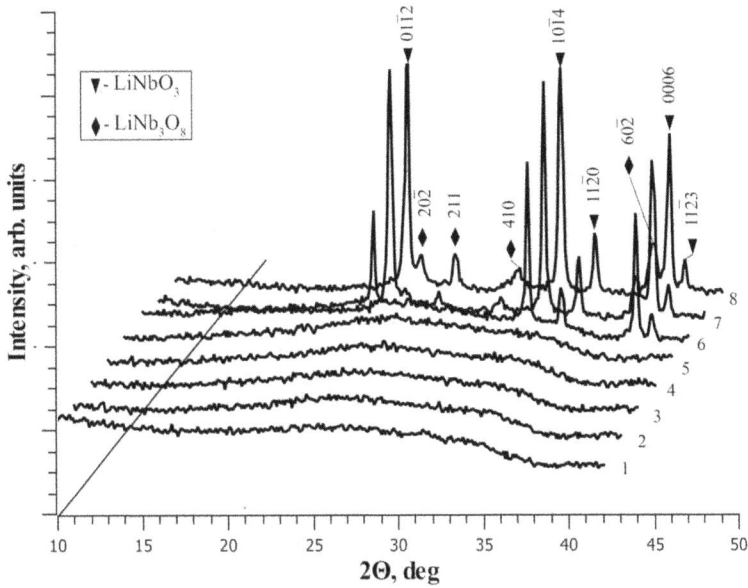

Figure 5.1. XRD patterns of an as-grown film (1) and films after TA at temperatures of 350 °C (2), 400 °C (3), 425 °C (4), 450 °C (5), 475 °C (6), 500 °C (7), and 525 °C (8). Reproduced from [16], with permission from Springer Nature.

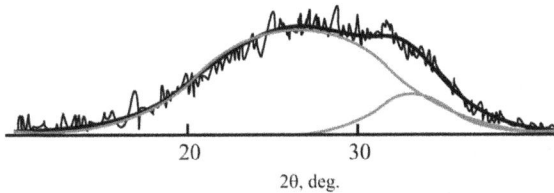

Figure 5.2. A fragment of the XRD pattern shown in figure 5.1 corresponding to the as-grown films. Two Gaussian curves approximating the experimental XRD spectra are shown in red. Reproduced from [17], with permission from Springer Nature.

In figure 5.2, a halo-like fragment of the XRD spectra, depicted in figure 5.1, spans the scattering angle range of 10°–40°, corresponding to the as-grown films. The halo profile observed in figure 5.2 can be modeled by the superposition of two Gaussian functions (normal distribution) with peaks corresponding to scattering angles of 26.5° and 33°, respectively. Analysis of this halo suggests the presence of two phases in the amorphous film, likely reflecting the structural units of crystalline niobium and lithium oxides—precursors to LN formation, akin to the process observed in the liquid phase.

Crystallization of the films starts at $T = 450$ °C, which can be seen in figure 5.1 as diffraction peaks associated with the crystal lattice of LN with the unit cell parameters of the $R3c$ space group ($a = b = 0.515$ nm, $c = 1.39$ nm). As the annealing temperature increases, the fraction of the crystalline phase in the film

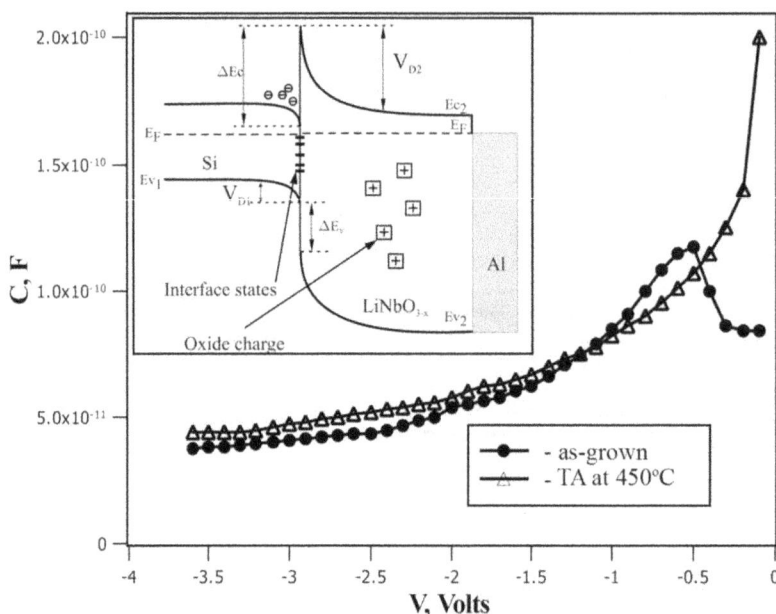

Figure 5.3. Typical C–V characteristics of the studied heterostructures. The inset shows a band diagram of the studied heterostructures. The labels V_{D1} and V_{D2} mark band bending in semiconductor 1 and semiconductor 2, respectively. The labels ΔE_c and ΔE_V mark conduction band and valence offsets, respectively, at the film–Si interface. Reproduced from [16], with permission from Springer Nature.

sharply rises; within the TA temperature range of 450 °C–475 °C, it increases from 45% to 100%. Beyond 500 °C, new reflections attributed to the LiNb$_3$O$_8$ phase emerge, although its presence in the film does not exceed 8%. The average size of a coherent scattering area, estimated using the Scherrer formula, increases from 200 to 300 nm as the temperature rises from 425 °C to 550 °C. Figure 5.3 illustrates the C–V characteristics of the as-grown Li–Nb–O/Si heterostructure (amorphous structure, curve 1) and the characteristics after TA at $T = 450$ °C (curve 2). Notably, the C–V characteristics of almost all samples closely resemble curve 1 in figure 5.3. The C–V curve of the as-grown heterostructure reveals two regions of capacitance modulation with voltage (below −0.5 V and above −0.5 V, respectively). This characteristic is attributed to the modulation of the space charge regions in the Si substrate and Li–Nb–O film under direct and reverse biases [8], describable within the framework of a model involving two back-to-back Schottky diodes in series [18]. This model implies the presence of a high density of interface states or an intermediate highly conductive layer at the Si–film interface, equivalent to a combined metal layer of two Schottky diodes.

The shift of C–V characteristics toward the negative voltage direction provides evidence for a positive fixed oxide charge in the amorphous films, aligning with findings in polycrystalline LN films, as reported in [7, 8]. The conduction band offset ΔE_c establishes a potential barrier for electrons in Si when the heterostructure is in an accumulation regime under zero bias (see the inset in figure 5.3). The observed

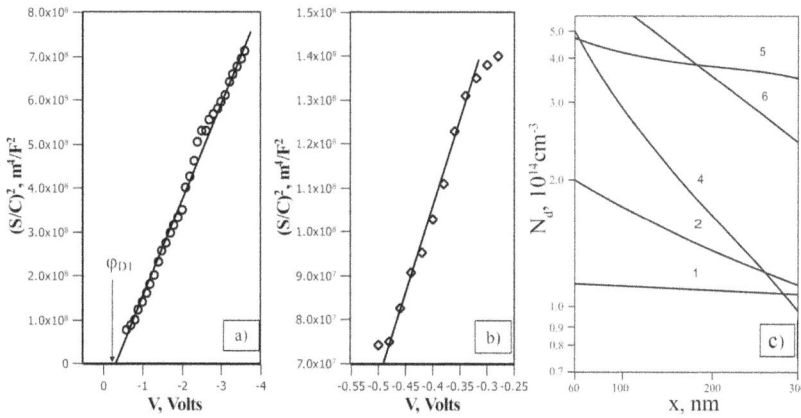

Figure 5.4. C–V characteristics of as-grown heterostructures in $(S/C)^2$–V coordinates ((a) and (b)) and donor distributions in the studied films (c) after TA at temperatures of 300 °C (1), 400 °C (2), 450 °C (4), 500 °C (5), and 600 °C (6). Here, curve (a) corresponds to a space charge region in Si, curve (b) corresponds to a space charge region in the film, and x is the distance from the film–substrate heterojunction. Reproduced from [16], with permission from Springer Nature.

C–V characteristic is typical for annealed samples with an amorphous–crystalline structure which is completely crystallized.

The only exceptions among the studied samples are the heterostructures with an amorphous–crystalline structure after TA at 400 °C and 450 °C. The C–V curves for these samples resemble those for Schottky diodes, lacking a section corresponding to the modulation of the space charge region in a film. In figures 5.4(a) and (b), the voltage dependence of the junction capacitance is presented in $(S/C)^2$–V coordinates, corresponding to the depletion regions in Si and Li–Nb–O films, respectively.

As per the theory of space charge effects in semiconductors [19], the slope of the $(S/C)^2$–V graph depends on the concentration of ionized donors in the semi-conductor. The linear section of this graph intercepts the horizontal axis at a point corresponding to the diffusion potential ϕ_D. The effective positive fixed charge in these films can be estimated by analyzing the shift of the experimental C–V curves relative to theoretical ones, following standard C–V analysis [20]. The results of such analysis for the studied heterostructures are presented in figures 5.4 and 5.5.

Ionized donor distributions for the studied films are depicted in figure 5.4(c). The C–V analysis reveals that the effective positive charge in the amorphous film changes non-monotonically with TA and reaches a minimum value at a temperature of 450 ° C (see figure 5.5(a)).

The TA of films in an oxygen-rich environment results in a reduction of oxygen vacancies, which are formed during RFMS due to the desorption of lithium oxide (Li_2O). According to a proposal made in [5], this desorption plays a crucial role in the nucleation of the LN and $LiNb_3O_8$ phases, leading to the formation of oxygen vacancies (i.e. positive charge) within the film. Amorphous films deposited on Si substrates undergo crystallization in LN with a notable oxygen deficit under TA at temperatures of up to 600 °C. Notably, TA does not significantly alter the oxygen

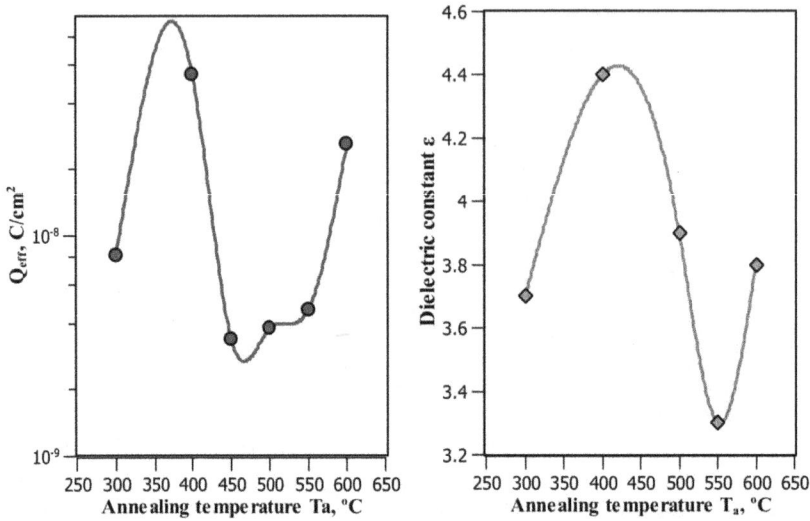

Figure 5.5. The density of the effective positive charge (a) and the effective dielectric constant (b) of amorphous Li–Nb–O films annealed at various temperatures.

content in the films. Consequently, the observed change in the density of effective positive charge following the TA of as-grown films can be elucidated by the redistribution of oxygen from the bulk to grain boundaries. The extent of oxygen atom redistribution is contingent upon the change in the specific concentration of grain boundaries formed during the film crystallization process. This mechanism is supported by the findings presented in [21]. The authors of this study explored the impact of TA on the structure and composition of the topmost layers of LN single crystals using x-ray photoelectron spectroscopy and coaxial impact collision ion scattering spectroscopy. Their findings revealed that, at a temperature of 400 °C, the topmost layers of the crystal exhibit a deficiency in lithium and oxygen, leading to the creation of corresponding vacancies due to the desorption of Li_2O. However, subsequent annealing at higher temperatures of up to 600 °C resulted in a deceleration of Li and O loss, which was attributed to the effusion of these elements from the bulk LN.

Figure 5.5(b) illustrates the variation in the dielectric constant of the analyzed films following annealing at various temperatures. This outcome aligns with the findings presented in [22], where the investigation focused on the structure and dielectric properties of amorphous films after TA. The authors of this study noted two peaks in the dielectric permittivity $\varepsilon(T)$ curve occurring at temperatures of 430 °C and 550 °C, along with a trough at $T = 490$ °C.

In our $\varepsilon(T)$ curve (refer to figure 5.5(b)), the initial peak is evident at around $T = 450$ °C, corresponding to the development of an amorphous–crystalline film with a 45% crystalline phase fraction. This crystalline transformation induces a shift in the dielectric constant. Some researchers [23] attribute the observed peak at the same temperature to an increase in hopping conductivity in films post-crystallization, which

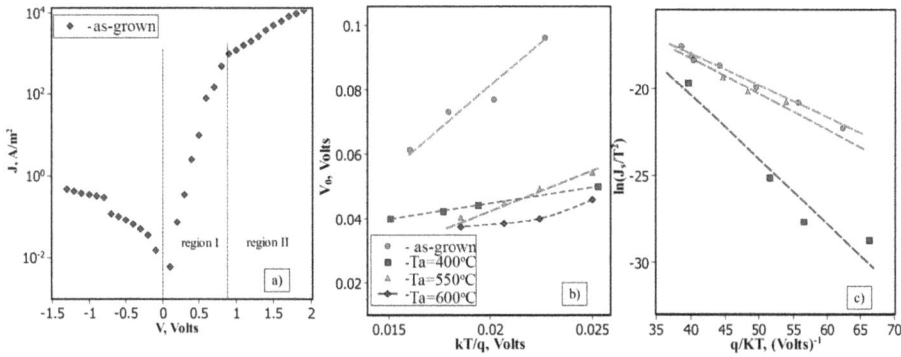

Figure 5.6. $I-V$ of the as-grown Li–Nb–O/Si heterostructure (a), temperature dependence of the parameter V_0 (b), and the dependence $\ln(J_s/T^2)-q/kT$ (c) derived from the $I-V$ characteristics of as-grown heterostructures (1) and heterostructures annealed at various temperatures T_a.

consequently influences the dielectric relaxation mechanism. They modeled the studied films using a resistor and a constant phase element (CPE) connected in parallel. To validate this hypothesis, we investigated the electrical conductivity of the studied heterostructures.

$C-V$ analysis of the heterostructures reveals that they operate in an accumulation regime under zero bias (refer to the inset in figure 5.3). Under forward bias, the $I-V$ characteristics are influenced by both the bulk properties of the film and the barrier properties at the film–substrate interface. The nonsymmetrical distribution of the $I-V$ characteristics in figure 5.6(a) underlines two distinct regions associated with the rapid and gradual increase in current with voltage (denoted as regions I and II, respectively). The first region pertains to the barrier properties of the film–Si heterojunction, while the second region is dictated by the bulk properties of the Li–Nb–O film.

According to the theory of semiconductor heterojunctions [24], the $I-V$ characteristics of the heterostructure depicted in figure 5.3 are described by the following expression:

$$J = J_s\left(\exp\left(\frac{qV_2}{n_2 \cdot kT}\right) - 1\right). \tag{5.2}$$

Here, J is the current density, q is the elementary charge, $V \approx V_2$ is the applied voltage, k is Boltzmann's constant, T is the temperature, and n_2 is an ideality factor that depends on the mechanism of charge transport through the heterostructure. The ideality factor characterizes the deviation of the $I-V$ characteristics from those of the ideal Schottky diode when $n = 1$. J_s is the saturation current density, described by the following expression:

$$J_s = B \cdot \exp\left(-\frac{qV_{D2}}{kT}\right). \tag{5.3}$$

Here, B is a voltage-independent constant and V_{D2} is the contact voltage. $I-V$ characteristics that obey equation (5.2) should be linear in $\ln(J)-V$ coordinates with

Table 5.1. The results of I–V analysis for the studied heterostructures.

Annealing temperature (°C)	Ideality factor n	Trap concentration N_t (cm^{-3})	Contact potential V_{D2} (V)
As-grown	5	6.1×10^{15}	0.2
400	1	4.7×10^{16}	0.3
550	—	3.5×10^{15}	—
600	2	2.3×10^{15}	0.2

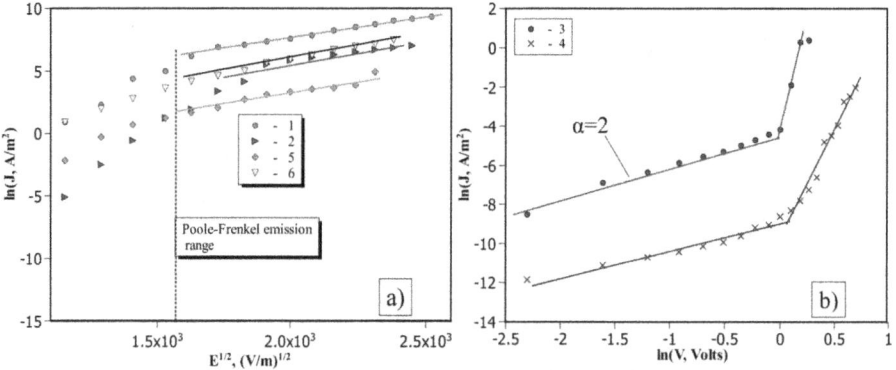

Figure 5.7. I–V characteristics of the studied heterostructures in Poole–Frenkel coordinates (a) and in double logarithmic coordinates (b). Annealing temperatures: 300 °C (1), 400 °C (2), 450 °C (3), 500 °C (4), 550 °C (5), and 600 °C (6).

a slope given by $d(\ln(J))/dV = 1/V_0 = q/nkT$ and an intercept giving $\ln(J_s)$. As suggested in [25], the temperature dependence $V_0 - kT/q$ determines the ideality factor. The temperature dependencies $V_0 - kT/q$ of the studied heterostructures, attributed to their I–V characteristics under low voltage (low electric fields $E < 2.5 \times 10^6$ V m^{-1}) are shown in figure 5.6(b). As follows from equation (5.3), the $\ln(J_s/T^2)$ $-q/kT$ graph should be linear with a slope of V_{D2}, which was observed for the majority of the heterostructures (figure 5.6(c)). The results of the I–V analysis are given in table 5.1.

Hence, in the presence of low electric fields, the transportation of charge in both as-grown heterostructures and those subjected to annealing at 400 °C and 600 °C is dictated by over-barrier emission. The temperature dependence of V_0 for the heterostructure post-TA at 550 °C (see figure 5.6(b)) indicates that within this range of electric fields, thermally assisted tunneling governs charge transport. Conversely, under intense electric fields, the I–V characteristics of the examined heterostructures (with the exception of those subjected to TA at 450 °C and 500 °C) are influenced by the bulk properties of the films. In $\ln(J)$–$E^{1/2}$ coordinates, these characteristics exhibit linear behavior, providing evidence of Poole–Frenkel emission in LN films (refer to figure 5.7(a)).

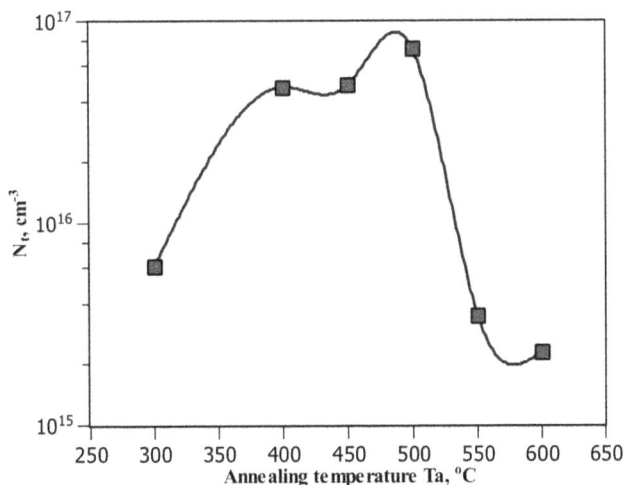

Figure 5.8. Trap concentration in as-grown film (1) and after TA at 400 °C (2), 450 °C (3), 500 °C (4), 550 °C (5), and 600 °C (6).

The trap concentration can be estimated through the magnitude of the electric field at which the linear section of the I–V characteristic in $\ln(J)$–$E^{1/2}$ coordinates begins [26].

The I–V characteristics of the studied heterostructures annealed at temperatures of 450 °C and 500 °C were linear in $\ln(J)$–$\ln(V)$ coordinates (see figure 5.7(b)), which indicates that space-charge-limited current (SCLC) dominates in LN films. In the framework of SCLC, the I–V characteristics obey the power law $J \sim V^{\alpha}$ with the typical 'quadratic law' ($\alpha = 2$). We determined the trap concentration from a fast-growing section of the graph with $\alpha > 2$ (see figure 5.7(b)). By analyzing the I–V characteristics of the studied heterostructures, we obtained the change of trap concentration in Li–Nb–O films during their TA, as shown in figure 5.8.

The precise nature of the trap states in the analyzed films remains unclear, and many researchers lean toward the notion that these states are localized at the grain boundaries. An illustrative example can be found in [27], where Si–LN hetero-structures were subjected to I–V analysis. The findings indicate a notably lower trap concentration in the textured LN films compared to their non-textured counterparts, which exhibit an increased concentration of grain boundaries in the bulk. The researchers reported that the I–V characteristics are influenced by both Poole–Frenkel emission and SCLC. In our investigation, the concentration peak in figure 5.8, observed at a temperature of approximately 470 °C, corresponds to the maximum of $\varepsilon(T)$, as depicted in figure 5.5(b). The trap concentration exhibits an upward trend in films with a crystalline phase fraction of up to 45%, resulting in an increase in hopping conductivity under specific conditions. This increase induces a shift in the dielectric relaxation mechanism, leading to the observation of a maximum in the effective dielectric constant for films exhibiting an amorphous–crystalline structure after TA at $T = 450$ °C.

This section has been reproduced with permission from [16].

5.2 Evolution of oxide charge during the crystallization of amorphous Li–Nb–O films

Certain applications, such as nonvolatile memory units and ferroelectric field-effect transistors (FeFETs), necessitate the creation of thin oxide films possessing a high dielectric constant ($\varepsilon = 200$–2000), minimal current leakage [9], and a substantial breakdown voltage ($E = 300$–$103\,\text{kV cm}^{-1}$ [28]). These parameters exhibit considerable sensitivity to the composition, point defects (Li and O) [29], and the charge localization centers at the film surface [30]. Additionally, charged defects in LN not only impact the electrical aspects but also influence its optical and ferroelectric properties [3, 29, 31–33]. Specifically, when a depletion layer is linked in series to a film, the high dielectric constant of the gate material leads to a depolarization field [34]. Consequently, achieving prolonged retention times of the stored polarization state in downsized FeFET devices becomes a formidable task. Furthermore, researchers at the forefront of ferroelectric memory design have revealed that charges within a ferroelectric material impede polarization reversal, a crucial aspect for effective memory application [34].

Conversely, some investigators have harnessed oxygen vacancies to fabricate memristive systems based on ferroelectric diodes utilizing LN [35]. Their study focused on the I–V characteristics of a Pt–LN–Pt metal–ferroelectric–metal capacitor that exhibited memristive behavior. Oxygen vacancies, concentrated at the bottom Pt–LN interface, played a pivotal role by 'memorizing' the switched state through electron trapping. Another research group reported rectifying filamentary resistive switching after the formation of LN films [36]. The phenomenon was elucidated by a model which posited that the local filament did not penetrate the entire LN thin film, leading to asymmetric contact barriers at the two interfaces. This was attributed to the high concentration of oxygen vacancies generated during TA in an Ar environment, which led to the formation of conductive filaments. Importantly, the filaments composed of oxygen vacancies did not traverse the entire LN film, suggesting potential applicability in high-density memories requiring uniform performance across different cells.

Despite the important role played by the oxide charge in ferroelectric memories, the origin and distribution of this charge in LN-based heterostructures during the crystallization of LN films are not yet clear. In this section, we investigate how the oxide charge evolves during the crystallization of amorphous Li–Nb–O films deposited onto Si substrates.

For this purpose, we sputtered films under the conditions given in table 5.2 [37]. Silicon wafers were cleaned by ion etching in an Ar plasma (2 min) before the RFMS process.

Figure 5.9 illustrates the structure and composition of both as-grown Li–Nb–O films and post-TA films deposited onto Si substrates under conditions corresponding to sample S4 from table 5.2.

As depicted in figures 5.9(a)–(c), thin Li–Nb–O films exhibit morphological inhomogeneity and possess an amorphous structure. The inhomogeneities manifest as anisotropic aggregates, roughly equivalent in size to the film itself (with lateral

Table 5.2. Fabrication regimes used to produce the studied samples.

Sample #	Reactive gas environment	Reactive gas pressure (Pa)	Magnetron power (W)	Substrate type	Substrate temperature	Film thickness d (nm)	Annealing duration and atmosphere
S1	Ar	1.5×10^{-1}	100	Si, n-type, $\rho = 4.5\,\Omega \cdot cm$	550 °C	3000	—
S2						300	—
S3						100	—
S4			50		Unheated	300	60 min in air
S5						100	
S6	Ar(60%) + O$_2$(40%)					100	

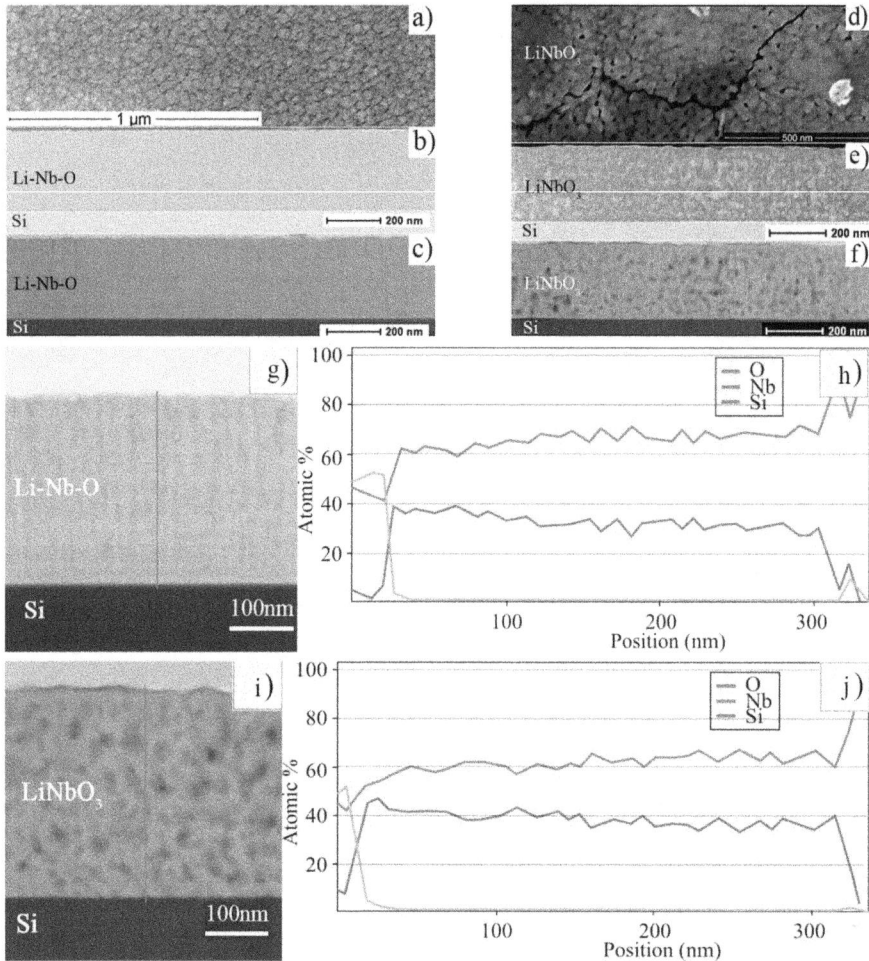

Figure 5.9. The structure and composition of both as-grown Li–Nb–O films and post-TA films deposited under regimes corresponding to sample S4 in table 1 (a)–(f). SEM patterns of surfaces (a and d) and cross-sectional bright-field (b) and (e) and dark-field (c), (f) TEM images of both as-grown Li–Nb–O films (a, b, and c) and after TA (d, e, and f). (g)–(j) cross-sectional dark-field TEM images and distributions of elements in as-grown Li–Nb–O films (g) and (h) and after TA (i) and (j). Reproduced from [37]. CC BY 4.0.

dimensions ranging from 15 to 50 nm and a vertical dimension of 250 nm). Analysis of scanning transmission electron microscopy (STEM) images indicates that the boundaries of these aggregates contain a higher concentration of light elements (oxygen and lithium) compared to the bulk of the film.

The energy-dispersive X-ray microanalysis (EDAM) results reveal a gradient in elemental composition (see figures 5.9(g) and (h)). The Nb–O ratio changes from 0.33 at the film–substrate interface to 0.5 at the film surface. The periodic change in the concentration of elements corresponds to separate film layers with a thickness of 15 nm, which correlates with the dimensions of the amorphous aggregates.

The annealed films demonstrate a crystalline structure with morphological inhomogeneity (the dispersed crystallites are 50 nm in size). As follows from the bright- and dark-field TEM images (see figures 5.9(e) and (f)), the elemental composition of adjacent crystallites differs greatly. The Nb–O ratio changes from 0.6 at the Si–LN interface to 0.8 at the LN film surface (see figures 5.9(i) and (j)).

In our preceding studies [1, 7], TA was found to diminish the positive effective charge (Q_{eff}) present in as-grown Si–LN heterostructures containing polycrystalline LN films. Notably, when Si–LN heterostructures fabricated by RFMS in an Ar + O_2 environment were annealed, Q_{eff} underwent a transformation, shifting from a positive to a negative state [7]. The precise origin of Q_{eff} (either within the bulk of the LN films or at the Si–LN interface) remains uncertain. In our earlier investigations focusing on heterostructures based on 'thick' LN films, Q_{eff} consistently exhibited a positive nature [1, 38]. Additionally, we documented that TA at 650 °C resulted in a reduction of Q_{eff}. Furthermore, Si–LN heterostructures grown by the RFMS technique in an Ar + O_2 gas mixture and subjected to annealing at 650 °C for 1 h exhibited a negative Q_{eff} [7, 38]. In the present study, we explore the evolution of Q_{eff} in as-grown heterostructures subjected to annealing at various temperatures.

The C–V characteristics of Si–LN heterostructures grown in an Ar environment with LN layers of varying thicknesses closely resembled those observed in metal–oxide–semiconductor (MOS) structures (refer to figure 5.10) [20].

All curves presented in figure 5.10 exhibit horizontal axis shifts in relation to the corresponding theoretical C–V characteristics. This phenomenon is attributed to the presence of the effective oxide charge, denoted by Q_{eff}, within the examined samples, as per the established theory of MOS heterostructures [20]. Figure 5.11 illustrates the dependency of Q_{eff} on the film thickness in the investigated Si–LN heterostructures.

As depicted in figure 5.11, the effective oxide charge exhibits a negative value for 'thin' films ($d < 200$ nm) of LN and a positive value for 'thick' films ($d > 200$ nm),

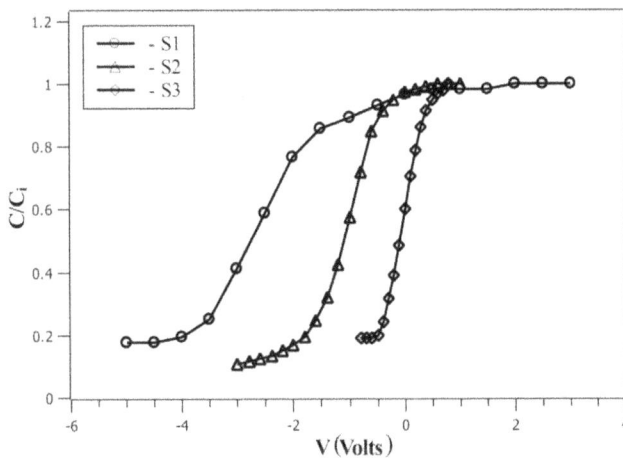

Figure 5.10. Typical normalized C–V characteristics of the studied heterostructures. Reproduced from [37]. CC BY 4.0.

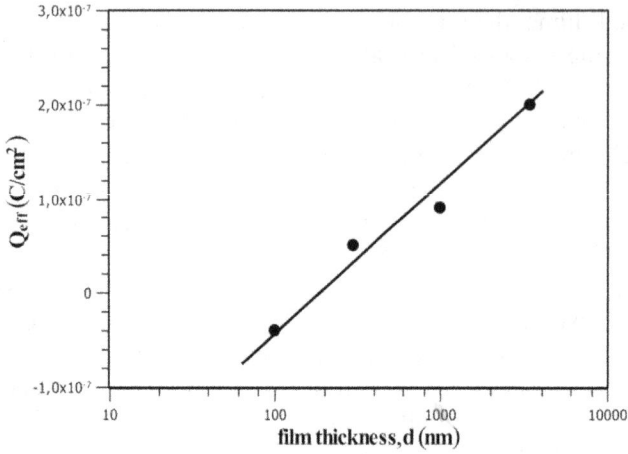

Figure 5.11. Effective oxide charge as a function of the film thickness in Si–LN heterostructures. Reproduced from [37]. CC BY 4.0.

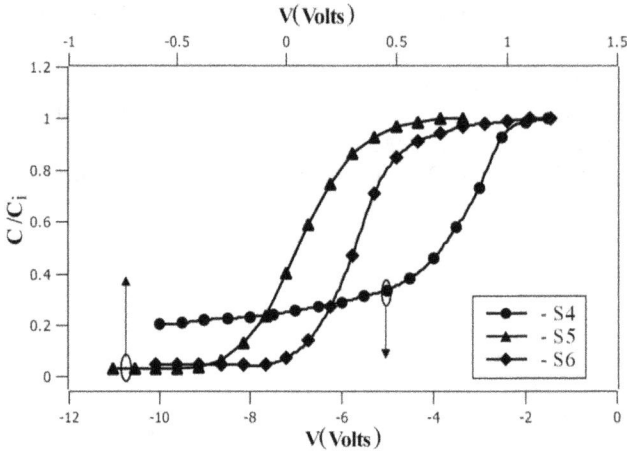

Figure 5.12. Normalized C–V characteristics of as-grown heterostructures annealed at $T_a = 400$ °C. C_i is the geometry capacitance of an LN layer. Reproduced from [37]. CC BY 4.0.

with an increase in magnitude as the thickness rises. It is noteworthy that the literature provides evidence for a substantial impact of LN film thickness on its electrical properties. Notably, the authors of [39] reported that 'thick' polycrystalline LN films, characterized by a small grain size, manifest a heightened concentration of oxygen defects and defect dipole complexes situated in proximity to the grain boundaries. These defects, carrying a positive charge, contribute to the observed $+ Q_{eff}$ for the thick LN films in our study.

Figure 5.12 shows the typical C–V characteristics of the studied samples.

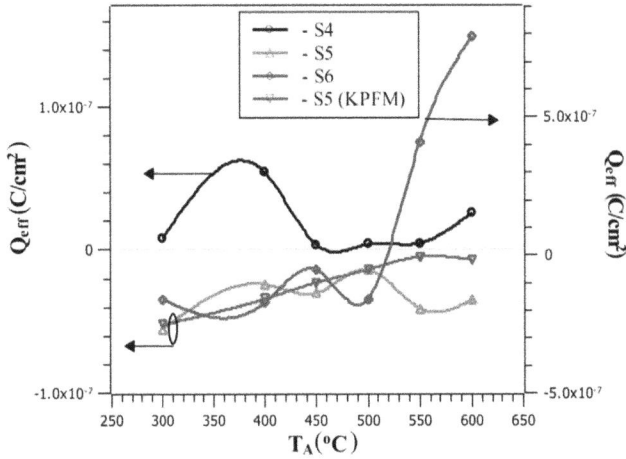

Figure 5.13. The effective oxide charge as a function of the annealing temperature for the studied heterostructures. Reproduced from [37]. CC BY 4.0.

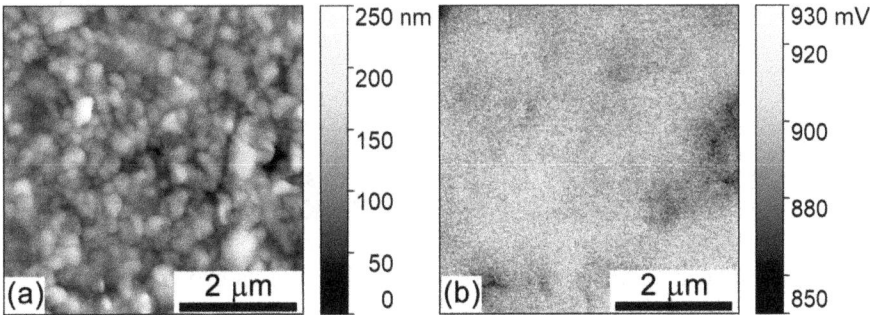

Figure 5.14. The surface topography (a) and surface potential (b) of sample S5 obtained by the KPFM method. Reproduced from [37]. CC BY 4.0.

Figure 5.13 shows the effective oxide charge Q_{eff} as a function of annealing temperature T_a for the studied samples, derived from their C–V characteristics at different T_a.

Furthermore, to validate the accurate interpretation of Q_{eff} derived from C–V measurements, Kelvin probe force microscopy (KPFM) measurements were conducted for sample S5. The resultant findings are illustrated in figure 5.14. The KPFM measurements of the examined sample reveal a nearly uniform surface potential of approximately 900 mV, with localized decreases of up to 850 mV, attributed to the presence of secondary phase inclusions. Additionally, KPFM scanning was performed at various distances, denoted by h, from the sample.

The graph in figure 5.15 illustrates the average surface potential given by the tip–surface distance, as acquired through the KPFM technique for the as-grown sample S5. By considering the linear relationship of $V_{surf}(h)$ as depicted in figure 5.15 and assuming a uniform charge distribution across the sample surface, the effective

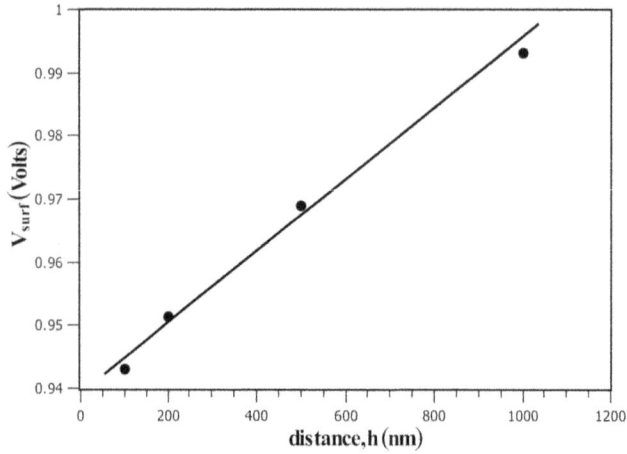

Figure 5.15. Average surface potential as a function of distance from the scanned surface for sample S5. Reproduced from [37]. CC BY 4.0.

charge was estimated using the following expression for the electric field of a charged surface:

$$-\frac{\mathrm{d}V}{\mathrm{d}h} = \frac{Q_{\mathrm{eff}}}{2\varepsilon\varepsilon_0} \tag{5.4}$$

where ε_0 represents the dielectric permittivity of free space. The Q_{eff} values obtained for various TA temperatures are depicted by the S5 (KPFM) curve in figure 5.13 and align with those derived from $C–V$ characteristics (refer to figure 5.11).

The implication derived from figure 5.13 is that the as-grown heterostructures fabricated in an Ar environment exhibit negative effective charges in the presence of 'thin' amorphous Li–Nb–O films (samples S5 and S6), while the samples based on 'thick' films demonstrate a positive effective charge. This observation is consistent with the data obtained for the polycrystalline films, as depicted in figure 5.11. Additionally, TA induces changes in the effective charge values. Notably, the effective charge varies oppositely for heterostructures with 'thick' and 'thin' films: as the magnitude of $|+Q_{\mathrm{eff}}|$ increases, the magnitude of $|-Q_{\mathrm{eff}}|$ decreases, and vice versa, as illustrated by samples S4 and S5 in figure 5.13. This phenomenon can be elucidated by postulating that the effective charge comprises two components: a negative charge $(-Q_{\mathrm{eff}})$, located near the Si–LN interface or at the film surface, and a positive charge $(+Q_{\mathrm{eff}})$, distributed within the bulk of the LN films. Analysis of various models suggests that charged antisite defects or oxygen vacancies are the most plausible sources of the positive oxide charge [32]. Li and O vacancies may arise in the deposited LN thin film due to its volatile nature and effusion from the film, as depicted in the following potential reactions [25]:

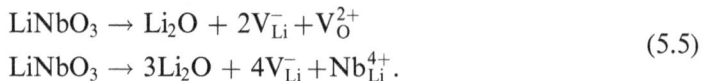

$$\begin{aligned} \mathrm{LiNbO_3} &\rightarrow \mathrm{Li_2O} + 2\mathrm{V_{Li}^-} + \mathrm{V_O^{2+}} \\ \mathrm{LiNbO_3} &\rightarrow 3\mathrm{Li_2O} + 4\mathrm{V_{Li}^-} + \mathrm{Nb_{Li}^{4+}}. \end{aligned} \tag{5.5}$$

On the other hand, according to the model proposed in [27], the structural defects can be redistributed as follows:

$$3O_O + 2V_{Li}^- + Nb_{Nb} \leftrightarrow \frac{3}{2}O_2 + Nb_{Li}^{4+} + 6e^-. \tag{5.6}$$

Therefore, the formation of the antisite defects Nb_{Li}^{4+} and the oxygen vacancies (V_O^{2+}) distributed in the bulk of a film are likely to be the origins of the positive charge in the as-grown Li–Nb–O films.

If oxygen is present in the reaction chamber, it fills the existing oxygen vacancies, and the positive charge declines. Another explanation can be based on the formation of the $V_{Li}^- - 1/2O_2$ complex, which produces a negative charge that partially compensates for the positive one. The following reaction describes the possible charge transfer that occurs if oxygen resides in the nearest position with a lithium vacancy [3]:

$$V_{Li}^- + \frac{1}{2}O_2 \Rightarrow V_{Li}^0 + \frac{1}{2}O_2^- \tag{5.7}$$

Consequently, in Li–Nb–O films grown in an Ar + O$_2$ gas mixture, the positive charge must be lower, which agrees with our results presented in figure 5.13.

The findings of previous research [26] have established that in films annealed within an oxygen atmosphere, there is a reduction in lithium concentration. This reduction leads to a slight decrease in lithium and oxygen vacancies, resulting in a positive charge in our particular case. According to these models, oxygen diffusion within a film during TA reduces the concentration of vacancies, subsequently decreasing the $+Q_{eff}$ value. Evidently, in heterostructures containing 'thin' LN films, the dominant charge is $-Q_{eff}$, which is associated with structural defects near the Si–LN interface. Conversely, in 'thick' LN films, the $+Q_{eff}$ can be attributed to oxygen vacancies or Nb_{Li}^{4+} defects that prevail over the interface-based negative charge. As previously mentioned, amorphous films deposited onto Si substrates crystallize in LN with a significant oxygen deficit under TA. This observation is corroborated by our findings on elemental composition (refer to figures 5.9(i) and (j)). TA in air significantly reduces the oxygen deficit in bulk thick Li–Nb–O films, while the alteration in interface-based charge does not contribute to the overall charge evolution. The negative component of effective charge in heterostructures based on 'thin' films exceeds its positive counterpart, and the change in $+Q_{eff}$ is observed as an inverse variation in $-Q_{eff}$ (which represents the net charge). As evidenced by our recent study, TA leads to the crystallization of amorphous Li–Nb–O films, forming polycrystalline LN films at a temperature of 470 °C [28]. This temperature closely aligns with the minimum net charge in Si–LN heterostructures (refer to figure 5.16 for samples S4 and S5).

TA at temperatures exceeding 520 °C induces the formation of the LiNb$_3$O$_8$ phase, which is characterized by a lithium deficit [16]. This, in turn, facilitates the generation of defects as described by equations (5.5) and (5.6). Consequently, the positive effective charge experiences a significant surge within LN films, as illustrated in

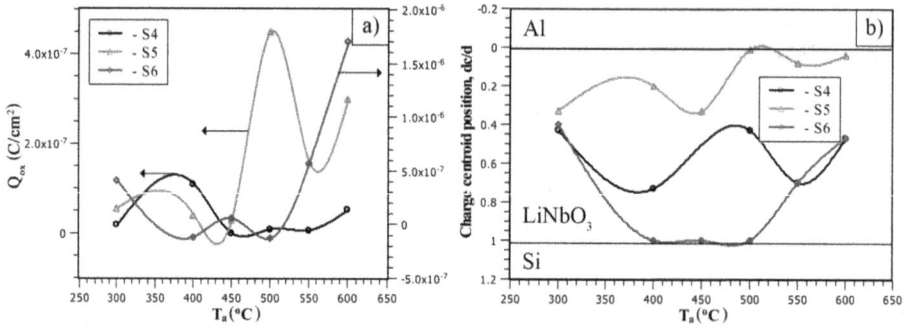

Figure 5.16. The oxide charge (a) and its centroid position (b) in the studied Si–LN–Al heterostructures as a function of the annealing temperature. Reproduced from [37]. CC BY 4.0.

figure 5.13, effectively compensating the $-Q_{eff}$. Exploiting this phenomenon holds promise for crafting Si–LN heterostructures devoid of effective oxide charge, which would therefore have potential applications in nonvolatile memory units. Notably, the charged domain walls, pivotal in domain wall memories, are capable of attracting ionic defects such as oxygen vacancies. This mechanism adversely impacts the smooth erasure and regeneration of domain walls [9].

The effective charge, by definition, comprises the oxide charge Q_{ox} and the interface charge Q_{ss}, the latter being linked to the density of interface states N_{ss}. This N_{ss} is quantified by the density of interface states per unit energy, D_{ss}, which can be derived from standard C–V analysis [30], expressed as:

$$N_{ss} = \int D_{ss}(E) \cdot f(E) \mathrm{d}E. \tag{5.8}$$

Here, $f(E)$ is the Fermi function.

We calculated the oxide charge as $Q_{ox} = Q_{eff} - Q_{ss}$. The oxide charge centroid position d_c can be determined through the following expression [40]:

$$\Delta V_{FB} = \frac{Q_{ox}}{C_i} \cdot \frac{d_c}{d}. \tag{5.9}$$

Here, ΔV_{FB} is the shift in the flat band voltage between the experimental and theoretical C–V curves and d is the film thickness.

The oxide charge Q_{ox} and its centroid position as a function of the annealing temperature are shown in figure 5.16.

As illustrated in figure 5.16, the as-grown amorphous Li–Nb–O films exhibit a positive Q_{ox}, with its centroid (the center of mass) positioned at the film's center, irrespective of the reactive gas environment. In samples produced in an Ar atmosphere, this charge fluctuates similarly to Q_{eff} (refer to figure 5.13) during TA. The centroid of the charge oscillates around the film's center in 'thick' films (sample S4) or approaches the film surface in 'thin' LN films (sample S5). In the case of films deposited in an Ar + O_2 gas mixture, Q_{ox} becomes negative and localizes near the film–substrate interface (see sample S6 in figure 5.16). This observation

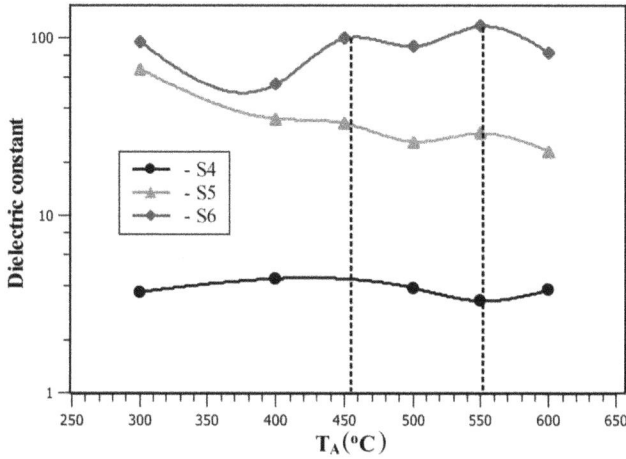

Figure 5.17. The dielectric constant as a function of the annealing temperature for the studied samples. Reproduced from [37]. CC BY 4.0.

substantiates our hypothesis that the negative component of Q_{eff} is situated at the Si–LN interface.

Figure 5.17 presents the dielectric constant ε of the investigated heterostructures annealed at different temperatures. As depicted, the ε values of samples S5 and S6 display two relatively broad peaks at temperatures of around 450 °C and 550 °C.

Similar peaks at 430 °C and 550 °C have been documented by other researchers [22, 41] for amorphous LN films. They corroborated, via XRD analyses and dielectric relaxation studies, that the peak at 550 °C corresponds to the recrystallization of the film, while the peak at 430 °C is attributed to Debye-like relaxation. Conversely, the dip in ε between the two peaks observed at 500 °C in figure 5.17 was also noted in [22]. This characteristic corresponds to the complete crystallization of LN films. Our recent work has unveiled this result at a temperature of 470 °C, which aligns with findings in [22], where a decrease in ε was linked to a change in the microscopic structure of the films at 490 °C. The dielectric constant values for single-crystal LN are reported as 31 along the a-axis and 78 along the c-axis at 25 °C and 100 kHz [42]. The closest ε to that value for single-crystal LN is observed for sample S5. As indicated in [43], the dielectric constant of polycrystalline LN films is greatly influenced by the grain size. Films with smaller grains exhibit a lower ε compared to LN films with larger grains.

This section has been reproduced from [37].

5.3 Transport properties and crystallization of Li–Nb–O films

As mentioned earlier, two primary synthesis approaches are proposed for complex oxides such as LN in thin film form. The first involves the crystallization of polycrystalline LN films with the desired properties, a popular method despite its inherent drawbacks. Epitaxial LN film growth under optimal RFMS conditions necessitates a relatively high substrate temperature, typically no less than 550 °C [5].

Due to the elevated pressures of lithium and lithium oxide vapors, the elemental composition of the growing film significantly deviates from the stoichiometry of LN, leading to the formation of the paraelectric phase $LiNb_3O_8$. Furthermore, researchers have observed varied crystallization behaviors of LN films on Si and SiO_2 substrates. Specifically, on SiO_2 substrates heated up to 530 °C, randomly oriented LN crystallites are synthesized, accompanied by the formation of a LiNbSiO alloy interlayer [44]. The presence of numerous vacancies in amorphous SiO_2 facilitates the acceptance of Li atoms. The reactivity of Li with SiO_2 and the migration rate of Li atoms both influence the degree of incorporation. The growth direction in LN films becomes randomized when the crystallization temperature on the SiO_2 substrate overlaps with the temperature corresponding to the interdiffusion of elements across the interface.

The alternative approach involves the crystallization of an amorphous Li–Nb–O film deposited on an unheated substrate [17]. Various post-deposition techniques, such as TA and PPT, have been employed to induce the crystallization of LN films [7, 45, 46]. However, the structure of amorphous films with compositions close to LN, the kinetics of their crystallization, as well as changes in texture, substructure, and surface morphology during crystallization and recrystallization, remain poorly explored and are reflected in only a few studies [5, 23, 47, 48].

It has been emphasized that the solid-phase crystallization process mitigates the impact of interfacial reactions, which is generally beneficial. Typically, amorphous LN films deposited at 300 °C undergo spontaneous transformation into a crystalline state at 460 °C, with the c-axis oriented towards the surface normal; this orientation is attributed to the minimal chemical potential. Concurrently, the interfacial region constrains the interdiffusion of atoms and limits interaction with the substrate. During crystallization, some Li atoms diffuse into the SiO_2 substrate, resulting in misorientation (i.e. orientation other than that along the c-axis). Upon crystallization of the LN film, each atom becomes anchored in a lattice position bonded to its neighbors. Other researchers have noted the absence of crystallization seeds at the LN–SiO_2 interface when SiO_2 is employed as a substrate [5]. They proposed the vacancy-mediated nucleation model, suggesting that the initial annealing event leads to the formation of nucleation centers. According to this model, vacancies generated during the incubation period stem from Li_2O evaporation, facilitating atomic movement and nucleating LN–$LiNb_3O_8$ complexes and phase separation. Once LN and $LiNb_3O_8$ seeds are formed, crystallization and phase conversion proceed. The emergence of Nb^{4+} defects in LN, which function as electronic traps with energies of 0.7 eV below the bottom of the LN conduction band, is attributed to oxygen and Li loss [21].

Despite the aforementioned advancements in structural and morphological studies of crystallization in amorphous LN films, their transport properties such as conductivity, the Hall effect, carrier concentration, and scattering have not received adequate investigation. Specifically, annealing as-grown amorphous powders at 450 °C promotes the formation of single-phase LN crystalline films; their DC conductivity depends on the crystalline phase quantity and increases with prolonged heat-treatment time [12]. Other research groups have reported that increasing the thickness of LN films from 45.3 to 192.9 nm results in a rise in charge carrier concentration [49]. Unfortunately, the underlying mechanisms of these effects remain undiscussed.

In our recent study, we investigated the crystallization process of Li–Nb–O amorphous films deposited onto Si substrates using the RFMS technique [16]. Through examination of their electrical properties, we demonstrated that quasi-amorphous Li–Nb–O films contain uniformly distributed donor centers. TA of these films leads to a concentration gradient from the surface towards the film–substrate interface, resulting in increased concentration compared to as-grown films. While the transverse DC conductivity and dielectric properties have been explored and discussed, the Hall effect and scattering phenomena were not addressed in the aforementioned study.

This section aims to investigate how the carrier mobility, concentration, and their transport evolve during the crystallization of Li–Nb–O films. This study logically continues our past research into the evolution of the electrical properties of amorphous LN films under TA.

The studied heterostructures were fabricated by depositing the Li–Nb–O system at a thickness of 300 nm onto oxidized silicon wafers (Si(001) n-type conductivity, $\rho = 4.5\,\Omega \cdot$ cm) using RFMS of the LN targets. The RFMS process was conducted under the following conditions: a vacuum of $\sim 10^{-1}$ Pa, a reactive Ar gas environment (0.75 Pa), and a power of 15 W cm^{-2} incident on the target. The Ar flow rate was 1.15 l h^{-1}. TA of the as-grown heterostructures was conducted in air at temperatures of $T_A = 400\,°$C, 450 °C, 500 °C, 550 °C, and 600 °C (for 1 h at each temperature). The studied samples were numbered S1,…,S5 according to the above set of temperatures.

Figure 5.18 shows the Hall coefficient (R_H) as a function of temperature, represented in Arrhenius coordinates, for the samples under study. It can be observed in figure 5.18 that R_H exhibits a negative value, indicating that electrons are the major charge carriers in the examined LN films. Additionally, at low temperatures, the Hall coefficient remains nearly independent of temperature, while its magnitude experiences an exponential decrease as temperature rises, commencing from the threshold temperature T_k. Notably, the threshold temperature T_k demonstrates an increase corresponding to the annealing temperature TA, as illustrated in figure 5.19. Previous research, as cited in our earlier works [16, 17], has established that TA initiates the crystallization process of amorphous LN films, which is typically completed within the temperature range of 450 °C–475 °C. Consequently, the linear segment displayed in figure 5.19 signifies the annealing of polycrystalline LN films, succeeded by their recrystallization at approximately 550 °C [50], and adheres to the following law: $T_k = 0.45 T_A - 24.6$.

For an n-type semiconductor, the Hall coefficient is given by the following formula [52]:

$$R_H = -\frac{1}{qn}r.$$

(5.10)

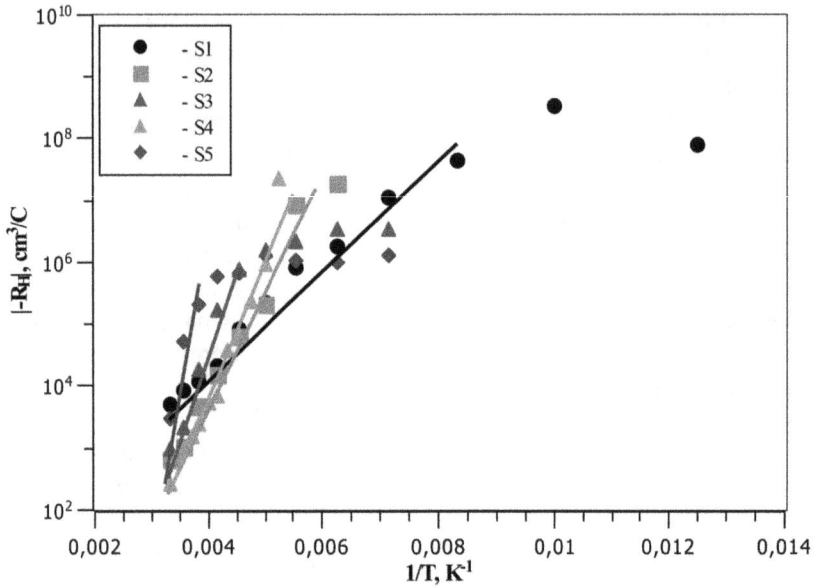

Figure 5.18. Temperature dependence of the Hall coefficient magnitude in Arrhenius coordinates. Reprinted from [51], Copyright (2022), with permission from Elsevier.

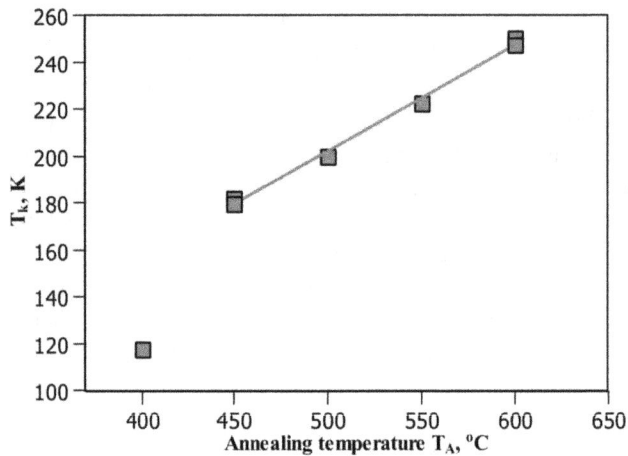

Figure 5.19. Threshold temperature T_k as a function of annealing temperature for the studied LN films.

Here, n is the concentration of electrons, q is the elementary charge, and r is a scattering factor. As a function of carrier energy, the factor r, determined by a prevailing scattering mechanism, equals the ratio of Hall and drift mobility. In bulk semiconductors, both thermal photon scattering and scattering by ionized impurities prevail. By contrast, in polycrystalline materials, other scattering mechanisms dominate (surface scattering, grain boundary scattering, and potential barrier scattering) [53]. Thus, in polycrystalline samples, the temperature dependence of the Hall coefficient is determined by the Hall mobility rather than the carrier concentration.

The presence of grain boundaries and interface-trapped charge, integral components of polycrystalline semiconductors, introduces complexities in the interpretation of Hall measurements.

Interface barriers, stemming from the interface charge, modulate band bending within the bulk of the grains. Additionally, interface states, which exhibit a high concentration at grain boundaries, serve to scatter or trap free carriers from the bulk of the grains. Complete depletion of the grains may occur under conditions where either the interface state density is relatively high or the bulk doping level is sufficiently low. Under such circumstances, films exhibit relatively high electrical resistivity, attributed to the significant thermal activation energy, E_a. Both the density per unit area and the energy of traps, corresponding to their activation energy, dictate the maximum trapped charge.

The carrier's mean free path, which is contingent upon the depletion layer width, may either surpass or fall short of the grain size. When the depletion regions only partially extend through the grains, the carrier concentration derived from Hall measurements accurately reflects their magnitude within the bulk of the grains. In such instances, Hall mobility adheres to the following principle [52]:

$$\mu_b = \mu_0 \exp\left(-\frac{\varphi_b}{kT}\right). \tag{5.11}$$

Here, φ_b is the band bending, and μ_0 is a parameter proportional to the average grain diameter. As demonstrated in [52], in a homogeneous material, the measured Hall coefficient represents the bulk grain properties.

When the depletion regions only partially extend over the grains, the Hall coefficient is almost temperature independent (as observed in our experiments at low temperatures), φ_b is very small, and the measured value of R_H is determined by the concentration of carriers in the neutral region of the grains. Due to an increase in grain size under the recrystallization process, as reported in [17], grain boundaries influence Hall mobility to a lesser extent, and the threshold temperature T_k rises, as observed in our experiment (see figure 5.19). The carrier concentration, estimated from the temperature-independent Hall coefficient as a function of the annealing temperature, is presented in figure 5.20.

The increase in carrier concentration with the annealing temperature, as depicted in figure 5.20, contradicts the findings presented in [54], where it decreases from 9×10^{12} to $4 \times 10^{12}\,\mathrm{cm}^{-3}$ with the substrate temperature, albeit with a simultaneous rise in mobility. This phenomenon was attributed to a reduction in film defectiveness.

The primary disparity between our results and those cited above lies in the synthesis methodology. LN films deposited onto heated substrates are synthesized in a polycrystalline form, whereas in our experiment, the as-grown 'quasi-amorphous' Li–Nb–O films undergo crystallization under TA [17]. Furthermore, recent demonstrations have indicated that these quasi-amorphous films comprise appropriate

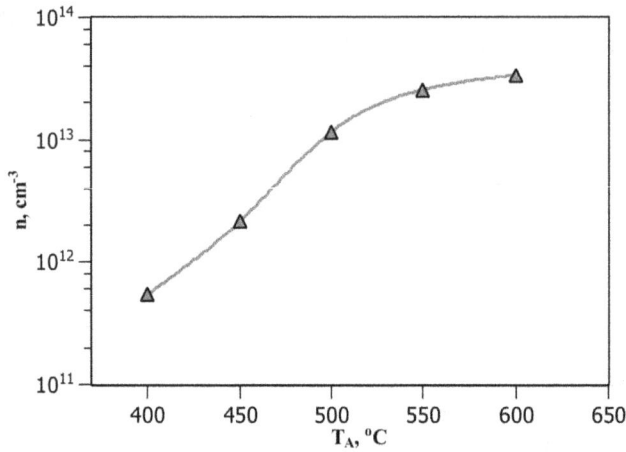

Figure 5.20. Carrier concentration as a function of the annealing temperature for the studied LN films.

nuclei of lithium oxide (Li_2O) and niobium oxide (Nb_2O_5), along with LN as a product of their reaction [17].

As per the model proposed in [5], annealing amorphous LN films in an O_2 atmosphere induces an incubation period before crystallization. Consequently, a larger volume fraction of $LiNb_3O_8$ is formed post-crystallization compared to annealing in a vacuum. Oxygen molecules permeate the films, impeding atom migration and retarding crystallization. The depletion of Li_2O generates vacancies that aid in the nucleation of LN and $LiNb_3O_8$ pairs. Additionally, some authors have suggested vacancy reactions that elucidate the generation of free electrons as the principal carriers in LN [55]:

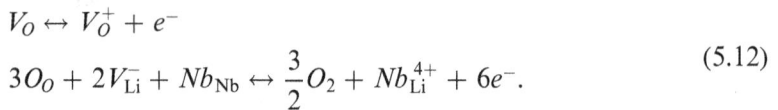

$$V_O \leftrightarrow V_O^+ + e^-$$
$$3O_O + 2V_{Li}^- + Nb_{Nb} \leftrightarrow \frac{3}{2}O_2 + Nb_{Li}^{4+} + 6e^-. \tag{5.12}$$

Here, V_{Li} and V_O are lithium and oxygen vacancies, respectively, and Nb_{Li} is an antisite defect at V_{Li}. Consequently, the vacancies produced under TA in LN, saturated when the annealing temperature reaches 550 °C, generate electrons according to figure 5.20. This result is in good agreement with our previous work, where the redistribution of shallow donors has been revealed in LN films under TA [16].

Figure 5.21 shows the typical temperature dependence of conductivity in Arrhenius coordinates for the studied samples.

As seen in figure 5.21, the conductivity has a very weak temperature dependence at low temperatures (attributed to thermally assisted tunneling [52]) and manifests an activation character at high temperatures. This thermally activated behavior can be affected by either the Hall mobility or the concentration of carriers in the film. The typical Hall mobility as a function of temperature for the studied samples is presented in figure 5.22 [51].

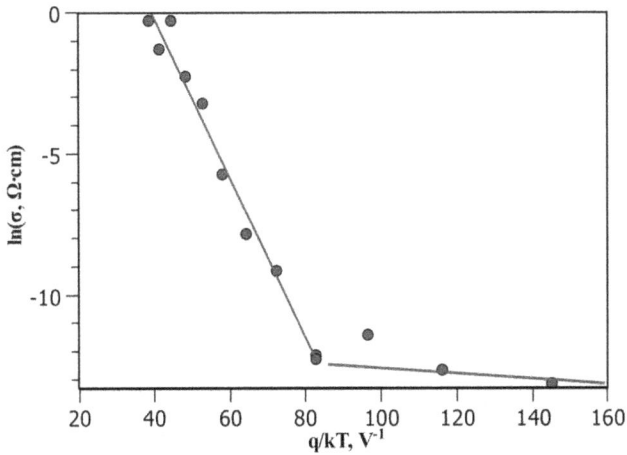

Figure 5.21. Temperature dependence of conductivity for sample S1 in Arrhenius coordinates.

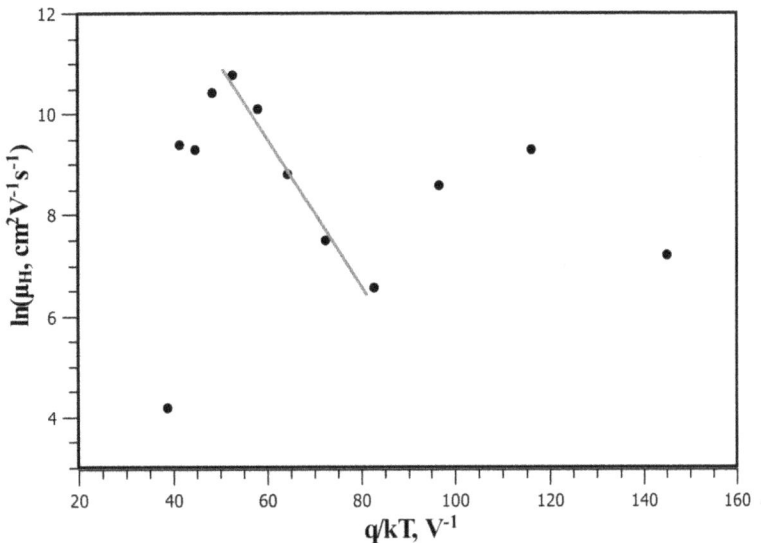

Figure 5.22. Temperature dependence of the Hall mobility for sample S1 in Arrhenius coordinates. Reprinted from [51], Copyright (2022), with permission from Elsevier.

The Hall mobility's activation characteristics (referenced by the linear segment in figure 5.22) align with the conductivity's analogous behavior discussed earlier. Three primary scattering mechanisms prevail in thin films: surface scattering, dislocation scattering, and grain boundary scattering [52]. Given that the thickness of our film (300 nm) surpasses the carrier mean free path, we disregard surface scattering in our analysis. Regarding grain boundary scattering, it is influenced by interface barriers arising from interface charges whose mobility is described by equation (5.11). In accordance with equation (5.11), the Arrhenius plot of μ_H should manifest as a

Table 5.3. The activation energies of the Hall coefficient (E_{aR}), the Hall mobility ($E_{a\mu}$), and conductivity ($E_{a\sigma}$) for the studied samples.

Sample	E_{aR} (eV)	$E_{a\mu}$ (eV)	$E_{a\sigma}$ (eV)
S1	0.30	0.15	0.30
S2	0.30	0.22	0.50
S3	0.45	0.45	1.0
S4	0.45	0.61	1.3
S5	0.53	0.53	0.50

straight line with a slope of φ_b, mirroring our experimental observations. Electrons do not contribute to charge transport when confined between grain boundaries with energies below φ_b. Conversely, electrons align with an applied magnetic field if their energies surpass φ_b. Consequently, the carrier concentration determined via Hall effect measurements diverges from bulk carrier concentration in this scenario. Table 5.3 outlines the activation energies of the Hall mobility ($E_{a\mu}$), the Hall coefficient (E_{aR}), and conductivity ($E_{a\sigma}$) derived from the examined samples.

Several authors have argued that hopping serves as the predominant charge transport mechanism in LN [56–58]. Notably, in our experimental findings, the activation energy of the Hall mobility is nearly half that of the conductivity, except for sample S5 (refer to table 5.3). This finding contrasts with the outcomes presented in [59] concerning bulk LN within the small-polaron theory framework, where the activation energy of μ_H amounts to only one-third of that associated with conductivity. Nonetheless, the activation energy for R_H, corresponding to sample S1, surpasses that of the Hall mobility by a factor of two, consistent with said theory. Moreover, since the Hall mobility can be expressed as a function of the Hall coefficient and conductivity, namely $\mu_H = R_H \sigma$, the activation energy of the Hall mobility ($E_{a\mu}$) is observed to be $E_{a\mu} = E_{a\sigma} - E_{aR}$, as evident in samples S2–S4 (see table 5.3).

Small polarons are postulated to form in LN through the trapping of electrons at niobium cations, contingent upon a sufficiently strong electron–lattice interaction. The Hall mobility within the small polaronic mechanism framework is governed by the following relationship [59]:

$$\mu_H \propto T^{-\frac{1}{2}} \exp\left(-\frac{W_H}{3kT}\right). \tag{5.13}$$

Here, W_H is the hopping activation energy. At sufficiently high temperatures, the pre-exponential term in equation (5.13) dominates. Thus, the Hall mobility peaks and then declines with temperature, as observed in our experiments (see figure 5.22).

As mentioned above, as-grown Li–Nb–O films (sample S1) exhibit a 'quasi-amorphous' structure, and the charge transport is limited by small-polaron hopping, determined by the bulk properties of the material. TA leads to the crystallization of the films, and the grain boundaries start to play a dominant role in charge transport.

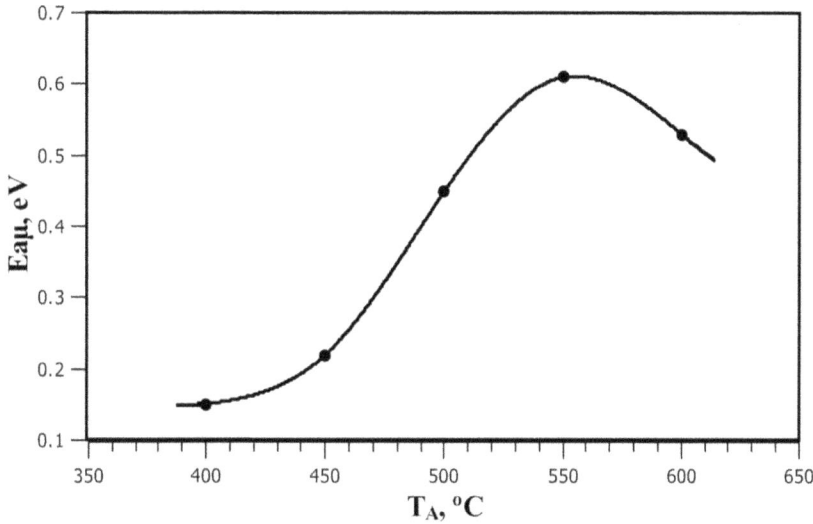

Figure 5.23. Activation energy of the Hall mobility as a function of annealing temperature for the studied LN films. Reprinted from [51], Copyright (2022), with permission from Elsevier.

Consequently, hopping polarons are not necessarily the major carriers in poly-crystalline LN films (samples S2–S4). Thus, the thermally activated part, like that shown in figure 5.22, might be interpreted in terms of grain boundary scattering. As demonstrated in [52], the effective Hall mobility is given by the following expression:

$$\frac{1}{\mu_H} = \frac{1}{\mu_b} + \frac{1}{\mu_1}. \tag{5.14}$$

Here, μ_1 and μ_b are the bulk single-crystal and the grain-boundary-limited mobility, respectively. Because μ_1 and μ_b change oppositely with temperature, the contributions of both mechanisms to the net mobility can be distinguished by analyzing its temperature dependence. Specifically, the section that declines with temperature in figure 5.22 is attributed to the second term in equation (5.14), whereas the rising section is described by the first term.

Figure 5.23 shows the activation energy $E_{a\mu}$ as a function of the annealing temperature given in table 5.3 for the studied samples.

It is evident from figure 5.23 that the activation energy increases with TA and peaks at a temperature of about 550 °C. Even though the crystallization temperature in a vacuum is 573 °C and a fully transformed state is reached at 753 °C [60], this behavior correlates with the change in trap concentration in Li–Nb–O amorphous films during the process of TA, which reaches a maximum at $T_A = 470$ °C, as investigated in our recent work [16]. Furthermore, according to the results for the crystallization of amorphous LN films, Nb^{4+} defects, which act as electronic traps, are produced at the grain boundaries due to the evaporation of Li_2O under TA [5]. Further recrystallization occurs at $T_A = 600$ °C, when the loss of Li_2O ceases and the number of grain boundaries (and thus the intergranular barriers) decreases.

This occurs because annealing replenishes lithium and oxygen vacancies through the effusion of these elements from the bulk of LN [21]. The intergranular barriers of φ_b limit the charge transport processes at this stage, and the corresponding activation energies are equal to the same value $E_{a\mu} = E_{a\sigma} = E_{aR} = \varphi_b$, as shown in table 5.3.

This section has been reproduced with permission from [51].

5.4 Charge phenomena at Si–LiNbO$_3$ heterojunctions

It is widely acknowledged that the performance of thin film devices can be constrained by electronic phenomena occurring at the film–substrate interface. Moreover, fabricating LN–substrate heterojunctions capable of modulating space charge areas and controlling charge injection poses a significant challenge for their application in electro-optic devices. Several authors have suggested the use of various buffer layers to promote the c-oriented growth of LN films on Si semi-conductor substrates. Specifically, ZnO-buffered Si (001) substrates were employed for LN film fabrication [61]. It has been demonstrated that a ZnO thickness of 100 nm in ZnO–Si heterojunctions ensures the formation of good interfaces with few charge traps. Conversely, when the ZnO thickness decreases to 15 nm, the dominant interface affecting the electrical properties of the heterojunction transitions from ZnO–Si to LN–Si, thus manifesting highly efficient injection properties. Additionally, the C–V characteristics of synthesized heterostructures were found to be influenced by the quality of the LN–Si heterojunction. In contrast to the findings in [61], our recent paper reported counterclockwise C–V curves. These occurred in LN–Si heterostructures due to the trapping of minority charges (holes) at border traps distributed in the SiO$_2$ layer that inevitably forms at the LN–Si interface during the RFMS process [38].

Another research group proposed the fabrication of an LN–Si heterojunction based on heavily Fe-doped c-oriented LN films fabricated by pulsed laser deposition. The resulting heterojunction yielded an excellent prototype device that exhibited rectifying I–V characteristics. These were explained in terms of the LN–Si band diagram [62]. Experimental results for the synthesis of epitaxial or c-oriented LN films suggest that the substrate temperature must be at least 550 °C under optimal RFMS conditions [63]. Due to the high lithium and lithium oxide vapor pressures, the elemental composition deviates considerably from the LN stoichiometry in as-grown films, leading to the formation of the undesired paraelectric phase LiNb$_3$O$_8$.

As an alternative to direct growth of polycrystalline LN films on various heated substrates, other techniques have been proposed. One possible efficient approach for fabricating LN–Si heterostructures with the required properties is to form amorphous films on unheated substrates, followed by their crystallization. This can be achieved through the design of optimal synthesis regimes [16, 17]. Despite extensive research into the structure and composition of LN films, only a few investigations have examined how the sputtering conditions, such as reactive gas composition, affect the electrical properties of LN–Si heterojunctions and influence the perform-ance of LN-based devices. Furthermore, to the best of our knowledge, the evolution

Table 5.4. Sputtering parameters of the studied samples.

Sample #	Reactive gas composition	Magnetron power (W)	Reactive gas pressure (Pa)	Substrate–target distance (cm)	Substrate position	Film thickness (nm)
S1	Ar	50	0.8	6	Offset from the target erosion zone	100
S2	Ar(80%) + O$_2$(20%)					200
S3	Ar(60%) + O$_2$(40%)					100

of the electrical properties of LN–Si heterojunctions during the crystallization of LN films grown in various reactive gas mixtures has not yet been studied.

This section aims to review the crystallization of amorphous LN films grown by RFMS in a reactive gas mixture with various oxygen contents, and to examine how the electrical properties of LN–Si heterojunctions evolve during this process. The novelty of this work lies in separating the contributions of the LN–Si heterojunction and the film to the electrical performance of the synthesized heterostructures during their crystallization. The primary objective of this study is to clarify the evolution of barrier properties in the synthesized heterojunctions and to propose optimal crystallization regimes, as reflected in our work [64].

The studied samples were fabricated by depositing the Li–Nb–O system onto unheated (001)Si wafers of n-type conductivity ($\rho = 4.5\ \Omega \cdot$ cm) by RFMS of the LN targets. The sputtering conditions corresponding to the studied samples are given in table 5.4. The substrates were located at an offset from the target erosion zone, thus preventing the intensive bombardment of the films by plasma particles [65]. Before the RFMS process, the silicon wafers were cleaned by ion etching in an Ar plasma (2 min). Subsequent TA was conducted in air for 1 h at the temperature range of $T = 300\ °C{-}600\ °C$.

The crystallization of as-grown amorphous films synthesized in an Ar reactive gas environment has been the subject of a recent investigation [16]. It has been observed that the proportion of the crystalline phase within the films increases from 45% to 100% as the annealing temperature is elevated from 450 °C to 475 °C. Subsequent annealing initiates a recrystallization process, leading to the formation of a large-block crystalline structure. In this section, we present a study of film compositions grown in an Ar + O$_2$ atmosphere. Figure 5.24 illustrates the Raman spectra of as-grown amorphous films fabricated under pure Ar (a) and Ar + O$_2$ gas environments (b), (c), followed by annealing at various temperatures.

For the films fabricated under a pure Ar atmosphere and annealed at 300 °C (refer to figure 5.24(a), curve 1), not all the peaks attributed to LN were detected; this result is unlike the Raman spectra of films annealed at higher temperatures (figure 5.24(a), spectra 2–6). This observation aligns with findings from our recent work, where the absence of certain peaks was attributed to the presence of an

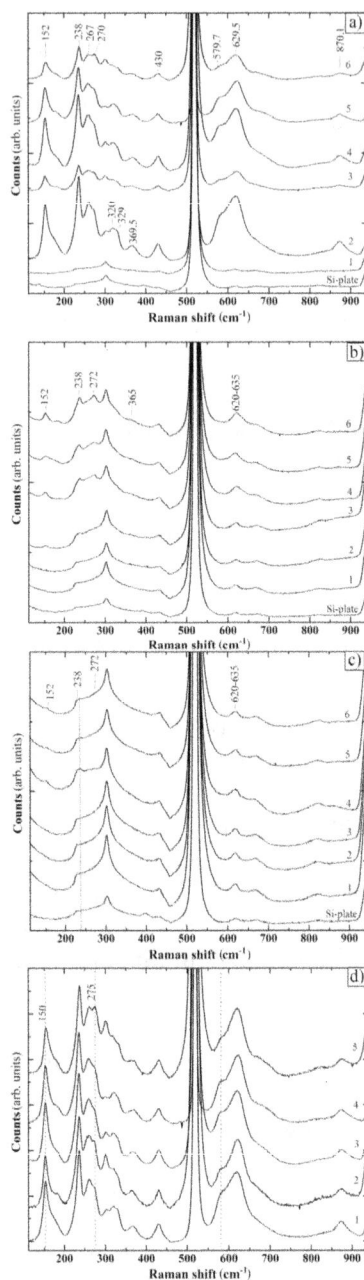

Figure 5.24. Raman spectra of Li–Nb–O films deposited on Si substrates by a RFMS technique and annealed at various temperatures T_A: 1—300 °C, 2—400 °C, 3—450 °C, 4—500 °C, 5—550 °C, 6—600 °C. (a) sample S1, (b) sample S2, (c) sample S3, (d) spectra of sample S1 normalized to the most intensive line (238 cm^{-1}) corresponding to LN (T_A: 1—400 °C, 2—450 °C, 3—500 °C, 4—550 °C, 5—600 °C). Reproduced from [64]. CC BY 4.0.

amorphous Li–Nb–O phase at this temperature [16]. For films annealed at temperatures ranging from 400 °C to 600 °C, the following Raman scattering peaks are presented in figure 5.24(a): 152, 238, 267, 270, 320, 329, 369.5, 430, 579.7, 629.5, and 870.1 cm^{-1}. According to reference [66], these observed spectra correspond to LN. The spectra in figure 5.24(a) can be conditionally divided into three wavenumber ranges: 100–450, 550–650, and 850–900 cm^{-1}, which correspond to the oscillation of Nb ions residing in the oxygen octahedra, oscillations of NbO_6 oxygen octahedra, and the Nb–O–Nb valence bridge oscillations of the oxygen ions in NbO_6 octahedra, respectively. Various oxidation states of LNs appear as adjacent peaks in the Raman spectra; one of the most intensive Raman modes attributed to LN is the line at 238 cm^{-1} [66]. Normalizing the Raman spectra based on the most intensive line corresponding to LN (see figure 5.24(d)) allows certain features to be discerned, particularly for $T_A = 600$ °C.

In figure 5.24(d), an increase in the intensity of the 275 cm^{-1} mode, accompanied by a shift in the 150 cm^{-1} peak, is observed as the annealing temperature rises. It is noteworthy that the 275 cm^{-1} mode, being the most prominent in the Raman spectra, is likely attributed to the $LiNb_3O_8$ phase [66]. Thus, the intensified presence of this peak can be linked to the concurrent formation of the $LiNb_3O_8$ and LN phases in the annealed films, as previously reported in our work [16] and by other researchers [14].

Regarding sample S2, an 'arm' appears at a wavenumber of 238 cm^{-1} in the Raman spectra for an annealing temperature of 450 °C, possibly indicating the onset of LN crystallization [16]. Furthermore, TA leads to the enhancement of this mode alongside the emergence of a 272 cm^{-1} peak associated with the $LiNb_3O_8$ phase. In the case of films produced under higher oxygen content (Ar(60%) + O_2(40%)) in a reactive gas environment (refer to figure 5.24(c) for sample S3), this mode is absent, suggesting the absence of the $LiNb_3O_8$ phase. This finding aligns with previous results, endorsing the Ar(60%) + O_2(40%) composition as the optimal reactive gas environment for the formation of single-phase LN films on Si [50].

Figure 5.25 illustrates the normalized C–V characteristics of as-grown Si–LN heterostructures [64].

As illustrated in figure 5.25, the C–V characteristics of the examined samples exhibit similarities to those observed in metal–insulator–semiconductor structures. Consequently, through the application of standard C–V analysis [20], several significant characteristics of the investigated heterostructures have been derived.

The dielectric constant of the analyzed films exhibits variation with respect to annealing temperature, as depicted in figure 5.26(a). Notably, a broad peak is evident in the $\varepsilon(T_A)$ curves at approximately 450 °C across all studied samples, irrespective of the composition of the reactive gas. Similar peaks have been reported by other researchers at temperatures of 430 °C and 550 °C for amorphous LN films [22, 41]. These studies suggest that the peak observed at 430 °C can be attributed to Debye-like relaxation, while the peak at 550 °C corresponds to film recrystallization.

As demonstrated in recent research [16], a temperature of 470 °C signifies complete crystallization of as-grown amorphous Li–Nb–O films deposited onto unheated Si substrates. Consequently, only polycrystalline films are present at

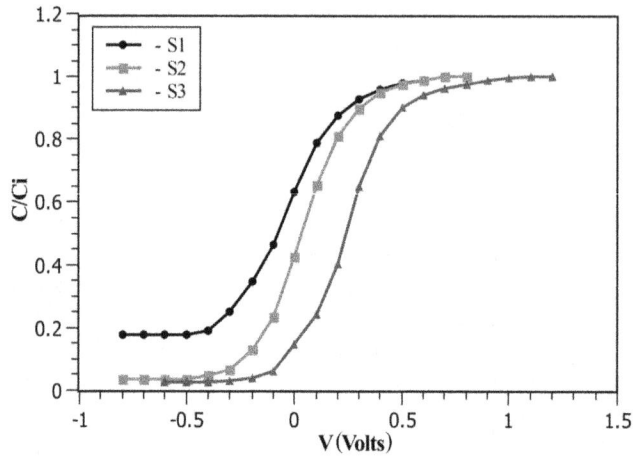

Figure 5.25. Normalized C–V characteristics of the as-grown studied samples. Reproduced from [64]. CC BY 4.0.

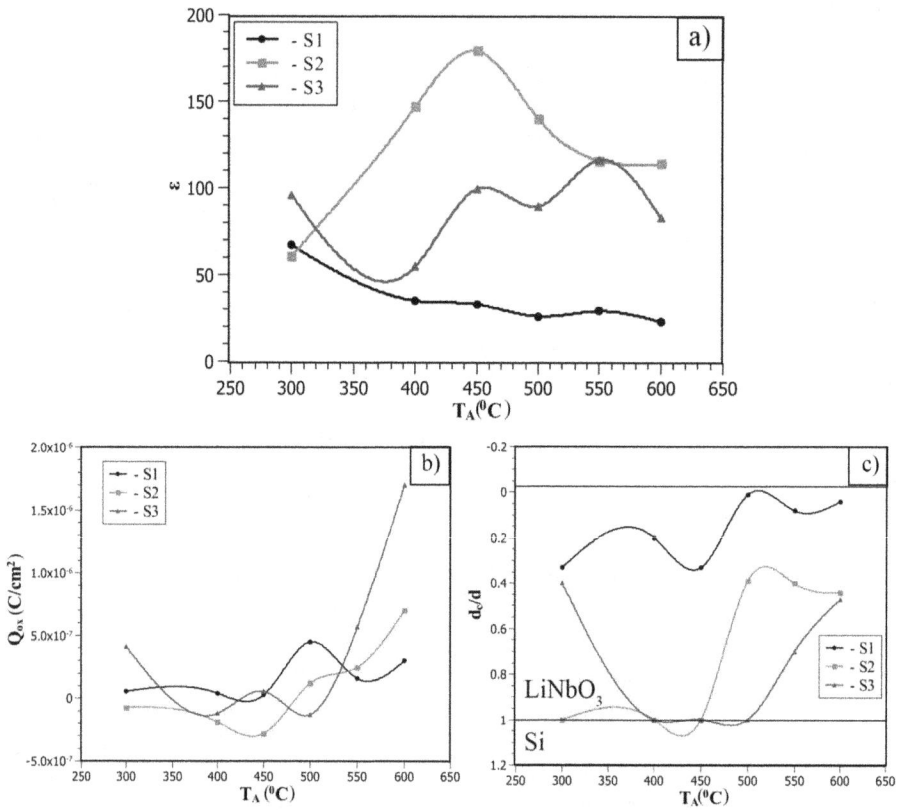

Figure 5.26. Dielectric constant ε (a), the fixed oxide charge Q_{ox} (b), and its relative centroid position (c) d_c as a function of annealing temperature T_A for the studied samples. Reproduced from [64].CC BY 4.0.

temperatures exceeding 470 °C. Notably, this temperature closely aligns with the crystallization temperature (473 °C) reported in prior literature [60]. As posited in another study [22], the Debye-like relaxation peak observed in our $\varepsilon(T_A)$ curves at 450 °C is linked to the transfer of Li ions to adjacent O octahedra within the micronetwork of niobium oxides existing in amorphous LN. This ion transfer demonstrates relaxation behavior. Given that LN films synthesized in an Ar + O_2 plasma exhibit lower Li deficiency [67], the concentration of Li in these films surpasses that of films produced in a pure Ar environment. Consequently, the relaxation peak at $T_a = 450$ °C is more pronounced for samples S2 and S3 compared to S1 (refer to figure 5.26(a)).

It is noteworthy that the dielectric constant of polycrystalline LN films, fabricated in both Ar(80%) + O_2(20%) and Ar(60%) + O_2(40%) gas mixtures and annealed at 550 °C ($\varepsilon = 120$), exceeds that of films deposited in a pure Ar atmosphere and annealed at the same temperature (refer to figure 5.26(a)). This elevated ε value holds particular promise for potential applications of LN-based heterostructures in nonvolatile memory units [34].

The C–V characteristics of the examined samples deviate from ideal C–V curves. This deviation is attributed to the effective charge, Q_{eff}, present in Si–LN heterostructures, comprising the fixed oxide charge, Q_{ox}, and the charge captured at interface states, Q_{ss}. By applying standard C–V analysis and segregating various components of Q_{eff} as detailed in [20, 40], we determined the fixed oxide charge Q_{ox} in the samples under investigation. The Q_{ox} values and the relative positions of its centroid (center of mass) d_c are depicted as a function of annealing temperature in figures 5.26(b) and (c).

From figures 5.26(b) and (c), it is evident that $+Q_{\mathrm{ox}}$ is situated at the center of the as-grown amorphous LN films, while $-Q_{\mathrm{ox}}$ is distributed at the film–substrate interface, consistent with previous findings [37]. TA has minimal impact on Q_{ox} in heterostructures fabricated in pure Ar and in an Ar(80%) + O_2(20%) gas mixture (samples S1 and S2 in figures 5.26(b) and (c)) up to the crystallization temperature ($T_C = 470$ °C). However, beyond T_C, the Q_{ox} values in heterostructures fabricated in an Ar(80%) + O_2(20%) atmosphere change sign from negative to positive, increase in magnitude, and reside in the middle of the LN film. In films grown in a pure Ar reactive environment, the $+Q_{\mathrm{ox}}$ migrates toward the film surface at temperatures below T_A. For heterostructures synthesized in an Ar(60%) + O_2(40%) gas mixture, the positive oxide charge in the as-grown films becomes negative at temperatures below T_A and is distributed at the film–substrate interface (refer to figure 5.26(c)). Upon crystallization of LN films (at $T_A > 500$ °C), the oxide charge becomes positive again, residing in the middle of the film.

Analysis of the $(S/C)^2$–V curves (not shown) revealed that the Si substrates become highly doped after the deposition of Li–Nb–O films. We calculated the donor impurity distribution $N_d(x)$ using the following formula [20]:

$$N_d(x) = -\frac{2}{q\varepsilon\varepsilon_0 S^2}\left(\frac{d}{dV}\left(\frac{S}{C(V)}\right)^2\right)^{-1}. \tag{5.15}$$

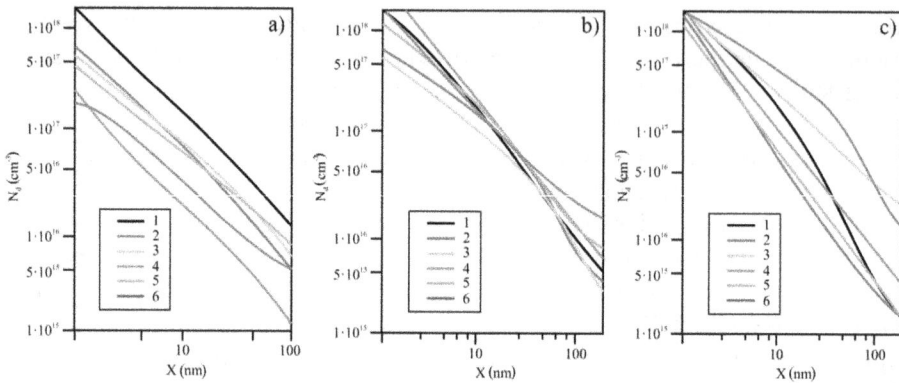

Figure 5.27. Donor distributions in Si substrates of the studied heterostructures (a—sample S1, b—sample S2, c—sample S3), annealed at various temperatures T_A (1—300 °C, 2—400 °C, 3—450 °C, 4—500 °C, 5—550 °C, 6—600 °C). X is the distance from the Si–LN interface toward the Si bulk. Reproduced from [64]. CC BY 4.0.

Here, ε and ε_0 are the dielectric constant of Si and the electric constant, respectively, and S is the contact area. The obtained donor distributions for the studied heterostructures annealed at various temperatures are shown in figure 5.27 [64].

In our prior investigations of polycrystalline LN films [1, 11], a similar doping effect was noted. The genesis of donors in Si is explained by the model proposed by reference [68], which revolves around the presence of *fast-diffusing gas-like molecular oxygen* within silicon. Our research indicates that thermal donor formation takes place at approximately 450 °C, driven by the diffusion of oxygen into Si at a rate of around 10^{-9} cm^2 s^{-1}. Figure 5.27 illustrates how N_d decreases to its nominal level within 200 nm of the substrate. Notably, under TA at temperatures up to 500 °C, there is a significant reduction in donor concentration solely in the heterostructures produced within a pure Ar environment (refer to figure 5.27(a)). This phenomenon can be ascribed to the cessation of oxygen diffusion from Si to LN films, which display a greater oxygen deficit compared to films grown in an Ar + O$_2$ gas mixture. Subsequent annealing results in a rise in N_d. The escalation observed in the examined heterostructures annealed at $T_A > 500$ °C (see figure 5.27) is attributable to the process of oxygen precipitation occurring during heat treatments within the temperature range of 550 °C–800 °C [68].

Figure 5.28 depicts a band diagram of the investigated heterostructures featuring a highly doped region in Si.

The highly doped area near the LN–Si interface serves as a reservoir for electrons, providing a good injection contact in some optoelectronic applications, such as tunable microring resonators [69].

The I–V characteristics of the studied samples are shown in figure 5.29.

As illustrated in figure 5.29, the examined samples exhibit rectifying I–V characteristics that arise from the barrier properties inherent in the LN–Si heterointerface (refer to figure 5.28). The saturation currents, denoted by J_s and observed

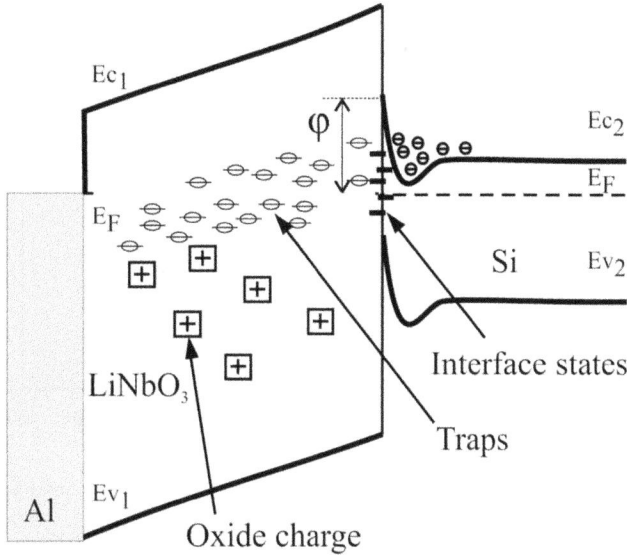

Figure 5.28. Band diagram of the studied Si–LN heterostructures. Reproduced from [64]. CC BY 4.0.

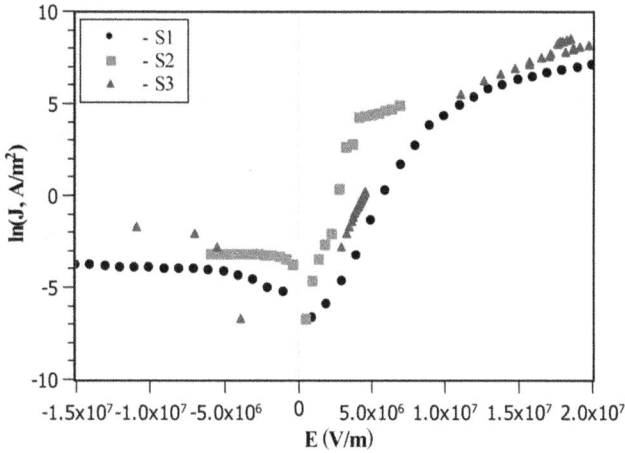

Figure 5.29. *I–V* characteristics of the studied samples at room temperature. Reproduced from [64]. CC BY 4.0.

under reverse bias (marked with '−' at the Al contacts), are influenced by the barrier height represented by φ. This relationship can be expressed as follows [70]:

$$J_s = \frac{4\pi q m}{h^3}(kT)^2 \exp\left(-\frac{q\varphi}{kT}\right). \tag{5.16}$$

Here, m is the effective mass of the electrons, h is Planck's constant, k is Boltzmann's constant, q is the elementary charge, T is the temperature, and φ is the average

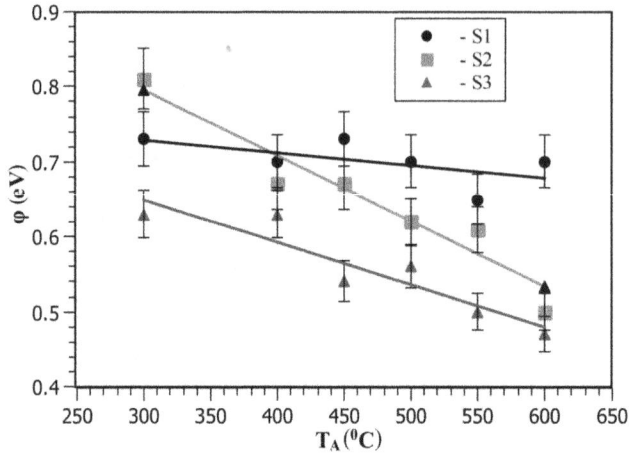

Figure 5.30. The potential barrier height, φ, at the Si–LN interface as a function of annealing temperature, T_A, for the studied samples. Reproduced from [64]. CC BY 4.0.

potential barrier height. Thus, the average potential barrier height can be determined from the slope of a linear section of the $\ln(J_s/T^2)-1/T$ graph. Figure 5.30 shows the evolution of the barrier height, φ, in the studied heterostructures during TA.

As depicted in figure 5.30, the barrier height within Si–LN heterostructures grown in an Ar environment is approximately 0.7 eV and remains independent of T_A. In contrast, the φ value decreases with increasing annealing temperature in samples fabricated using an Ar + O_2 reactive gas mixture. While the barrier height in as-grown heterostructures deposited in an Ar(80%) + O_2(20%) atmosphere is higher compared to those of samples fabricated in an Ar(60%) + O_2(40%) mixture or a pure Ar environment, TA results in a decrease in the φ value to approximately 0.5 eV for samples S2 and S3. The observed decline in the barrier height for the studied heterostructures fabricated in an Ar + O_2 gas mixture correlates with the evolution of the oxide charge Q_{ox} (refer to figure 5.31). The highest $-Q_{ox}$, located at the LN–Si interface, corresponds to the highest φ value observed for as-grown heterostructures fabricated in an Ar(80%) + O_2(20%) gas mixture (sample S2) [64].

The direct branches of the I–V characteristics ('+' at the top Al contact) are linear in $\ln J$ vs. $\ln V$ coordinates (see figure 5.32), suggesting that SCLC affects the charge transport in the studied samples.

The absence of translational symmetry and the high defect concentration in an amorphous material result in a high density of localized states within the bandgap. In such cases, it is more probable for distributed states rather than monoenergetic traps to exist within the bandgap. According to the SCLC theory, the specific energy distribution of these traps can be determined from the experimental I–V curves, as proposed in [71].

A detailed analysis of the I–V characteristics has revealed that the traps exhibit an exponential distribution within the bandgap of the as-grown amorphous films. This distribution can be described by the following expression [72]:

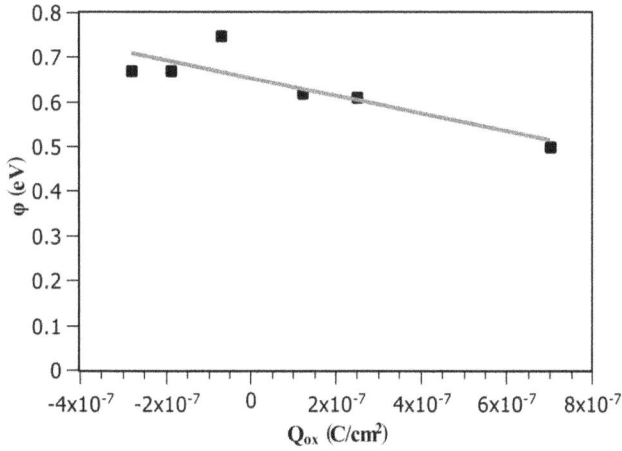

Figure 5.31. Correlation between the barrier height, φ, and the oxide charge, Q_{ox}, in LN–Si heterostructures grown in an Ar(80%) + O$_2$(20%) gas mixture. Reproduced from [64]. CC BY 4.0.

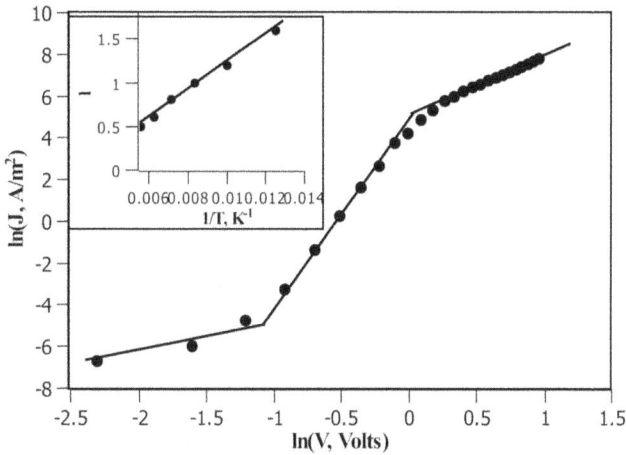

Figure 5.32. Typical I–V characteristic in $\ln(J)$ vs. $\ln(V)$ coordinates of the studied as-grown samples recorded at room temperature. The inset shows $I(T)$ as a function of temperature (see further comments in the text). Reproduced from [64]. CC BY 4.0.

$$N_t(E) = \frac{N_t}{kT_c} \exp\left(-\frac{qE}{kT_c}\right). \tag{5.17}$$

Here, k denotes Boltzmann's constant, q represents the elementary charge, T_c signifies a characteristic temperature greater than the measurement temperature, and E stands for the energy measured from the bottom of the conduction band. Consequently, the voltage dependence of the SCLC can be expressed as [20]:

$$I \propto V^{l+1} \tag{5.18}$$

where the exponent $(l = T_c/T)$ is a temperature-dependent parameter. It has been observed in our experiments (see inset in figure 5.32) that l must be a linear function of $1/T$. By deriving l from the slope of the experimental I–V characteristics, we have obtained the energy distribution of traps $N_t(E)$ within the bandgap of LN films. The evolution of $N_t(E)$ under the annealing process for the studied samples is depicted in figure 5.33.

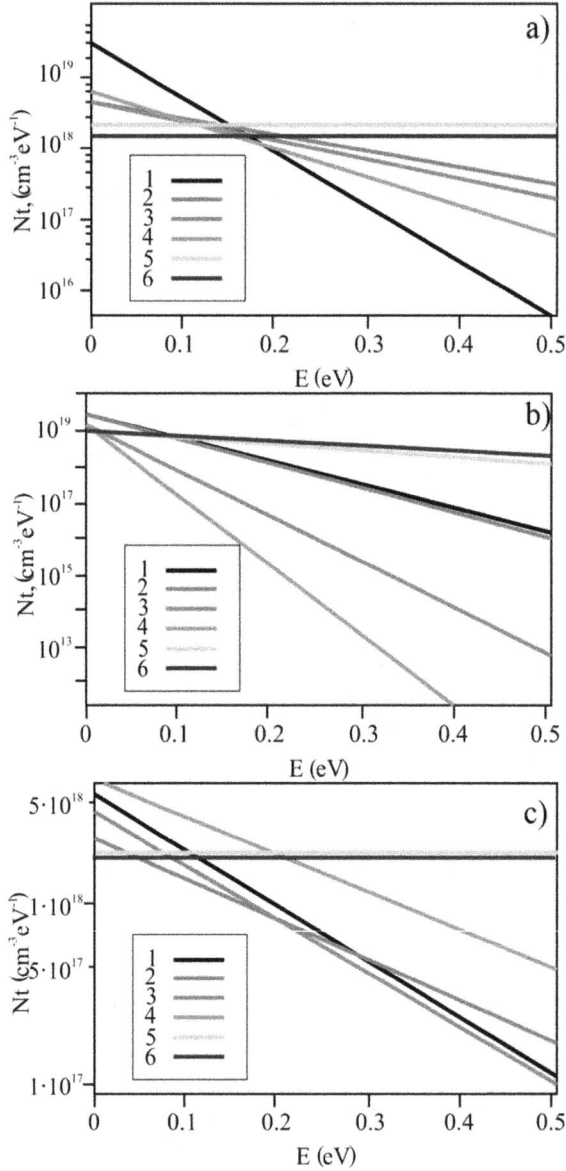

Figure 5.33. Trap distribution in the studied samples (a—sample S1, b—sample S2, c—sample S3), annealed at various temperatures T_A (1—300 °C, 2—400 °C, 3—450 °C, 4—500 °C, 5—550 °C, 6—600 °C). Reproduced from [64]. CC BY 4.0.

Figure 5.33 illustrates the comparative 'smoothness' of the exponential distribution in the as-grown films produced from an Ar(60%) + O$_2$(40%) gas mixture in contrast to others. TA induces changes in the shape of the exponential distribution across all samples during the crystallization process. Notably, recrystallization at annealing temperatures exceeding 500 °C changes the trap distribution to a uniform pattern. Particularly noteworthy is the significant reduction in trap density observed in LN films sputtered in an Ar(80%) + O$_2$(20%) gas mixture and annealed at 500 °C.

This section has been reproduced from [64].

5.5 Charge phenomena in NiSi$_2$–LiNbO$_3$ heterostructures

As demonstrated earlier, the structural, optical, and electrical properties of LN-based heterostructures are greatly influenced by the sputtering conditions [3, 73–75]. Incorporating various buffer layers is recognized as one of the most efficient methods of modifying the interface. This approach enhances the desired characteristics of LN-based heterostructures on silicon while also preventing interfacial reactions [76–78].

The increasing degree of integration and emerging commutation issues in microelectronics have driven the widespread application of transition-metal silicides for diverse purposes. Among these, nickel silicide (NiSi$_2$) stands out for its compatibility with existing MOS technology, alongside its high thermal and chemical stability, resistance to electromigration, and superior electrical conductivity. Extensive studies have delved into the kinetics, synthesis mechanisms, phase formation, and electrical and mechanical properties of NiSi$_2$ through conventional methods reliant on thermal treatment of nickel on silicon [79].

The advent of ferroelectric and ferromagnetic thin film heterostructures for applications in spintronics, optoelectronics, and nanoelectronics poses new technological challenges. The envisioned creation of multiferroic-based heterostructures on a single substrate for signal processing purposes holds significant practical interest.

The synthesis of epitaxial and defect-free LN films faces challenges, particularly in the formation of high-angle boundaries and multi-orientation nucleation during the growth of thick films ($d > 1\,\mu$m). Our previous research has shown that c-oriented LN films produced on Si substrates by the RFMS technique do not exhibit a single orientation throughout their thickness [63, 65]. Additionally, we have observed the potential growth of epitaxial LN films on (111)Ag, with the orientation relation (0001)LN||(111)Ag, inducing the formation of a mosaic substructure of LN [65].

To address these challenges, we have proposed an approach where an intermediate epitaxial NiSi$_2$ layer forms on the surface of (111)Si, effectively suppressing the formation of high-angle boundaries. This strategy promotes the singly oriented nucleation and growth of LN films due to the 60° symmetry and the low lattice mismatch between NiSi$_2$ and LN. Furthermore, the NiSi$_2$ layer acts as a barrier, preventing the diffusion of Si atoms into LN films, thereby preserving their dielectric properties.

Employing NiSi$_2$ as an intermediate (buffer) layer yields several benefits: (1) the inhibition of Si atom diffusion from the substrate, resulting in LN ferroelectric films with minimal conductivity; (2) the promotion of biaxial texture formation in defect-free epitaxial LN films; (3) the integration of epitaxial LN films into existing MOS technologies, utilizing the LN–Ni–NiSi$_2$–Si heterosystem as a functional element for spintronics, optoelectronics, and nanoelectronics.

While TA remains a conventional post-growth treatment method, enhancing the structural, electrical, and other properties of as-grown LN films [5, 7, 80], alternative competitive and efficient techniques, such as PPT, have been proposed. PPT stands out as an ultrafast, energy-saving technique, reducing thermal processing times and thus cutting production time significantly. The primary advantage of this method lies in its ability to generate localized heating through pulses of light, each lasting less than 1 ms. This enables the processing of thin films in ambient atmospheres without the need for vacuum or inert environments. PPT initiates recrystallization processes, leading to the formation of a nanocrystalline structure in the subsurface layers of semiconductors [81–83].

However, to the best of our knowledge, despite the promising prospects for contemporary optoelectronics, the study of LN–NiSi$_2$ heterostructures has not yet been undertaken. The present section endeavors to discuss the results for the synthesis of LN–NiSi$_2$ heterostructures with the requisite electrical properties for their successful application in optoelectronics and nonvolatile memory units.

Nickel silicide layers, approximately 0.4 μm thick, were fabricated through the thermal evaporation and condensation of Ni onto (111)Si wafers heated to 550 °C, as detailed in [79]. Thin LN films were deposited onto Si–NiSi$_2$ substrates using the RFMS method, both in a pure Ar atmosphere and in an Ar + O$_2$ reactive gas environment, using the parameters outlined in table 5.5. The substrates were positioned above the target erosion zone to facilitate the directional growth of LN films under the influence of reactive plasma, as demonstrated in our previous studies [50, 65].

Table 5.5. RFMS synthesis regimes used to produce LN–NiSi$_2$ heterostructures.

Sample#	Reactive gas environment	Magnetron power (W)	Substrate temperature, T_{sub} (°C)	Substrate–target distance (cm)	LN film thickness (nm)	PPT
S1	Ar	100	550	5	300	—
S2	Ar(90%) + O$_2$(10%)				440	—
S3	Ar(90%) + O$_2$(10%)				1000	+
S4	Ar(60%) + O$_2$(40%)				650	—
S5	Ar(60%)+O$_2$(40%)				650	+

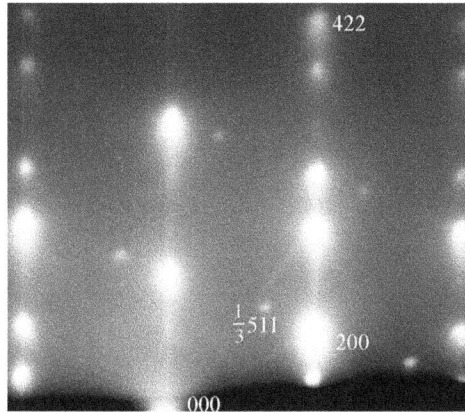

Figure 5.34. Reflection RHEED pattern of a NiSi$_2$ layer with a thickness of 100 nm grown on (111)Si wafers. Reprinted from [84], Copyright (2020), with permission from Elsevier.

The thickness of the films was determined using scanning electron microscopy (SEM, Tescan MIRRA 3) by examining cross-sections of the samples. Subsequent PPT was conducted by irradiating the sample with powerful radiation (spectral range: $\lambda = 0.2$–$1.2\,\mu$m) in air, delivered in pulses lasting 10^{-2} s each, for a total duration of 1.0 s (equivalent to a supplied radiation energy of $E_I \sim 120\,\mathrm{J\,cm}^{-2}$). The structure of the synthesized coatings was analyzed using the reflection high-energy electron diffraction (RHEED) method (EG-100M, acceleration voltage of 100 kV).

The electrical properties of the as-grown heterostructures and those subjected to PPT were evaluated through I–V characteristics and Hall measurements (ECOPIA HMS-5500) within a temperature range of $T = 80$–300 K.

Figure 5.34 shows the RHEED pattern of a NiSi$_2$ layer, 100 nm in thickness, cultivated on (111)Si wafers. This layer serves as a substrate for subsequent LN film deposition [84]. As evidenced by figure 5.34, the NiSi$_2$ layer exhibits a crystallographic orientation parallel to the lattice of the (111)Si substrate. The resulting film presents a granular structure characterized by multiple-twinning NiSi$_2$ crystallites. Notably, a distinct diffraction contrast, evidenced by streaks in the crystallographic directions of $\langle 022 \rangle$, $\langle 422 \rangle$, and $\langle 422 \rangle$, as well as superstructural reflexes 1/3 and 2/3 (511), signifies that the multiple twinning yields a highly dispersed NiSi$_2$ crystal structure.

We have previously reported the structural properties of LN films grown under sputtering conditions similar to those used for samples S1 and S4 (see table 5.5) [7, 8]. Here, our focus shifts to the structure of the films fabricated in an Ar(90%) + O$_2$(10%) gas mixture. We deposited Li–Nb–O films onto unheated Si–NiSi$_2$ substrates, ensuring the formation of amorphous films on silicon substrates [17]. This was followed by TA and PPT to investigate the film crystallization under these conditions. Figure 5.35 illustrates the RHEED patterns of as-grown Li–Nb–O films, each with a thickness of 300 nm, grown on NiSi$_2$, and the films after TA and PPT.

As depicted in figure 5.35(a), the as-grown films display an amorphous–nano-crystalline structure consistent with the stoichiometry of LN, aligning with our

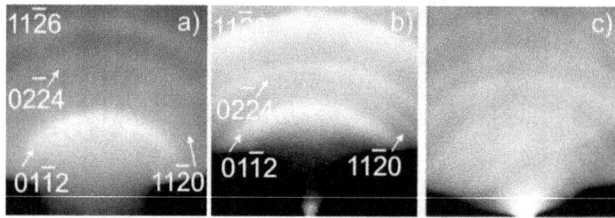

Figure 5.35. RHEED patterns of as-grown Li–Nb–O films with a thickness of 300 nm grown on NiSi$_2$ in an Ar (90%) + O$_2$(10%) gas mixture (a) and the films after TA (b) and PPT (c). Reprinted from [84], Copyright (2020), with permission from Elsevier.

recent findings for Li–Nb–O films crystallized on silicon [17]. Figure 5.35(b) indicates that TA induces the crystallization of as-grown films into polycrystalline LN structures without a preferred orientation dictated by the Si–NiSi$_2$ substrate.

In contrast, the RHEED patterns of films after PPT (figure 5.35(c)) reveal halo-like rings corresponding to the same Bragg angles as the diffraction maxima attributed to the LN crystal lattice. Consequently, the films after PPT exhibit an amorphous–nanocrystalline structure, with LN nanocrystals embedded in an amorphous matrix.

Numerous authors have presented diverse findings regarding the conductivity type of NiSi$_2$, which ranges from p-type to mixed conduction. Notably, the significant outcomes of additional analyses suggest that the metallic nature of NiSi$_2$ does not primarily stem from Ni electrons (as seen in metallic Ni) but rather from s- and p-like free electrons, rendering it similar to a compound noble metal. This proposition implies that NiSi$_2$ could be approximated as a metallic, simple cubic Si phase stabilized by Ni atoms centered in every other cube. Various reports by researchers indicate that NiSi$_2$ can exhibit either p- or n-type conductivity. Hall measurements performed on our samples have indicated that the synthesized NiSi$_2$ layers demonstrate n-type conductivity, with a carrier concentration of $n = 6.1 \times 10^{20}$ cm^{-3}, a value consistent with previous reports. Furthermore, the synthesized NiSi$_2$ exhibits metal-like conductivity, with resistivity dependent on temperature, as shown in figure 5.36 [84].

The metallic temperature dependence of resistivity, as depicted in figure 5.36, can be approximated by the following well-known expression:

$$\rho = \rho_r(1 + \alpha(t - t_r)). \tag{5.19}$$

Here, ρ_r represents the room temperature resistivity, α denotes the temperature coefficient of resistivity (TRC), and t_r is the room temperature. Fitting the experimental results to equation (5.19) yields the following values: $\alpha = 2.48 \times 10^{-3}$ °C^{-1} and $\rho_r = 2.26 \times 10^{-4}\,\Omega \cdot$ cm, which is eight times higher than that reported by other researchers [85]. Consequently, we regard the synthesized NiSi$_2$ layers as degenerate semiconductors with n-type conductivity. Hall measurements of the as-grown LN films (sample S2) revealed that electrons at a concentration of 1.0×10^{16} cm^{-3} at room temperature are the major carriers in the studied films.

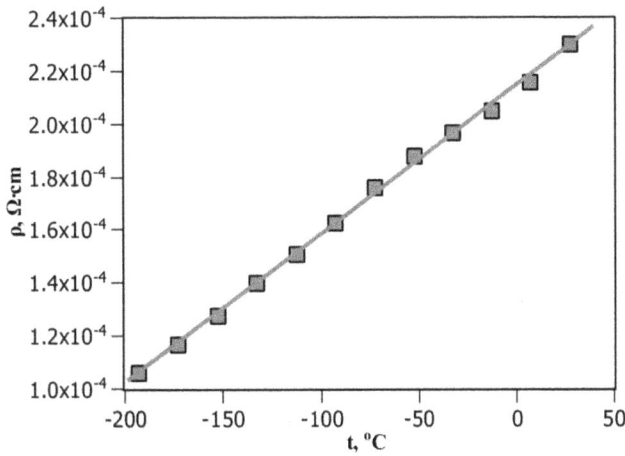

Figure 5.36. Resistivity of synthesized NiSi$_2$ films as a function of temperature.

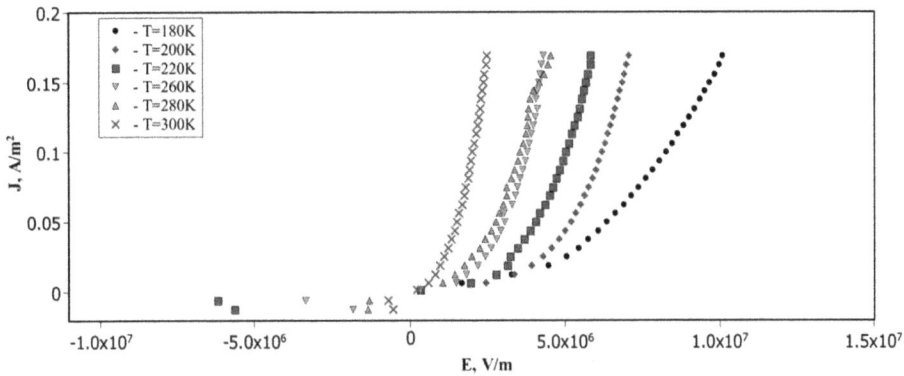

Figure 5.37. I–V characteristics of sample S1 at various temperatures.

The I–V characteristics of sample S1, measured at various temperatures, are illustrated in figure 5.37 [84].

As illustrated in figure 5.37, the I–V characteristics exhibit a rectifier nature. For further analysis, it would be beneficial to utilize a band diagram of the studied Al–LN–NiSi$_2$ heterostructures. All the fundamental parameters necessary to construct an idealized band diagram are provided in table 5.6.

Figure 5.38 illustrates a simplified band diagram of the investigated heterostructures [84]. Initially, before the materials are brought into contact, the band diagram is as depicted in figure 5.38(a). In this scenario, we assume that polycrystalline LN films harbor deep donor centers and traps, characterized by the concentrations N_d and N_t, respectively. Recent studies have shown that LN films fabricated by the RFMS technique in an Ar atmosphere possess a positive oxide charge, with a centroid situated approximately at the center of LN films [37, 90]. Therefore, the band diagram of the examined heterostructures, featuring LN films grown in an Ar

Table 5.6. The basic parameters of materials in Al–LN–NiSi$_2$ heterostructures.

	NiSi$_2$	LN	Al
Electron affinity, χ work function, φ_m (eV)	4.75 [86]	1.5 [87]	4.1 [88]
Bandgap energy, E_g (eV)	0.99 [89]	4.2 [73]	
Carrier concentration (cm^{-3})	6.1×10^{20}	1.0×10^{16}	
Type of conductivity	n-type	n-type	

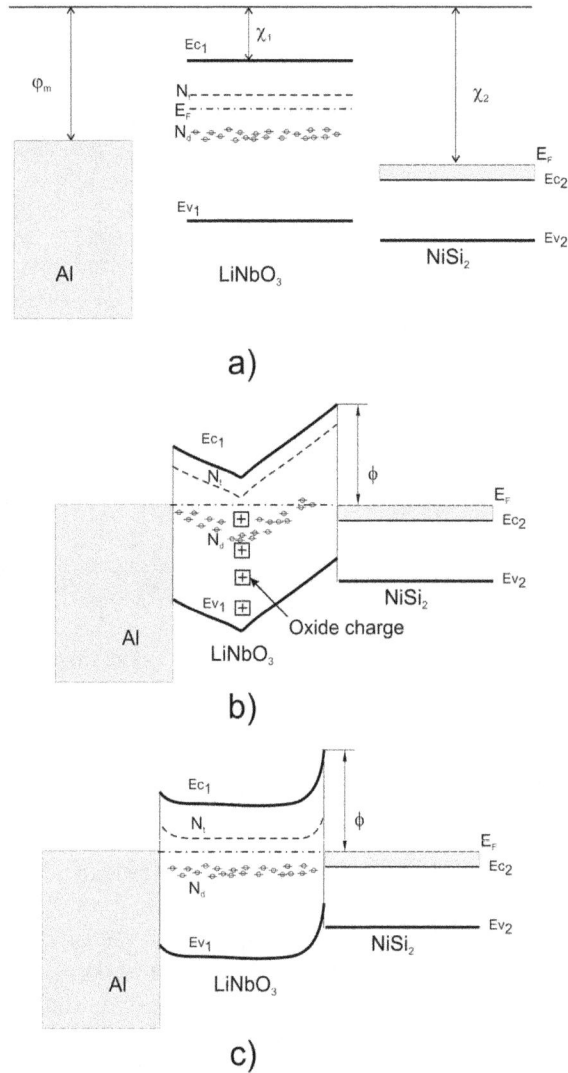

Figure 5.38. Band diagram of Al–LN–NiSi$_2$ heterostructures: (a) before all materials are brought into contact, (b) as-grown heterostructures based on LN films that have a positive oxide charge and are fabricated in an Ar atmosphere, (c) as-grown heterostructures based on LN films fabricated in an Ar + O$_2$ gas mixture. Reprinted from [84], Copyright (2020), with permission from Elsevier.

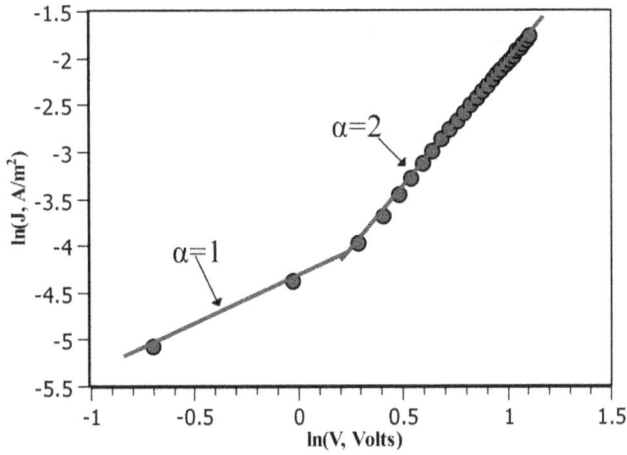

Figure 5.39. *I–V* characteristics (under forward bias) of sample S1 in double logarithmic coordinates.

atmosphere and exhibiting a positive oxide charge (referred to as sample S1), is presented in figure 5.38(b).

In figure 5.38(b), the presence of an oxide charge renders the band diagram highly asymmetrical, resulting in different barriers (φ) for electrons on the different sides of the LN films. This asymmetry manifests as a rectifying character in the *I–V* characteristics, as depicted in figure 5.37. Analysis indicates that under forward bias ('+' at the Al contact), the *I–V* characteristics are linear in double logarithmic coordinates (see figure 5.39), which is typical for SCLC; they follow a power law $I \sim V^\alpha$ with a quadratic section corresponding to $\alpha = 2$ [71]. According to SCLC theory, this section is attributed to the presence of monoenergetic shallow traps, with a concentration of $N_t = 2.0 \times 10^{15}\,\text{cm}^{-3}$ estimated from the experimental *I–V* curve, as described in [71]. The slope of the *I–V* characteristics increases with temperature, suggesting that traps are not monoenergetic but are rather distributed over different energies. This distribution can be inferred from the experimental *I–V* characteristic using the following expression [71]:

$$N(E) = \frac{CV}{qkT}\left(\frac{V}{I}\frac{\mathrm{d}I}{\mathrm{d}V} - 1\right)^{-1}. \tag{5.20}$$

Here, $N(E)$ represents the number of traps per unit energy range, C is the capacitance of the sample, k is the Boltzmann constant, T is the temperature, and q is the elementary charge. Differential analysis of the experimental *I–V* curves reveals a uniform energy distribution of traps with a density of $N(E) = 3.0 \times 10^{16}\,\text{cm}^{-3}\,\text{eV}^{-1}$ [84].

Figure 5.40 illustrates the *I–V* characteristics of Al–LN–NiSi$_2$ heterostructures based on as-grown LN films fabricated in an Ar(90%) + O$_2$(10%) gas mixture (sample S2).

As can be seen in figure 5.40, the *I–V* characteristics of sample S2 exhibit greater symmetry relative to the polarity of the applied voltage compared to those of sample

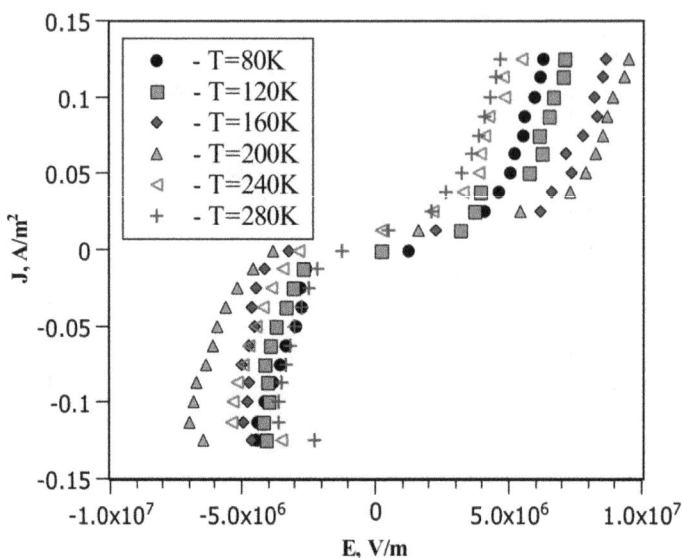

Figure 5.40. *I–V* characteristics of sample S2 measured at various temperatures.

S1, indicating a change in the band diagram. Previous research indicates that the positive oxide charge decreases with the addition of oxygen to the reaction chamber [37, 90]. Consequently, the band diagram of Al–LN–NiSi$_2$ heterostructures fabricated in an Ar(90%) + O$_2$(10%) gas mixture (sample S2) can be represented as shown in figure 5.38(c), resulting in more symmetrical *I–V* curves.

Analysis reveals that charge transport in the examined sample follows the SCLC theory due to the presence of traps uniformly distributed in the bandgap of LN, with a density of $N(E) = 4.0 \times 10^{17}\,\text{cm}^{-3}\text{eV}^{-1}$ estimated from the experimental *I–V* characteristics described above using equation (5.20). The resistivity of LN films in both S1 and S2 samples, estimated from the ohmic sections of their *I–V* characteristics under low electric fields, was $\rho = 8.0 \times 10^9\,\Omega \cdot \text{cm}$, consistent with our previous results for LN films grown in a pure Ar environment [63].

Figure 5.41 illustrates the *I–V* characteristics of Al–LN–NiSi$_2$ heterostructures based on LN films. These films were grown in an Ar(90%) + O$_2$(10%) atmosphere and subjected to PPT treatment (sample 3).

It is evident from figure 5.41 that LN films after PPT display ohmic-type *I–V* characteristics, with the resistivity showing a metallic-like temperature dependence. This behavior adheres to equation (5.19) for the values $\alpha = 1.6 \times 10^{-3}\,°\text{C}^{-1}$ and $\rho_r = 4.2 \times 10^3\,\Omega \cdot \text{cm}$.

Figure 5.42 illustrates the *I–V* characteristics of Al–LN–NiSi$_2$ heterostructures with as-grown LN films fabricated in an Ar(60%) + O$_2$(40%) gas mixture (sample S4).

Under forward bias (with '+' applied to the Al contact) within the range of low temperatures and high electric fields ($E > 10^7\,\text{V}\,\text{m}^{-1}$), the *I–V* characteristics of

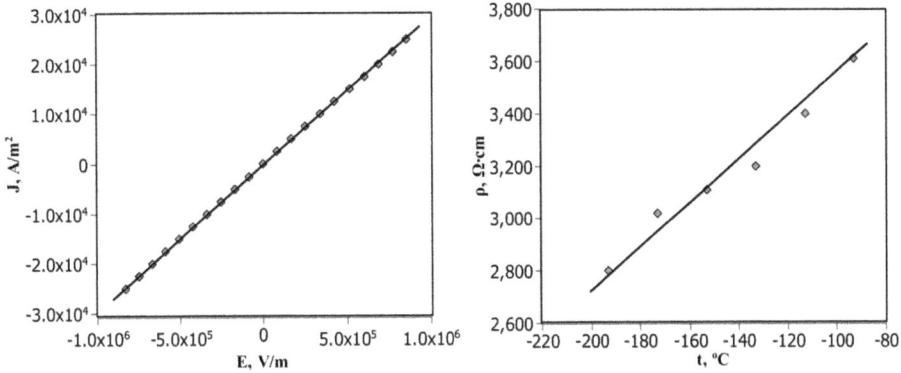

Figure 5.41. (a) $I\text{--}V$ characteristics and (b) resistivity as a function of temperature for sample S3.

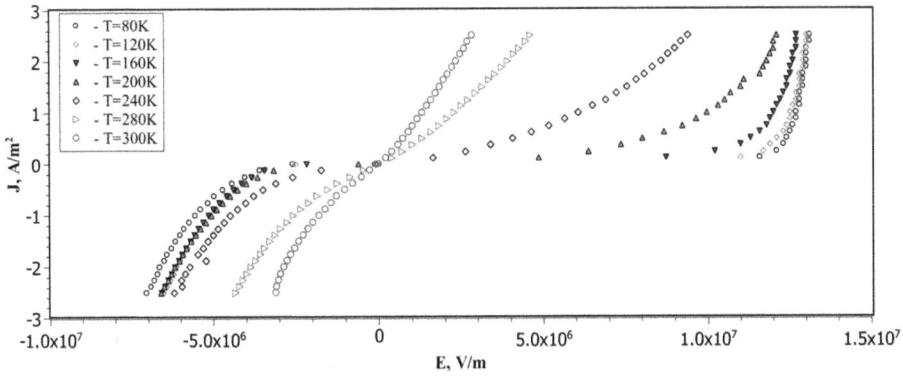

Figure 5.42. $I\text{--}V$ characteristics of sample S4 at various temperatures.

sample S4 were linear in $\ln J\text{--}V$ coordinates (see figure 5.43(a)) and can be described by the following expression:

$$J = J_s \exp\left(\frac{V}{V_o}\right). \tag{5.21}$$

Here, J_s represents the saturation current density, V denotes the applied voltage, and V_0 is a temperature-dependent parameter.

Following a classification proposed in [25], the temperature dependence of V_0 depicted in figure 5.43(b) is attributed to the thermionic-field emission charge transport mechanism. Within this framework, V_0 becomes temperature independent at low temperatures, approaching a horizontal section with an intercept corresponding to V_{00}, as seen in figure 5.43(b), and described by the following formula [91]:

$$V_{00} = \frac{h}{4\pi} \sqrt{\frac{N_d}{m^* \varepsilon \varepsilon_0}}. \tag{5.22}$$

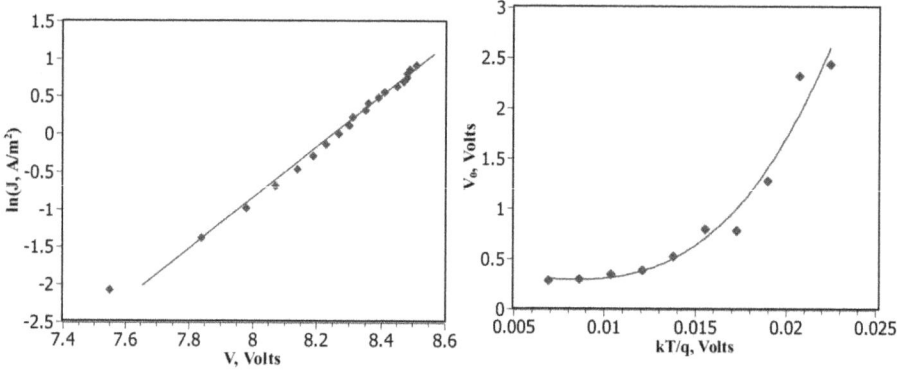

Figure 5.43. Forward I–V characteristics at $T = 80$ K (a) and a plot of V_0 versus kT/q in equation (5.21) for sample S4.

Here, N_d represents the donor concentration, h stands for Planck's constant, m^* denotes the effective mass of carriers, ε signifies the dielectric constant of the material, and ε_0 represents the electric constant. The donor concentration, estimated experimentally from figure 5.43(b) using equation (5.22), is $N_d = 7 \times 10^{15}$ cm^{-3} [84]. As proposed in [91], the potential barrier height at the LN–NiSi$_2$ interface was estimated from the slope of a linear graph $\ln(J_s \cosh(q V_{00}/kT)/T)$ versus $1/V_0$ (not shown), yielding a value of $\varphi = 2.4$ eV. Primarily, disordered and polycrystalline materials exhibit traps within the bandgap, participating in the charge transport process. Consequently, tunneling most likely occurs over these traps rather than to the conduction band in LN. If a trap continuum is situated in the bandgap of LN with an energy E_t below its conduction band, it can be estimated as $E_t = \chi_2 - \chi_1 - \varphi = 0.85$ eV (see figure 5.38), which closely aligns with values obtained in our previous works for Si–LN heterostructures [63], attributed by some authors to lithium vacancies [92].

At higher temperatures ($T = 240$–300 K), the I–V characteristics of sample S4 exhibited symmetry relative to the electric field polarity, as illustrated in figure 5.42. One possible charge transport mechanism accounting for this phenomenon is Poole–Frenkel emission. For disordered materials such as amorphous substances or materials with defects, the I–V characteristics can be elucidated within the framework of Poole–Frenkel emission, which is described by equation [26]:

$$J = AT^2E \exp\left(\frac{E_a}{kT}\right)\sinh\left(\frac{\beta E^{1/2}}{kT}\right).$$

(5.23)

Here, E represents the applied field, A is a constant, T denotes temperature, E_a signifies the average potential barrier height between defect centers, and β stands for the Poole–Frenkel coefficient, expressed by equation (1.16) (see chapter 1).

In equation (5.23), the I–V characteristics are expected to exhibit linearity in $\ln(J/ET^2)$–$E^{1/2}$ coordinates, with a slope proportional to β, as observed in our experiments (refer to figure 5.44). The effective dielectric constant ε of LN films, estimated

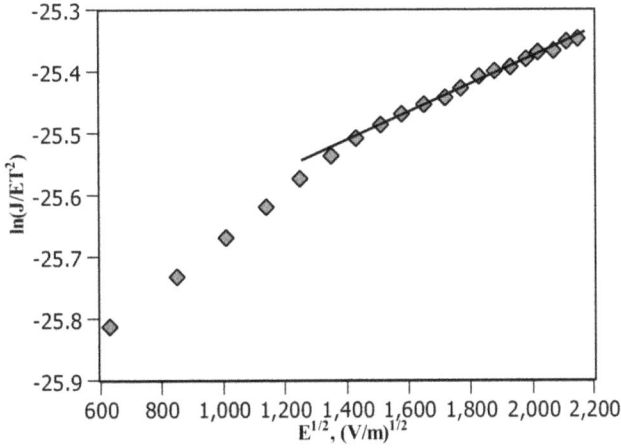

Figure 5.44. Forward $I-V$ characteristics of sample S4 at room temperature in Poole–Frenkel coordinates.

from the experimentally determined Poole–Frenkel coefficient, is 89, closely resembling that of bulk LN along the nonpolar axis ($\varepsilon = 83$) [63].

When the Poole–Frenkel effect prevails, the trap concentration can be estimated using the following expression [26]:

$$N_t = \left(\frac{q\sqrt{E_k}}{\beta}\right)^3. \tag{5.24}$$

Here, q represents the elementary charge, E_k signifies the electric field corresponding to the beginning of a linear section of the $\ln(J/ET^2)$–$E^{1/2}$ graph, and β is determined by equation (1.16) (see chapter 1). The resistivity and trap concentration associated with LN films, derived from the $I-V$ characteristics of sample S4 in Poole–Frenkel coordinates (refer to figure 5.44), are $\rho = 2.5 \times 10^9\,\Omega \cdot \text{cm}$ and $N_t = 3 \times 10^{18}\,\text{cm}^{-3}$, respectively. A relatively weak temperature dependence of the reverse $I-V$ characteristics, observed under high electric fields (refer to figure 5.42, when for negative biases $E > 5 \times 10^6\,\text{V}\,\text{m}^{-1}$), can be attributed to non-activated hopping conductivity over the defect centers distributed near the Fermi level in the bandgap of LN films [93]. The defect center concentration in sample S4, estimated from the experimental $I-V$ characteristics as described in [93], is $N_t = 7 \times 10^{18}\,\text{cm}^{-3}$, which is close to that determined above from the forward branches.

The PPT of LN–NiSi$_2$ heterostructures fabricated in an Ar(60%) + O$_2$(40%) gas mixture transforms their $I-V$ characteristics (not shown) into ohmic-type ones as presented in figure 5.41. These heterostructures have a metallic-like temperature dependence, approximated by equation (5.19) with $\alpha = 1.0 \times 10^{-3}\,°\text{C}^{-1}$ and $\rho_r = 4.9 \times 10^3\,\Omega \cdot \text{cm}$. It is noteworthy that the obtained parameters closely resemble those of LN–NiSi$_2$ heterostructures fabricated in an Ar(90%) + O$_2$(10%) reactive gas mixture and subjected to PPT (see sample S3). Thus, despite the obvious advantages associated with the PPT of LN films in terms of their structural properties, this post-growth treatment technique adversely affects the electrical properties of LN–NiSi$_2$ heterostructures, significantly reducing their resistivity. The resulting metallic-like conductivity of LN films induced by PPT constitutes a limiting factor for many potential applications of LN-based heterostructures.

This section has been reproduced with permission from [84].

5.6 Summary and discussion

1. Oxygen vacancies play a crucial role in the generation of positive charges within these films. The amorphous as-grown Li–Nb–O films harbor donor centers with a concentration of approximately $N_d = 1 \times 10^{14} \, \mathrm{cm}^{-3}$, uniformly distributed throughout the film. TA induces a concentration gradient from the surface towards the film–substrate interface, increasing the number of donor centers compared to that of the as-grown films. The concentration of grain boundaries experiences an increase in nanocrystalline films with a crystalline phase fraction of up to 45%, diminishing as a large-block structure forms.

2. The trap concentration in the films varies with annealing temperature, reaching a minimum at $T = 600 \, ^\circ\mathrm{C}$, which is attributed to the formation of the large-block structure. Charge transport in the investigated heterostructures is dictated by the barrier properties of the LN–Si heterojunction under low applied electric fields ($E < 2.5 \times 10^6 \, \mathrm{V\,m^{-1}}$). Conversely, under strong electric fields ($E > 2.5 \times 10^6 \, \mathrm{V\,m^{-1}}$), charge transport is influenced by the bulk properties of the LN films, involving Poole–Frenkel emission and SCLC.

3. The concentration of traps distributed within the films is dependent on the annealing temperature and attains its maximum value at $T_a = 450 \, ^\circ\mathrm{C}$, aligning with the formation of the amorphous–crystalline phase with a crystalline phase fraction of 45%. The evolution of electric properties is steered by structural transformations in the crystallization process, aiming to minimize the effective positive oxide charge in LN films annealed at $T = 450 \, ^\circ\mathrm{C}$.

4. All as-grown films exhibit a net effective oxide charge comprising two components: a negative $-Q_{\mathrm{eff}}$, predominant in 'thin' ($d < 200 \, \mathrm{nm}$) Li–Nb–O films and a positive $+Q_{\mathrm{eff}}$, prevailing in 'thick' ($d > 200 \, \mathrm{nm}$) films. The $-Q_{\mathrm{eff}}$ is attributed to structural defects near the substrate–film interface, while the $+Q_{\mathrm{eff}}$ is determined by a deficit of Li and O (vacancies) within the bulk of Li–Nb–O films.

5. TA of as-grown films induces changes in the vacancy concentration due to Li$_2$O desorption, thereby altering $+Q_{\mathrm{eff}}$. At a temperature of 470 $^\circ\mathrm{C}$, the as-grown Li–Nb–O films crystallize, forming Si–LN heterostructures with minimal net oxide charge. Heterostructures grown in an Ar(60%) + O$_2$(40%) reactive gas mixture exhibit a lower $+Q_{\mathrm{eff}}$ compared to those fabricated in pure Ar. TA at a temperature of about 520 $^\circ\mathrm{C}$ results in the formation of the second phase, LiNb$_3$O$_8$, increasing $+Q_{\mathrm{eff}}$ and

compensating $-Q_{\text{eff}}$ entirely. This process can potentially be used for the fabrication of thin film Si–LN heterostructures free from the effective oxide charge.

6. Electrons are the major carriers in the as-grown quasi-amorphous Li–Nb–O films, and small-polaron hopping is the main charge transport mechanism. TA in air causes the crystallization of the studied films, increasing the carrier concentration from 6×10^{11} to $6 \times 10^{14}\,\text{cm}^{-3}$ due to the generation of Nb_{Li} antisite defects produced by the evaporative loss of Li_2O from the films.

7. Two concurrent transport processes contribute to the net Hall mobility: grain-boundary-limited mobility and bulk single-crystal-limited mobility, the latter of which prevails when the grains become larger upon recrystallization.

8. The presence of oxygen in the reaction chamber promotes the formation of $-Q_{\text{ox}}$, the highest value of which is observed in films fabricated in an Ar $(80\%) + \text{O}_2(20\%)$ gas mixture. Additionally, the barrier height φ at the LN–Si interface is inversely correlated with Q_{ox}, decreasing with positive Q_{ox} and increasing with negative Q_{ox}.

9. TA significantly reduces $-Q_{\text{ox}}$ in heterostructures fabricated in an Ar + O_2 gas mixture and decreases the φ value. The lowest barrier height is observed in LN–Si heterostructures fabricated under an Ar/O_2 ratio of 60/40. The most significant reduction in trap density occurs upon annealing these films at $T_a = 500\,°\text{C}$.

10. The Al–LN–NiSi$_2$ heterostructures fabricated in an Ar environment exhibit pronounced rectifying I–V characteristics attributed to a nonsymmetrical potential barrier structure at the Al–LN and LN–NiSi$_2$ interfaces. This structure arises from the presence of positive oxide charge in LN films and is influenced by electronic traps with a concentration of $2.0 \times 10^{15}\,\text{cm}^{-3}$ within the bandgap of LN. The trap concentration, which affects charge transport in the heterostructures, increases with the oxygen content in the reaction chamber.

As an efficient post-growth treatment technique, PPT enhances the structural properties of LN films, promoting their recrystallization while dramatically reducing their resistivity and transforming their conductivity to resemble that of a metal.

References

[1] Sumets M, Ievlev V, Kostyuchenko A and Kuz'mina V 2014 Influence sputtering conditions on electrical characteristics of Si-LiNbO$_3$ heterostructures formed by radio-frequency magnetron sputtering *Mol. Cryst. Liq. Cryst.* **603** 202–15

[2] Wang X, Ye Z, Li G and Zhao B 2007 Influence of substrate temperature on the growth and optical waveguide properties of oriented LiNbO$_3$ thin films *J. Cryst. Growth* **306** 62–7

[3] Simões A Z, Zaghete M A, Stojanovic B D, Gonzalez A H, Riccardi C S, Cantoni M and Varela J A 2004 Influence of oxygen atmosphere on crystallization and properties of LiNbO$_3$ thin films *J. Eur. Ceram. Soc.* **24** 1607–13

[4] Kiselev D A, Zhukov R N, Bykov A S, Voronova M I, Shcherbachev K D, Malinkovich M D and Parkhomenko Y N 2014 Effect of annealing on the structure and phase composition of thin electro-optical lithium niobate films *Inorg. Mater.* **50** 419–22

[5] Akazawa H and Shimada M 2007 Mechanism for LiNb3O$_8$ phase formation during thermal annealing of crystalline and amorphous LiNbO$_3$ thin films *J. Mater. Res.* **22** 1726–36

[6] Sumets M, Kostyuchenko A, Ievlev V and Dybov V 2016 Electrical properties of phase formation in LiNbO$_3$ films grown by radio-frequency magnetron sputtering method *J. Mater. Sci., Mater. Electron.* **27** 7979–86

[7] Sumets M, Kostyuchenko A, Ievlev V, Kannykin S and Dybov V 2015 Influence of thermal annealing on structural properties and oxide charge of LiNbO$_3$ films *J. Mater. Sci., Mater. Electron.* **26** 7853–9

[8] Sumets M, Ievlev V, Kostyuchenko A, Vakhtel V, Kannykin S and Kobzev A 2014 Electrical properties of Si-LiNbO$_3$ heterostructures grown by radio-frequency magnetron sputtering in an Ar + O$_2$ environment *Thin Solid Films* **552** 32–8

[9] Hwang C S and Mikolajick T 2019 Ferroelectric memories ed B Magyari-Köpe and Y Nishi *Advances in Non-Volatile Memory and Storage Technology* 2nd edn (Woodhead Publishing) pp 393–441

[10] Gudkov S I, Baklanova K D, Kamenshchikov M V, Solnyshkin A V and Belov A N 2018 Electrical conductivity and barrier properties of lithium niobate thin films *Phys. Solid State* **60** 743–6

[11] Ievlev V, Sumets M, Kostyuchenko A, Ovchinnikov O, Vakhtel V and Kannykin S 2013 Band diagram of the Si-LiNbO$_3$ heterostructures grown by radio-frequency magnetron sputtering *Thin Solid Films* **542** 289–94

[12] Graça MPF, Prezas P R, Costa M M and Valente M A 2012 Structural and dielectric characterization of LiNbO$_3$ nano-size powders obtained by Pechini method *J. Sol–Gel Sci. Technol.* **64** 78–85

[13] Fakhri M A, Hashim U, Salim E T and Salim Z T 2016 Preparation and charactrization of photonic LiNbO$_3$ generated from mixing of new raw materials using spry pyrolysis method *J. Mater. Sci., Mater. Electron.* **27** 13105–12

[14] Akazawa H and Shimada M 2006 Precipitation kinetics of LiNbO$_3$ and LiNb$_3$O$_8$ crystalline phases in thermally annealed amorphous LiNbO$_3$ thin films *Phys. Status Solidi* **203** 2823–7

[15] Huang S 1984 *Structure and Structure Analysis of Amorphous Materials* (Oxford: Clarendon)

[16] Sumets M, Ievlev V, Dybov V, Kostyuchenko A, Serikov D, Kannykin S, Kotov G and Belonogov E 2019 Electrical properties of amorphous films and crystallization of Li–Nb–O system on silicon *J. Mater. Sci., Mater. Electron.* **30** 15662–9

[17] Ievlev V M, Belonogov E K, Dybov V A, Kannykin S V, Serikov D V, Sitnikov A V and Sumets M P 2019 Synthesis of lithium niobate during crystallization of amorphous Li–Nb–O film *Inorg. Mater.* **55** 1237–41

[18] van Opdorp C and Kanerva HKJ 1967 Current–voltage characteristics and capacitance of isotype heterojunctions *Solid-State Electron.* **10** 401–21

[19] Böer K W 2010 *Introduction to Space Charge Effects in Semiconductors* (Berlin: Springer)

[20] Sze S M and Kwok K N 2006 *Physics of Semiconductor Devices* 3rd edn (New York: Wiley)

[21] Tabata K, Choso T and Nagasawa Y 1998 The topmost structure of annealed single crystal of LiNbO$_3$ *Surf. Sci.* **408** 137–45

[22] Kitabatake M, Mitsuyu T and Wasa K 1984 Structure and dielectric properties of amorphous LiNbO$_3$ thin films prepared by a sputtering deposition *J. Appl. Phys.* **56** 1780–4

[23] Kim S H, Yang Y S, Kim H J, Kang M S and Jang M S 1995 Thermal and dielectric properties of amorphous LiNbO$_3$ *Jpn. J. Appl. Phys.* **34** 4842–8

[24] Sharma B L and Purohit R K 1974 *Semiconductor Heterojunctions* (Oxford: Pergamon)

[25] Saxena A N 1969 Forward current–voltage characteristics of Schottky barriers on n-type silicon *Surf. Sci.* **13** 151–71

[26] Hill R M 1971 Poole–Frenkel conduction in amorphous solids *Philos. Mag.* **23** 59–86

[27] Joshi V, Roy D and Mecartney M L 1995 Nonlinear conduction in textured and non textured lithium niobate thin films *Integr. Ferroelectr.* **6** 321–7

[28] Haertling G H 1991 Ferroelectric thin films for electronic applications *J. Vac. Sci. Technol.* A **9** 414–20

[29] Bredikhin S, Scharner S, Klingler M, Kveder V, Red'kin B and Weppner W 2000 Nonstoichiometry and electrocoloration due to injection of Li+ and O$_2$− ions into lithium niobate crystals *J. Appl. Phys.* **88** 5687–94

[30] Sanna S, Hölscher R and Schmidt W G 2014 Temperature dependent LiNbO3(0 0 0 1): surface reconstruction and surface charge *Appl. Surf. Sci.* **301** 70–8

[31] Sweeney K L and Halliburton L E 1983 Oxygen vacancies in lithium niobate *Appl. Phys. Lett.* **43** 336–8

[32] Smyth D M 1983 Defects and transport in LiNbO$_3$ *Ferroelectrics* **50** 93–102

[33] Cao E, Zhang Y, Qin H, Zhang L and Hu J 2013 Vacancy-induced magnetism in ferroelectric LiNbO$_3$ and LiTaO$_3$ *Physica* B **410** 68–73

[34] Coll M *et al* 2019 Towards oxide electronics: a roadmap *Appl. Surf. Sci.* **482** 1–93

[35] Li H, Xia Y, Xu B, Guo H, Yin J and Liu Z 2010 Memristive behaviors of LiNbO$_3$ ferroelectric diodes *Appl. Phys. Lett.* **97** 012902

[36] Pan X *et al* 2016 Rectifying filamentary resistive switching in ion-exfoliated LiNbO$_3$ thin films *Appl. Phys. Lett.* **108** 032904

[37] Sumets M, Ievlev V, Belonogov E, Dybov V, Serikov D, Kotov G and Turygin A 2020 Oxide charge evolution under crystallization of amorphous Li–Nb–O films *J. Sci. Adv. Mater. Devices* **5** 256–62

[38] Sumets M, Ievlev V, Kostyuchenko A, Dybov V, Kotov G and Sidorkin A 2018 Charge phenomena at the Si/LiNbO$_3$ heterointerface after thermal annealing *Ceram. Int.* **44** 15058–64

[39] Simões A Z, González AHM, Ries A, Zaghete M A, Stojanovic B D and Varela J A 2003 Influence of thickness on crystallization and properties of LiNbO$_3$ thin films *Mater. Charact.* **50** 239–44

[40] Nagai K, Sekigawa T and Hayashi Y 1985 Capacitance–voltage characteristics of semi-conductor–insulator–semiconductor (SIS) structure *Solid-State Electron.* **28** 789–98

[41] Glass A M, Lines M E, Nassau K and Shiever J W 1977 Anomalous dielectric behavior and reversible pyroelectricity in roller-quenched LiNbO$_3$ and LiTaO$_3$ glass *Appl. Phys. Lett.* **31** 249–51

[42] Nassau K, Levinstein H J and Loiacono G M 1966 Ferroelectric lithium niobate. 2. Preparation of single domain crystals *J. Phys. Chem. Solids* **27** 989–96

[43] Vest R W and Wu RCR 1990 The electrical properties and epitaxial growth of LiNbO/sub 3/ films by the MOD process *Proc. 1990 IEEE 7th Int. Symp. Appl. Ferroelectr. (Piscataway, NJ)* (IEEE) 170–6

[44] Akazawa H and Shimada M 2004 Correlation between interfacial structure and c-axis-orientation of $LiNbO_3$ films grown on Si and SiO_2 by electron cyclotron resonance plasma sputtering *J. Cryst. Growth* **270** 560–7

[45] Ievlev V M, Sumets M P and Kostyuchenko A V 2012 Effect of thermal annealing on electrical properties of Si-$LiNbO_3$ *Mater. Sci. Forum* **700** 53–7

[46] Sumets M, Belonogov E, Dybov V, Serikov D, Kannykin S, Kostyuchenko A and Ievlev V 2021 Pulsed photon treatment effect on the optical bandgap of $LiNbO_3$ films grown by radio-frequency magnetron sputtering method *J. Mater. Sci., Mater. Electron.* **32** 4290–9

[47] Akazawa H 2014 Nucleation and crystallization of Li_2O–Nb_2O_5 ternary compound thin films co-sputtered from $LiNbO_3$ and Li_2O targets *Thin Solid Films* **556** 74–80

[48] Hongye S, Guoxiang L, Bing L, Shifen H and Huafu W 1990 Crystallization processes in amorphous LiNbO3 *Ferroelectrics* **101** 11–8

[49] Vakulov Z *et al* 2018 Size effects in LiNbO3 thin films fabricated by pulsed laser deposition *J. Phys. Conf. Ser.* **1124** 022032

[50] Sumets M, Ievlev V, Dybov V, Kostyuchenko A, Serikov D, Kannykin S and Belonogov E 2019 Synthesis and properties of multifunctional Si–$LiNbO_3$ heterostructures for non-volatile memory units *J. Mater. Sci., Mater. Electron.* **30** 16562–70

[51] Sumets M, Ievlev V, Dybov V, Serikov D, Belonogov E and Grebennikov A 2022 Transport properties and crystallization of Li–Nb–O system on silicon *Mater. Sci. Semicond. Process.* **142** 106519

[52] Orton J W and Powell M J 1980 The Hall effect in polycrystalline and powdered semiconductors *Rep. Prog. Phys.* **43** 1263

[53] Kireev P 1978 *Semiconductor Physics* 2nd edn (Moscow: Mir Publishers)

[54] Vakulov Z E, Zamburg E G, Golosov D A, Zavadskiy S M, Miakonkikh A V, Clemente I E, Rudenko K V, Dostanko A P and Ageev O A 2018 Effect of substrate temperature on the properties of $LiNbO_3$ nanocrystalline films during pulsed laser deposition *Bull. Russ. Acad. Sci. Phys.* 2017 8112 **81** 1476–80

[55] Wilkinson A P, Cheetham A K and Jarman R H 1993 The defect structure of congruently melting lithium niobate *J. Appl. Phys.* **74** 3080

[56] Dhar A and Mansingh A 1990 Polaronic hopping conduction in reduced lithium niobate single crystals *Philos. Mag.* B **61** 1033–42

[57] Dhar A, Singh N, Singh R K and Singh R 2013 Low temperature dc electrical conduction in reduced lithium niobate single crystals *J. Phys. Chem. Solids* **74** 146–51

[58] Shandilya S, Tomar M, Sreenivas K and Gupta V 2009 Purely hopping conduction in c-axis oriented $LiNbO_3$ thin films *J. Appl. Phys.* **105** 094105

[59] Nagels P 1980 Experimental hall effect data for a small-polaron semiconductor ed C R Chien and C L Westgate *The Hall Effect and Its Applications* (New York: Springer) pp 253–80

[60] Nakano H and Suyama Y 2010 *In-situ* TEM observation of crystallization process for $LiNbO_3$ and NaNbO3 *Adv. Sci. Technol.* **63** 47–51

[61] Hao L, Li Y, Zhu J, Wu Z, Deng J, Liu X and Zhang W 2013 Fabrication and electrical properties of $LiNbO_3$/ZnO/n-Si heterojunction *AIP Adv.* **3** 042106

[62] Li W, Cui J, Zheng D, Wang W, Wang S, Song S, Liu H, Kong Y and Xu J 2019 Fabrication and characteristics of heavily Fe-doped $LiNbO_3$/Si heterojunction *Materials* **12** 2659

[63] Sumets M 2018 *Lithium Niobate-Based Heterostructures: Synthesis, Properties and Electron Phenomena* (Bristol: IOP Publishing)

[64] Sumets M, Dybov V, Serikov D, Belonogov E, Seredin P, Goloshchapov D, Grebennikov A and Ievlev V 2020 Effect of reactive gas composition on properties of Si/LiNbO$_3$ hetero-junctions grown by radio-frequency magnetron sputtering method *J. Sci. Adv. Mater. Devices* **5** 512–9

[65] Sumets M, Kostyuchenko A, Ievlev V, Kannykin S and Dybov V 2015 Sputtering condition effect on structure and properties of LiNbO$_3$ films *J. Mater. Sci., Mater. Electron.* **26** 4250–6

[66] Bartasyte A, Plausinaitiene V, Abrutis A, Stanionyte S, Margueron S, Boulet P, Kobata T, Uesu Y and Gleize J 2013 Identification of LiNbO$_3$, LiNb$_3$O$_8$ and Li$_3$NbO$_4$ phases in thin films synthesized with different deposition techniques by means of XRD and Raman spectroscopy *J. Phys. Condens. Matter* **25**

[67] Gordillo-Vázquez F J and Afonso C N 2002 Influence of Ar and O$_2$ atmospheres on the Li atom concentration in the plasma produced by laser ablation of LiNbO$_3$ *J. Appl. Phys.* **92** 7651

[68] Gosele U and Tan T Y 1982 Oxygen diffusion and thermal donor formation in silicon *Appl. Phys. A: Solids Surf.* **28** 79–92

[69] Chen L, Xu Q, Wood M G and Reano R M 2014 Hybrid silicon and lithium niobate electro-optical ring modulator *Optica* **1** 112–8

[70] Simmons J G 1965 Richardson-Schottky effect in solids *Phys. Rev. Lett.* **15** 967–8

[71] Rose A 1955 Space-charge-limited currents in solids *Phys. Rev.* **97** 1538–44

[72] Lampert M A and Mark P 1970 *Current injection in solids* (New York: Academic) http://cds.cern.ch/record/112986 (accessed 13 April 2017)

[73] Sumets M, Ovchinnikov O, Ievlev V and Kostyuchenko A 2017 Optical band gap shift in thin LiNbO$_3$ films grown by radio-frequency magnetron sputtering *Ceram. Int.* **43** 13565–8

[74] Lee T-H, Hwang F-T, Lee C-T and Lee H-Y 2007 Investigation of LiNbO$_3$ thin films grown on Si substrate using magnetron sputter *Mater. Sci. Eng.* B **136** 92–5

[75] Akazawa H 2009 Target-quality dependent crystallinity of sputter-deposited LiNbO$_3$ films: observation of impurity segregation *Thin Solid Films* **517** 5786–92

[76] Hao L, Li Y, Zhu J, Wu Z, Wang J, Liu X and Zhang W 2014 Enhanced memory characteristics by interface modification of ferroelectric LiNbO$_3$3 films on Si using ZnO buffers *J. Alloys Compd.* **599** 108–13

[77] Shih W C and Sun X Y 2010 Preparation of C-axis textured LiNbO$_3$ thin films on SiO$_2$/Si substrates with a ZnO buffer layer by pulsed laser deposition process *Phys. B Condens. Matter.* **405** 1619–23

[78] He J and Ye Z 2003 Highly C-axis oriented LiNbO$_3$ thin film on amorphous SiO$_2$ buffer layer and its growth mechanism *Chin. Sci. Bull.* **48** 2290

[79] Ievlev V M, Shvedov E V, Soldatenko S A, Kushchev S B and Gorozhankin Y V 2006 Silicide formation during heat treatment of thin Ni–Pt and Ni–Pd solid-solution films and Pt/Ni bilayers on (111)Si *Inorg. Mater.* **42** 151–9

[80] Fakhri M A, Salim E T, Hashim U, Abdulwahhab A W and Salim Z T 2017 Annealing temperature effect on structural and morphological properties of nano photonic LiNbO$_3$ *J. Mater. Sci., Mater. Electron.* **28** 16728–35

[81] Bhosle V, Tiwari A and Narayan J 2006 Electrical properties of transparent and conducting Ga doped ZnO *J. Appl. Phys.* **100** 033713

[82] Khang Y and Kuk Y 1996 Multiple phase structures of on Si(001): an atomic view *Phys. Rev. B Condens. Matter Mater. Phys.* **53** 10775–80

[83] Belonogov E, Grebennikov A, Dybov V, Kannykin S, Kostyuchenko A, Kuschev S, Serikov D, Soldatenko S and Sumets M 2020 Photon treatment effect on the surface and figure of merit of thermoelectric material $Bi_2Te_{3-x}Se_x$ *Mater. Renew. Sustain. Energy* **9** 1–7

[84] Sumets M, Belonogov E, Ievlev V, Dybov V, Serikov D, Polovinkin A and Grebennikov A 2020 Synthesis and properties of $NiSi_2$–$LiNbO_3$ heterostructures fabricated by radio-frequency magnetron sputtering *Surf. Interfaces* **21** 100797

[85] Colgan E G, Mäenpää M, Finetti M and Nicolet M-A 1983 Electrical characteristics of thin Ni_2Si, $NiSi$, and $NiSi_2$ layers grown on silicon *J. Electron. Mater.* **12** 413–22

[86] Bucher E, Schulz S, Lux-Steiner M C, Munz P, Gubler U and Greuter F 1986 Work function and barrier heights of transition metal silicides *Appl. Phys.* A **40** 71–7

[87] Yang W-C, Rodriguez B J, Gruverman A and Nemanich R J 2004 Polarization-dependent electron affinity of $LiNbO_3$ surfaces *Appl. Phys. Lett.* **85** 2316–8

[88] Kiejna A and Wojciechowski K F 1981 Work function of metals: relation between theory and experiment *Prog. Surf. Sci.* **11** 293–338

[89] Lee I H, Lee J, Oh Y J, Kim S and Chang K J 2014 Computational search for direct band gap silicon crystals *Phys. Rev.* B **90**

[90] Sumets M, Belonogov E, Dybov V, Serikov D, Kostyuchenko A, Ievlev V and Kotov G 2019 Effective charge in $LiNbO_3$ films fabricated by radio-frequency magnetron sputtering method *Phys. Solid State* **61** 2367–70

[91] Padovani F A and Stratton R 1966 Field and thermionic-field emission in Schottky barriers *Solid-State Electron.* **9** 695–707

[92] Donnerberg H, Tomlinson S M, Catlow CRA and Schirmer O F 1989 Computer-simulation studies of intrinsic defects in $LiNbO_3$ crystals *Phys. Rev.* B **40** 11909–16

[93] Sumets M 2017 Charge transport in $LiNbO_3$-based heterostructures *J. Nonlinear Opt. Phys. Mater.* **26** 1750011

IOP Publishing

Lithium Niobate-Based Heterostructures (Second Edition)
Synthesis, properties, and electron phenomena
Maxim Sumets

Chapter 6

Bonus chapter: multifunctional Si–LiNbO$_3$ heterostructures for nonvolatile memory units

This bonus chapter of the monograph presents well-structured results on the potential applications of synthesized LiNbO$_3$ (LN)-based heterostructures as foundational elements for nonvolatile memory units. It proposes optimal radio-frequency magnetron sputtering (RFMS) conditions and subsequent thermal annealing (TA) for the fabrication of multifunctional Si–LN heterostructures. These heterostructures meet all the necessary parameters for successful application in field-effect transistors and nonvolatile memory units that utilize electro-optic properties.

Ferroelectric complex oxides find extensive applications in optoelectronic devices, memory units, and signal processing circuits due to their piezoelectric, electro-optic, acousto-optic, and pyroelectric properties. Moreover, promising multiferroic materials, which involve the coupling of light with magnetism and ferroelectricity, have been reported. The authors of [1] observed remarkable white-light-controlled ferromagnetism and ferroelectricity in multiferroic BiFeO$_3$ nanosheets at room temperature (300 K). The integration of thin ferroelectric films with existing silicon technology garnered significant interest in the 1980s.

LN stands out as a ferroelectric material due to its high Curie temperature, wide bandgap, and high electro-optic coefficients. LN exhibits higher speeds and data rates exceeding $40\,\mathrm{Gb\,s^{-1}}$ compared to other ferroelectrics [2]. Additionally, its environmentally friendly nature distinguishes it from other materials such as PbNbO$_3$ [3, 4].

Over the last few decades, numerous integrated devices have been developed and fabricated on various substrates such as Si [5, 6], SiO$_2$ [7], and Al$_2$O$_3$ [8]. These devices span a wide range, including the new generation of ferroelectric random-access memories (FRAMs), high-bandwidth electro-optic modulators, circuits with planar waveguides, multiplexers and demultiplexers, gigahertz-bandwidth devices, optical amplifiers, and frequency converters [9].

doi:10.1088/978-0-7503-6305-1ch6

Microelectronics predominantly rely on semiconductors, typically (001)Si wafers. However, the incompatibility between these wafers and most oxides presents challenges for the epitaxial growth of functional oxides on Si wafers. Ferroelectric oxides in polycrystalline form are already used in commercial memories and are poised to have a significant impact on new ferroelectric memories [10].

Multifunctional integrated heterostructures based on LN, such as Si–LN heterostructures, show immense promise due to a combination of effects such as optical amplification and switching, nonvolatile data storage, and broadband (gigahertz) frequency transformation [11]. Synthesizing Si–LN heterostructures with properties akin to those of bulk LN is a challenging technological endeavor. Specifically, it is critical to ensure that the ferroelectric polar axis aligns with the out-of-plane direction to achieve effective switching in nonvolatile memory devices. However, the thermal expansion mismatch between the materials in complex oxides on (001)Si tends to orient the axis in-plane [10].

Another challenge arises from domain wall memory applications of ferroelectric films. The presence of charged domain walls, crucial for such memories, can attract ionic defects such as oxygen vacancies, impeding the clean erasure and regeneration of domain walls [12]. Addressing these challenges may involve growing textured or epitaxial LN films with low defect concentrations or forming amorphous films followed by controlled crystallization using optimal synthesis techniques suitable for LN films.

Table 6.1 summarizes the primary requirements for Si–LN heterostructures, their associated applications, limiting factors, and potential solutions.

Table 6.1. The main requirements for Si–LN heterostructures associated with their fields of applications, limiting factors, and possible solutions.

Potential application of Si–LN heterostructures	Nonvolatile memory units, ferroelectric field-effect transistors (FeFETs), optical resonators.			
Physical properties	Ferroelectric domain structure	Dielectric properties	Remnant polarization P_r and coercive field E_c	Surface roughness
Requirements	Single domain structure or preferable orientation $\langle 0001 \rangle$ LN (in the case of poly-domain films) for effective switching [13]	High dielectric constant ($\varepsilon = 200$–2000), low leakage current [12], high breakdown voltage ($E = 300$–10^3 kV cm^{-1} [14]), tuned-up conductivity type	High magnitude of P_r (20–$30\,\mu$C cm^{-2} [14]), low values of E_c driving voltage (flat band voltage, less than 5 V[14]), effective switching [10]	To reduce the optical losses (<1 dB cm^{-1} [9]) the roughness should be around 4 nm for LN films 400 nm thick [15]
Limiting factors	Different growth rates for grains with Z^- and Z^+	Sensitivity to composition and point defects (Li	Formation of internal fields [21, 22], charged defects [23],	Roughness greatly depends on the

	domain orientations [16], (affected by Li content [17], grain boundaries and defects [18])	and O) [19], a high concentration of charge localization centers at a surface [20]	mechanical strain; E_c depends on the film's thickness and defect concentration [24]	deposition regime [25]
Possible solutions	Deposition of single-phase, oriented or epitaxial LN films	Reduction of the charged center concentration in a film; ionic doping: Fe, Cu—n-type [26], Zr^{4+}, Hf^{4+} [21], N [27]—p-type	TA, careful selection of the synthesis regime [25, 28]	Careful selection of the reactive gas composition, TA

This chapter is dedicated to the synthesis of multifunctional Si–LN heterostructures with electro-optic properties suitable for the fabrication of nonvolatile memory units. To achieve the desired outcomes outlined in table 6.1, we address the following tasks:

1. Synthesis of polycrystalline single-phase textured LN films with ferroelectric properties comparable to those of bulk LN
2. Minimization of oxide charge, coercive field, and surface roughness of LN films in Si–LN heterostructures

Various processes, such as chemical vapor deposition (CVD) [29, 30], liquid-phase epitaxy (LPE) [31], pulsed laser deposition [32], the sol–gel method [33], and excimer laser ablation [34], have been employed for the deposition of high-quality LN films. However, not all of these methods are suitable for depositing thin LN films onto semiconductor substrates for integrated electronics.

In this study, similar to our previous research [35], films were fabricated using the RFMS method, which is one of the most efficient deposition techniques for complex oxides and preserves their elemental composition. The choice of RFMS is motivated by several advantages:

- Lower working pressure, providing optimal free paths for ions in the space charge area.
- The ability to regulate ion energy over a wide range through magnetic fields.
- Relatively high deposition rate.
- Sputtering coefficients that are independent of the material's melting temperature.
- Capability to deposit films over wide substrate areas.

One major challenge associated with all synthesis processes is the deviation from LN stoichiometry (deficits) related to lithium and oxygen. Addressing this issue requires careful selection of the reactive gas composition used in the reaction chamber. Previous studies [36, 37] have shown that the presence of oxygen as a reactive gas component increases the partial lithium pressure in the chamber. However, the use of pure oxygen as a reactive gas does not allow the formation of single-phase LN films [38]. Hence, significant efforts have been made to develop an optimal Ar/O_2 ratio in the reactive gas environment, considering other RFMS parameters and the required properties of the synthesized films. In this study, we investigate how the addition of oxygen to the reactive gas mixture affects the film properties.

TA in vacuum or gas environments is a popular technique for reducing defect concentration and lowering mechanical stress in synthesized films. Typically, pure oxygen or air is used as the atmosphere for TA [39]. It has been found that the presence of oxygen greatly influences the crystallization process of LN films [40, 41]. Oxygen helps prevent the evaporation of volatile compounds such as Li_2O, controls stoichiometry, and improves electrical parameters. Many researchers have reported the formation of amorphous LN films on unheated substrates [40, 42, 43], which transition to polycrystalline films after TA. The film's roughness exhibits a nonlinear dependence on the TA temperature. When oriented LN films undergo TA, the final result is determined by the annealing temperature and duration. It has been emphasized that the O–Nb ratio approaches stoichiometry after the TA of LN films in an air environment [44]. Based on the results presented in this monograph, our aim was to develop optimal technological regimes for the fabrication of Si–LN heterostructures for nonvolatile memory unit applications.

Si–LN heterostructures were formed through RFMS of a single-crystal LN target in an Ar environment and in an $Ar + O_2$ reactive gas mixture, following the specified regimes outlined in table 6.2. Silicon wafers with n-type conductivity ($\rho = 4.5 \, \Omega \cdot cm$) underwent cleaning by ion etching in an Ar plasma (2 min) before the RFMS process and served as substrates.

Subsequent TA of the as-grown heterostructures was performed in air at temperatures ranging from 600 °C to 700 °C for 1 h. The film's thickness was determined using atomic force microscopy (AFM, Solver P47) on cleaved as-grown heterostructures. A controlled film was deposited alongside the studied films onto a

Table 6.2. The synthesis regimes used in this work.

	Regime #			
	1	2	3	4
Reactive gas environment		Ar		$Ar(60\%) + O_2(40\%)$
Magnetron power (W)			100	
Reactive gas pressure (Pa)			0.15	
Substrate–target distance (cm)	5	5	10	5
Substrate position	Over the target erosion zone		Offset from the target erosion zone	

substrate satellite partially shaded by glue. After deposition, the glue was dissolved by alcohol, leaving a step. The height of this step, which equalled the film thickness, was measured using AFM.

The composition and structure of the films were analyzed using x-ray diffraction (XRD, ARL X'TRA Thermo Techno), while their morphology was examined through AFM. The electrical properties of the as-grown heterostructures and those after TA were assessed through high-frequency ($f = 10^5$ Hz) capacitance–voltage (C–V) characteristics. Top contacts ($S = 4.5 \times 10^{-6}$ m^2) were created via the thermal evaporation and condensation of aluminum in a vacuum ($P = 1 \times 10^{-4}$ Pa). Bottom contacts were formed by applying an In/Ga eutectic alloy to the (001)Si wafer to establish ohmic contacts.

The ferroelectric properties of the investigated heterostructures were analyzed using the Sawyer–Tower method [45].

Figure 6.1 illustrates the *in situ* XRD patterns of both the as-grown Li–Nb–O film and its transformation during TA. As depicted in the figure, the as-grown films exhibit an amorphous structure which transitions into a crystalline state at 475 °C, as evidenced by the appearance of diffraction peaks corresponding to the crystal lattice of LN (space group R3c, with lattice parameters $a = b = 0.515$ nm and $c = 1.39$ nm). With an increase in annealing temperature from 450°C to 475°C, the proportion of the crystalline phase rapidly increases from 45% to 100%. Beyond 500 °C, additional reflections attributed to the LiNb$_3$O$_8$ phase emerge due to oxygen deficiency, although this phase comprises no more than 8% of the total phase

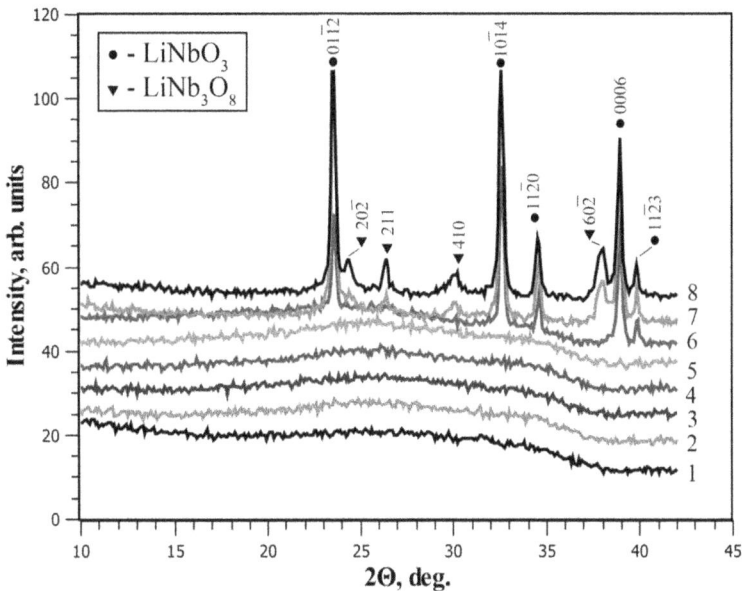

Figure 6.1. *In situ* XRD patterns acquired during the TA process of as-grown Li–Nb–O films (1) and film annealed at different temperatures: 350 °C (2), 400 °C (3), 425 °C (4), 450 °C (5), 475 °C (6), 500 °C (7), and 525 °C (8). Reproduced from [46], with permission from Springer Nature.

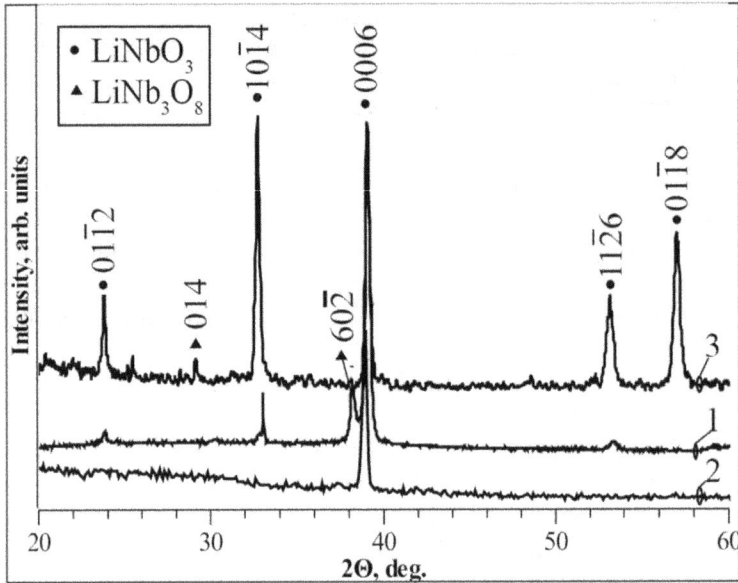

Figure 6.2. XRD patterns of films fabricated by the RFMS method under different regimes: 1—regime 1, 2—regime 2, 3—regime 3 (see table 6.2). Reproduced from [46], with permission from Springer Nature.

composition. Notably, LN films crystallized at 475 °C exhibit no discernible texture, a trend persisting with higher annealing temperatures.

Consequently, a substrate temperature of $T_s = 550$ °C was selected as the optimal condition for the formation of polycrystalline films in this study. Figure 6.2 showcases the XRD patterns of films grown under the influence of plasma components (the plasma effect). Spectrum 3 is taken from a sample produced without the plasma effect, while spectrum 1 represents its maximum impact.

Figure 6.2 reveals that in the absence of the plasma effect, films with random grain orientation (curve 3) are produced, whereas a moderate plasma effect (curve 2) induces the formation of single-phase textured $\langle 0001 \rangle$-LN films. Beyond the region of strong plasma influence, thermodynamic equilibrium conditions prevail, favoring the natural growth texture $\langle 0001 \rangle$ for LN films. In the target erosion zone (under an intense plasma effect), the stochastic nature of grain nucleation dominates, suppressing the $\langle 0001 \rangle$texture. Moreover, an increase in plasma effect intensity results in the formation of films composed of two phases: LN and the paraelectric $LiNb_3O_8$ phase.

According to the generally accepted model, during film growth under plasma influence, bombarding ions generate an appropriate density of mobile defects on the film surface [47]. However, to minimize defect generation in the subsurface layers, these ions, predominantly negative oxygen ions [47], should not possess extremely high energy. Therefore, only careful selection of the substrate–target distance and relative positioning creates optimal ion-assist conditions, facilitating the oriented growth of $\langle 0001 \rangle$-LN films (refer to figure 6.2).

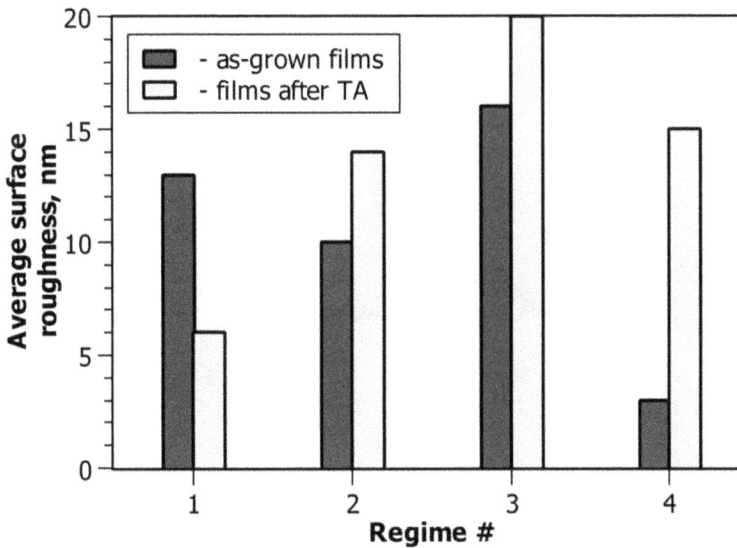

Figure 6.3. Average surface roughness of films fabricated under various regimes (see table 6.2).

Analysis of the XRD spectra of films fabricated in an Ar(60%) + O$_2$(40%) reactive gas mixture (refer to regime 4 in table 6.2) revealed that textured ⟨0001⟩ LN films are formed under the plasma effect during the RFMS process [46].

The grain size, as estimated by Selyakov–Scherrer analysis, remained consistent at approximately 40–50 nm across all films, irrespective of the RFMS regime employed. Analysis of the films' morphology indicated that LN films produced under regime 4 exhibit minimal surface roughness (refer to figure 6.3), a factor of significant importance in mitigating losses in optoelectronic devices.

Our results demonstrate that TA of the textured LN films leads to the disappearance of texture and the formation of the second phase, LiNb$_3$O$_8$, with a twofold increase in grain size, up to 100 nm [46]. Surface roughness increases during TA, except in films grown over the target erosion zone (see figure 6.3, regime 1).

Figure 6.4 shows the ferroelectric hysteresis loops of the films fabricated under regimes given in table 6.2.

As indicated in figure 6.4, all films exhibit ferroelectric properties, the parameters of which are listed in table 6.3. According to table 6.4, the remnant polarization of films grown under regimes 1, 2, and 4 is close to that of bulk LN ($P_r = 71\,\mu C\,cm^{-2}$) [48], which is desirable for nonvolatile memory units. Structural analysis of these three regimes shows the formation of textured ⟨0001⟩LN films.

Films with arbitrary grain orientation (regime 3) exhibit the lowest remnant polarization and higher coercive field, making them unsuitable for practical applications (see table 6.1). Notably, films synthesized in an Ar + O$_2$ gas environment (regime 4) have a minimal built-in field. The presence of built-in fields in ferroelectric films can be attributed to oxide charge (either bulk or interfacial). To investigate this effect, we studied the electrical properties of synthesized Si–LN

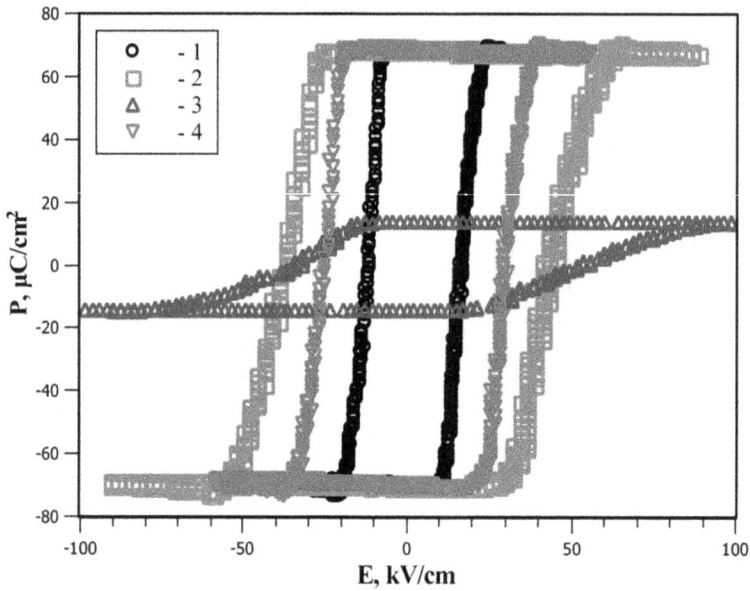

Figure 6.4. *P–E* hysteresis loops of films grown under the different RFMS regimes given in table 2: 1—regime 1, 2—regime 2, 3—regime 3, 4—regime 4. Reproduced from [46], with permission from Springer Nature.

Table 6.3. Hysteresis loop parameters of the studied films.

Regime #	Coercive field, E_c (kV cm^{-1})		Remnant polarization, P_r (μC cm^{-2})		Vertical shift, ΔP_r (μC cm^{-2})		Built-in field, E_i (kV cm^{-1})	
	As-grown	After TA	As-grown	After TA	As-grown	After TA	As-grown	After TA
1	14.0	68	69	12	−1	−6	2.8	2.4
2	39.5	69	69	37.8	−1	−7	2.8	1.2
3	44.0	—	14	—	−0.5	—	9.7	—
4	27.3	69	69	17.4	−1	−7	2.0	−2.7

Table 6.4. The results of *C–V* analysis for Si–LN–Al heterostructures fabricated under the different regimes given in table 6.2.

	Regime 1		Regime 2		Regime 4	
	As-grown	After TA	As-grown	After TA	As-grown	After TA
Dielectric constant of LN film ε	32	—	28	25	29	12
Flat band voltage, V_{FB} (volts)	−4.8	—	−5.6	−0.1	−3	2

Effective charge density, Q_{eff} (C cm^{-2})	$+9 \times 10^{-7}$	—	$+8 \times 10^{-7}$	$+5.5 \times 10^{-9}$	$+5.1 \times 10^{-8}$	-4.0×10^{-8}
Charged state density, N_{eff} (cm^{-2})	5.7×10^{12}	—	5.0×10^{12}	3.3×10^{10}	3.2×10^{11}	2.4×10^{11}
Position of the charge centroid (in units of film thickness d_c/d)	0.96	—	0.36	0.97	—	0.71

heterostructures through their $C-V$ characteristics. We revealed that the normalized high-frequency $C-V$ characteristics of the studied heterostructures were similar to those of metal–insulator–semiconductor heterostructures [46]. The common parameters, estimated through the standard $C-V$ analysis for the studied heterostructures, are given in table 6.4.

As shown in table 6.4, the dielectric constant of the as-grown films is close to that of bulk LN ($\varepsilon = 30$ [49]). Additionally, the effective charge density is nearly the same for films grown under regimes 1 and 2. According to [25], this charge is attributed to the plasma effect, which causes intense bombardment of the film. This effect leads to stochastic film growth, defect generation, and the formation of a paraelectric LiNb$_3$O$_8$ phase. Consequently, the positive oxide charge in LN films is not caused by the intense plasma effect or the presence of the LiNb$_3$O$_8$ phase. A similar magnitude of positive oxide charge was observed in LN films grown by the laser ablation method and was explained by the presence of Li vacancies [50].

Table 6.4 shows that films deposited in an Ar + O$_2$ gas mixture (regime 4) have the lowest density of effective positive charge. Therefore, the presence of oxygen in the reactive chamber is crucial for reducing this charge. Additionally, the location of the charge centroid indicates that the charge is not localized at the film–substrate interface. It has been demonstrated that sputtering parameters significantly influence defect generation in LN films. Specifically, it is recommended to conduct the RFMS process under a pressure of 1.3 Pa in an Ar(80%) + O$_2$(20%) reactive gas mixture. The authors of [43] suggest that these conditions facilitate the formation of LN films with minimal oxygen vacancy concentration. Conversely, some authors have reported that the presence of O$_2$ in the reactive chamber affects the plasma composition, increasing the concentration of Li ions [37]. An analysis of various models suggests that the most plausible origin of the positive oxide charge is charged antisite defects, such as Nb$_{Li}^{4\cdot}$ (where the dot denotes the positive charge) or oxygen vacancies [51]. According to these models, oxygen reduces the concentration of vacancies and, thus, the positively charged centers in the film. Our results revealed that the $P-E$ hysteresis loops of Si–LN heterostructures were deformed after TA [46]. All parameters derived from the $P-E$ loops are provided in table 6.3.

As indicated in table 6.3, TA does not affect the remnant polarization but causes all loops to shift downward by the same magnitude, regardless of the sputtering regime. This could indicate a preferred polarization orientation in the films after TA. Furthermore, TA of the films grown in an Ar(60%) + O$_2$(40%) gas mixture (regime

4) results in an increased coercive field and the emergence of a relatively strong built-in field in the opposite direction (see table 6.3). We investigated this effect using C–V analysis.

As demonstrated earlier, LN films deposited in pure Ar exhibit a positive effective charge, irrespective of the films' orientation, the plasma effect, and the presence of the $LiNb_3O_8$ phase (regimes 1 and 2). Therefore, we conducted a further comparison of C–V characteristics for films grown under regimes 2 and 4, both before and after TA. Our analysis of C–V curves (see table 6.4) revealed that TA reduces the effective positive charge in the films, resulting in a smaller shift along the horizontal axis [46]. The significant decrease in the effective positive charge compared to the as-grown heterostructures can be attributed to oxygen diffusion during TA. The diffusing oxygen neutralizes the positively charged oxygen vacancies in LN films, which aligns with previously reported results [52].

The results obtained in this section for the decrease in positive charge are particularly important for the effective switching of Si–LN heterostructures (see table 6.1). Moreover, for Si–LN heterostructures fabricated in pure Ar after TA, the nearly flat band regime occurs. This regime is ideal from a practical application standpoint, where space charge effects are detrimental. Specifically, gate materials with high dielectric constants result in a depolarization field if a depletion layer is connected in series [10]. Consequently, achieving long retention times of the stored polarization state in scaled-down FeFET devices is very challenging. Additionally, some authors at the forefront of ferroelectric memory design have noted that a preexisting charge in a ferroelectric material negatively impacts polarization reversal, which is crucial for the effective use of such materials in memory applications [10].

Furthermore, it is noteworthy that TA causes a sign change in the oxide charge from positive to negative for Si–LN heterostructures fabricated in an Ar(60%) + O_2(40%) atmosphere (see table 6.4, regime 4). This phenomenon can be explained by the formation of $V'_{Li} - 1/2O_2$ complexes. Oxygen diffuses into the film and fills the oxygen vacancies. If oxygen resides in a position near a lithium vacancy, charge transfer may occur according to the following reaction [41]:

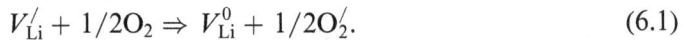

$$V'_{Li} + 1/2O_2 \Rightarrow V^0_{Li} + 1/2O'_2. \qquad (6.1)$$

The decrease in the built-in field in Si–LN heterostructures during TA (see table 6.3) can also be explained by the decline in positive charge.

To sum up, based on the results obtained in the present work, we propose the optimal RFMS regimes for Si–LN heterostructures according to the desired properties and requirements for their successful application (see table 6.1). The proposed regimes are provided in table 6.5.

Based on the results obtained in this study, the following conclusions can be drawn:

1. During the RFMS process under moderate plasma conditions, single-phase textured $\langle 0001 \rangle$ LN films are formed on Si substrates. The crystallization of amorphous films begins at 475 °C.

Table 6.5. Optimal RFMS regimes for Si–LN heterostructures.

Substrate: Si	Reactive gas environment	
	Ar	Ar + O_2 (60/40)
Magnetron power (W)	100	
Substrate temperature	550 °C	
Working pressure (Pa)	0.15	
Target-substrate distance (cm)	4–5 (offset the target erosion zone)	
TA	At $T = 650$ °C (60 min)	none
Film properties	Two-phase (LN + $LiNb_3O_8$) with an arbitrary grain orientation (average grain size 100 nm). Average surface roughness is 15 nm. Electrical properties: see tables 6.3 and 6.4 (regime 2)	Single-phase textured $\langle 0001 \rangle$ LN films with minimal surface roughness (3 nm). Average grain size is 40 nm. Electrical properties: see tables 6.3 and 6.4 (regime 4)

2. The presence of oxygen along with argon in the reactive chamber leads to the formation of highly oriented single-phase $\langle 0001 \rangle$ LN films, which minimizes both surface roughness and positive oxide charge.

3. Optimal RFMS conditions and subsequent TA methods have been proposed for the fabrication of multifunctional Si–LN heterostructures, meeting all the parameters required for their successful application in field-effect transistors and nonvolatile memory units that utilize electro-optic properties. These parameters include:
   - Single-phase textured $\langle 0001 \rangle$ LN films with minimal surface roughness (3 nm)
   - Flat band regime
   - Low coercive field
   - Remnant polarization and dielectric constant close to those of bulk LN
   - Lowest density of effective positive charge

This chapter has been reproduced with permission from [46].

References

[1] Sun B, Han P, Zhao W, Liu Y and Chen P 2014 White-light-controlled magnetic and ferroelectric properties in multiferroic $BiFeO_3$ square nanosheets *J. Phys. Chem.* C **118** 18814–9

[2] Noguchi K, Mitomi O and Miyazawa H n.d. Low-voltage and broadband $Ti:LiNbO_3$ modulators operating in the millimeter wavelength region *Optical Fiber Communications, OFC* (Optical Society of America) 205–6

[3] Dadami S T, Matteppanavar S, Shivaraja I, Rayaprol S, Angadi B and Sahoo B 2016 Investigation on structural, Mössbauer and ferroelectric properties of $(1-x)PbFe_{0.5}Nb_{0.5}O_3-(x)BiFeO_3$ solid solution *J. Magn. Magn. Mater.* **418** 122–7

[4] Matteppanavar S, Rayaprol S, Anupama A V, Sahoo B and Angadi B 2015 On the room temperature ferromagnetic and ferroelectric properties of $Pb(Fe_{1/2}Nb_{1/2})O_3$ *J. Supercond. Nov. Magn.* **28** 2465–72

[5] Lee Y S, Kim G-D, Kim W-J, Lee S-S, Lee W-G and Steier W H 2011 Hybrid $Si-LiNbO_3$ microring electro-optically tunable resonators for active photonic devices *Opt. Lett.* **36** 1119

[6] Mercante A J, Yao P, Shi S, Schneider G, Murakowski J and Prather D W 2016 110 GHz CMOS compatible thin film $LiNbO_3$ modulator on silicon *Opt. Express* **24** 15590

[7] Akazawa H and Shimada M 2004 Correlation between interfacial structure and c-axis-orientation of $LiNbO_3$ films grown on Si and SiO_2 by electron cyclotron resonance plasma sputtering *J. Cryst. Growth* **270** 560–7

[8] Huang C H-J 1995 Hybrid integrated optical devices utilizing thin films of lithium niobate *Integr. Ferroelectr.* **6** 355–62

[9] Wessels B W 2007 Ferroelectric epitaxial thin films for integrated optics *Annu. Rev. Mater. Res.* **37** 659–79

[10] Coll M *et al* 2019 Towards oxide electronics: a roadmap *Appl. Surf. Sci.* **482** 1–93

[11] Bartasyte A, Margueron S, Baron T, Oliveri S and Boulet P 2017 Toward high-quality epitaxial $LiNbO_3$ and $LiTaO_3$ thin films for acoustic and optical applications *Adv. Mater. Interfaces* **4** 1600998

[12] Hwang C S and Mikolajick T 2019 Ferroelectric memories ed B Magyari-Köpe and Y Nishi *Advances in Non-Volatile Memory and Storage Technology* 2nd edn (Woodhead Publishing) pp 393–441

[13] Rost T A, Lin H and Rabson T A 1991 Ferroelectric switching of a field-effect transistor with a lithium niobate gate insulator *Appl. Phys. Lett.* **59** 3654

[14] Haertling G H 1991 Ferroelectric thin films for electronic applications *J. Vac. Sci. Technol.* A **9** 414–20

[15] Fork D K, Armani-Leplingard F and Kingston J J 1996 Application of electroceramic thin films to optical waveguide devices *MRS Bull.* **21** 53–8

[16] Kingston J J, Fork D K, Leplingard F and Ponce F A 1994 C– outgrowths in C+ thin films of $LiNbO_3$ on Al_2O_3-c *MRS Proc.* **341** 289

[17] Feigelson R S 1996 Epitaxial growth of lithium niobate thin films by the solid source MOCVD method *J. Cryst. Growth* **166** 1–16

[18] Kadota M, Ogami T, Yamamoto K, Tochishita H and Negoro Y 2010 High-frequency lamb wave device composed of MEMS structure using $LiNbO_3$ thin film and air gap *IEEE Trans. Ultrason. Ferroelectr. Freq. Control* **57** 2564–71

[19] Bredikhin S, Scharner S, Klingler M, Kveder V, Red'kin B and Weppner W 2000 Nonstoichiometry and electrocoloration due to injection of Li^+ and O^{2-} ions into lithium niobate crystals *J. Appl. Phys.* **88** 5687–94

[20] Sanna S, Hölscher R and Schmidt W G 2014 Temperature dependent $LiNbO_3$(0 0 0 1): surface reconstruction and surface charge *Appl. Surf. Sci.* **301** 70–8

[21] Pei Z, Hu Q, Kong Y, Liu S, Chen S and Xu J 2011 Investigation on p-type lithium niobate crystals *AIP Adv.* **1** 032171

[22] Hao L *et al* 2012 Integration and electrical properties of epitaxial $LiNbO_3$ ferroelectric film on n-type GaN semiconductor *Thin Solid Films* **520** 3035–8

[23] Paturzo M *et al* 2005 On the origin of internal field in lithium niobate crystals directly observed by digital holography *Opt. Express* **13** 5416

[24] Chandra P, Dawber M, Littlewood P B and Scott J F 2004 Scaling of the coercive field with thickness in thin-film ferroelectrics *Ferroelectrics* **313** 7–13

[25] Sumets M, Kostyuchenko A, Ievlev V, Kannykin S and Dybov V 2015 Sputtering condition effect on structure and properties of LiNbO₃ films *J. Mater. Sci., Mater. Electron.* **26** 4250–6

[26] Volk T and Wöhlecke M 2009 *Lithium Niobate: Defects, Photorefraction and Ferroelectric Switching* (Berlin: Springer)

[27] Li W *et al* 2019 P-type lithium niobate thin films fabricated by nitrogen-doping *Materials* **12** 819

[28] Sumets M, Ievlev V, Kostyuchenko A and Kuz'mina V 2014 Influence sputtering conditions on electrical characteristics of Si-LiNbO₃ heterostructures formed by radio-frequency magnetron sputtering *Mol. Cryst. Liq. Cryst.* **603** 202–15

[29] Sakashita Y and Segawa H 1995 Preparation and characterization of LiNbO₃ thin films produced by chemical-vapor deposition *J. Appl. Phys.* **77** 5995

[30] Curtis B J and Brunner H R 1975 The growth of thin films of lithium niobate by chemical vapour de position *Mater. Res. Bull.* **10** 515–20

[31] Kondo S, Miyazawa S, Fushimi S and Sugii K 1975 Liquid-phase-epitaxial growth of single-crystal LiNbO₃ thin film *Appl. Phys. Lett.* **26** 489

[32] Chaos J, Perea A, Gonzalo J, Dreyfus R, Afonso C and Perrière J 2000 Ambient gas effects during the growth of lithium niobate films by pulsed laser deposition *Appl. Surf. Sci.* **154** 473–7

[33] Nashimoto K, Cima M J, McIntyre P C and Rhine W E 1995 Microstructure development of sol–gel derived epitaxial LiNbO₃ thin films *J. Mater. Res.* **10** 2564–72

[34] Shibata Y, Kaya K, Akashi K, Kanai M, Kawai T and Kawai S 1993 Epitaxial growth of LiNbO3 films on sapphire substrates by excimer laser ablation method and their surface acoustic wave properties *Jpn. J. Appl. Phys.* **32** L745–7

[35] Iyevlev V, Kostyuchenko A and Sumets M 2011 Fabricatoin, substructure and properties of LiNbO₃ films *Proc. SPIE—Int. Soc. Opt. Eng.* 77471J–8

[36] Tsirlin M 2004 Influence of gas phase composition on the defects formation in lithium niobate *J. Mater. Sci.* **39** 3187–9

[37] Gordillo-Vázquez F J and Afonso C N 2002 Influence of Ar and O₂ atmospheres on the Li atom concentration in the plasma produced by laser ablation of LiNbO₃ *J. Appl. Phys.* **92** 7651

[38] Ogale S B, Nawathey-Dikshit R, Dikshit S J and Kanetkar S M 1992 Pulsed laser deposition of stoichiometric LiNbO₃ thin films by using O₂ and Ar gas mixtures as ambients *J. Appl. Phys.* **71** 5718

[39] Fakhri M A, Salim E T, Hashim U, Abdulwahhab A W and Salim Z T 2017 Annealing temperature effect on structural and morphological properties of nano photonic LiNbO₃ *J. Mater. Sci., Mater. Electron.* **28** 16728–35

[40] Kiselev D A, Zhukov R N, Bykov A S, Voronova M I, Shcherbachev K D, Malinkovich M D and Parkhomenko Y N 2014 Effect of annealing on the structure and phase composition of thin electro-optical lithium niobate films *Inorg. Mater.* **50** 419–22

[41] Simões A Z, Zaghete M A, Stojanovic B D, Gonzalez A H, Riccardi C S, Cantoni M and Varela J A 2004 Influence of oxygen atmosphere on crystallization and properties of LiNbO₃ thin films *J. Eur. Ceram. Soc.* **24** 1607–13

[42] Simões A Z, Zaghete M A, Stojanovic B D, Riccardi C S, Ries A, Gonzalez A H and Varela J A 2003 LiNbO₃ thin films prepared through polymeric precursor method *Mater. Lett.* **57** 2333–9

[43] Shandilya S, Tomar M and Gupta V 2012 Deposition of stress free c-axis oriented LiNbO₃ thin film grown on (002) ZnO coated Si substrate *J. Appl. Phys.* **111** 10–6

[44] Edon V, Rèmiens D and Saada S 2009 Structural, electrical and piezoelectric properties of LiNbO$_3$ thin films for surface acoustic wave resonators applications *Appl. Surf. Sci.* **256** 1455–60

[45] Sawyer C B and Tower C H 1930 Rochelle salt as a dielectric *Phys. Rev.* **35** 269–73

[46] Sumets M, Ievlev V, Dybov V, Kostyuchenko A, Serikov D, Kannykin S and Belonogov E 2019 Synthesis and properties of multifunctional Si–LiNbO$_3$ heterostructures for non-volatile memory units *J. Mater. Sci., Mater. Electron.* **30** 16562–70

[47] Ellmer K and Welzel T 2012 Reactive magnetron sputtering of transparent conductive oxide thin films: role of energetic particle (ion) bombardment *J. Mater. Res.* **27** 765–79

[48] Wemple S H, DiDomenico M and Camlibel I 1968 Relationship between linear and quadratic electro-optic coefficiens in LiNbO$_3$, LiTaO$_3$, and other oxygen-octahedra ferro-electrics based on direct measurement of spontaneous polarization *Appl. Phys. Lett.* **12** 209–11

[49] Nassau K, Levinstein H J and Loiacono G M 1966 Ferroelectric lithium niobate. 2. Preparation of single domain crystals *J. Phys. Chem. Solids* **27** 989–96

[50] Hao L Z, Zhu J, Luo W B, Zeng H Z, Li Y R and Zhang Y 2010 Electron trap memory characteristics of LiNbO$_3$ film/AlGaN/GaN heterostructure *Appl. Phys. Lett.* **96** 032103

[51] Smyth D M 1983 Defects and transport in LiNbO$_3$ *Ferroelectrics* **50** 93–102

[52] Bollmann W 1977 The origin of photoelectrons and the concentration of point defects in LiNbO$_3$ crystals *Phys. Status Solidi* **40** 83–91

Lithium Niobate-Based Heterostructures (Second Edition)
Synthesis, properties, and electron phenomena
Maxim Sumets

Appendix A

Cell parameters and powder x-ray diffraction data for $LiNbO_3$ [1], $LiNb_3O_8$ [2], Li_3NbO_4 [3], and Nb_2O_5 [4].

$LiNbO_3$

Space Group: $R3c$

2Θ (°)	d-Spacing (Å)	Intensity	h	k	l
23.715	3.7488	999	0	1	2
32.713	2.7353	349	1	0	4
34.83	2.5737	204	1	1	0
38.968	2.3094	34	0	0	6
40.073	2.2482	66	1	1	3
42.573	2.1218	105	2	0	2
48.529	1.8744	154	0	2	4
53.249	1.7188	209	1	1	6
54.844	1.6726	4	2	1	1
56.133	1.6372	120	1	2	2
56.995	1.6144	62	0	1	8
61.112	1.5152	107	2	1	4
62.448	1.4859	86	3	0	0
64.693	1.4397	2	1	2	5
68.557	1.3676	37	2	0	8
71.203	1.3232	32	1	0	10
73.536	1.2869	22	2	2	0
73.792	1.283	12	2	1	7
76.111	1.2496	28	3	0	6
76.834	1.2396	3	2	2	3
77.437	1.2315	2	1	3	1
78.524	1.2171	47	3	1	2
79.256	1.2077	40	1	2	8
81.775	1.1768	17	0	2	10
82.831	1.1644	30	1	3	4

(Continued)

2Θ (°)	d-Spacing (Å)	Intensity	h	k	l
83.686	1.1547	6	0	0	12
86.034	1.1291	1	3	1	5
86.508	1.1241	35	2	2	6
88.864	1.1003	16	0	4	2

Cell parameters: $a = 5.147$ Å, $b = 5.147$ Å, $c = 13.856$ Å, $\alpha = 90.0°$, $\beta = 90.0°$, and $\gamma = 120.0°$.

LiNb₃O₈

$LiNb_3O_8$

Space Group: $P2_1/a$

2Θ (°)	d-Spacing (Å)	Intensity	h	k	l
12.15	7.2842	11.46	2	0	0
14.58	6.0760	22.18	−2	0	1
19.87	4.4686	13.95	2	0	1
21.46	4.1407	29.88	2	1	0
21.62	4.1095	39.21	0	1	1
22.94	3.8760	8.41	−2	1	1
23.9	3.7234	15.5	−4	0	1
24.34	3.6564	26.63	−2	0	2
24.44	3.6421	27.24	4	0	0
25.02	3.5591	11.53	0	0	2
26.68	3.3416	6.64	2	1	1
29.4	3.0380	6.91	−4	0	2
29.85	2.9933	2	−4	1	1
30.21	2.9581	100	−2	1	2
30.29	2.9506	99.26	4	1	0
30.73	2.9097	5.23	4	0	1
30.77	2.9059	1.06	0	1	2
31.08	2.8773	19.91	2	0	2
34.48	2.6009	2.39	−4	1	2
35.68	2.5165	23.69	0	2	0
35.95	2.4979	26.25	2	1	2
36.15	2.4846	3.13	−2	0	3
37.02	2.4281	3.19	6	0	0
38.14	2.3597	10.38	−6	0	2
38.63	2.3310	1.49	−4	0	3
38.73	2.3250	3.38	−2	2	1

39.74	2.2684	2.57	−6	1	1
40.49	2.2279	2.38	−2	1	3
40.63	2.2205	7.46	−3	2	1
40.81	2.2109	7.98	−1	1	3
40.94	2.2046	6.42	5	1	1
41.17	2.1927	2	2	2	1
42.1	2.1462	2.9	0	1	3
43.4	2.0850	4.44	−4	2	1
43.49	2.0808	1.27	2	0	3
43.66	2.0730	7.31	−2	2	2
43.72	2.0704	7.54	4	2	0
44.07	2.0547	2.88	0	2	2
44.36	2.0421	2.94	4	1	2
46.61	1.9487	4.25	6	1	1
46.88	1.9380	2.83	−4	2	2
47.27	1.9229	1.33	2	1	3
47.72	1.9059	1.15	−8	0	1
47.79	1.9034	2.48	4	2	1
48.03	1.8942	13.51	2	2	2
48.45	1.8789	4.58	−6	1	3
49.88	1.8282	6.12	−4	0	4
50.09	1.8210	5.9	8	0	0
51.26	1.7824	2.14	−8	1	1
51.34	1.7795	13.46	0	0	4
51.5	1.7746	13.57	6	0	2
51.7	1.7681	3.18	−2	2	3
51.88	1.7623	1	4	0	3
52.36	1.7473	3.33	6	2	0
52.4	1.7461	3.64	−8	1	2
52.41	1.7457	3.38	−2	1	4
53.21	1.7213	32.06	−6	2	2
53.31	1.7183	20.91	−4	1	4
53.51	1.7124	21.55	8	1	0
53.59	1.7101	1.91	−4	2	3
54.71	1.6777	1.17	0	1	4
54.86	1.6736	1.14	6	1	2
56.27	1.6349	1.43	2	3	0
56.34	1.6329	2.11	0	3	1
57.31	1.6076	1.66	−6	1	4
57.47	1.6036	1.5	2	2	3
60.74	1.5248	12.92	−2	3	2
60.79	1.5238	12.67	4	3	0
60.98	1.5193	1.27	−8	2	1
62.58	1.4844	1.17	−4	0	5
62.83	1.4791	4.76	−4	2	4
63.01	1.4753	4.71	8	2	0

(*Continued*)

A-3

(Continued)

2Θ (°)	d-Spacing (Å)	Intensity	h	k	l
64.03	1.4542	10.33	−8	1	4
64.09	1.4530	11.77	0	2	4
64.16	1.4515	10.24	−10	1	2
64.22	1.4503	12.11	6	2	2
64.27	1.4493	15.64	2	3	2
64.56	1.4435	1.08	4	2	3
64.8	1.4387	1.96	4	0	4
65.33	1.4283	1.1	6	1	3
66.46	1.4067	1.04	−6	2	4
66.95	1.3976	1.22	8	1	2
67.74	1.3833	11.12	4	1	4
68.49	1.3699	1.85	0	1	5
70.25	1.3399	1.02	−10	0	4
72.71	1.3005	1.78	−8	2	4
72.84	1.2985	1.97	−10	2	2
73.27	1.2920	1.47	−6	3	3
74.17	1.2785	1.22	−4	2	5
75.57	1.2582	2.71	0	4	0
75.75	1.2556	1.63	−10	2	3
76.42	1.2463	1.2	−8	3	2
77.17	1.2361	1.59	−4	3	4
77.33	1.2339	1.7	8	3	0
78.47	1.2188	2.56	−6	0	6
78.84	1.2140	2.73	12	0	0
80.32	1.1953	5.51	−2	1	6
80.59	1.1920	1.02	−4	4	1
80.65	1.1914	5.74	10	1	2
80.78	1.1898	1.1	−2	4	2
80.82	1.1893	1.09	4	4	0
81.36	1.1827	1.47	−10	2	4
81.6	1.1799	4.14	−12	0	4
83.94	1.1528	2.15	2	4	2
85.52	1.1355	1	−8	1	6
86.41	1.1260	3.17	−8	3	4
86.53	1.1248	3.2	−10	3	2
87.07	1.1192	1.47	2	0	6
87.27	1.1171	1.48	8	0	4
87.95	1.1103	5.86	−6	4	2
89.31	1.0969	4.4	−6	2	6
89.67	1.0934	4.66	12	2	0
89.76	1.0925	1.97	2	1	6
89.8	1.0921	4.18	4	3	4
89.96	1.0906	2.08	8	1	4

Cell parameters: $a = 15.2620\,\text{Å}$, $b = 5.0330\,\text{Å}$, $c = 7.4570\,\text{Å}$, $\alpha = 90.0°$, $\beta = 107.3°$, and $\gamma = 90.0°$.

Li₃NbO₄

Space Group: $I23$

2Θ (°)	d-Spacing (Å)	Intensity	h	k	l
14.86	5.9602	94.81	1	1	0
21.08	4.2145	4.47	2	0	0
25.89	3.4411	100	2	1	1
33.62	2.6655	13.74	3	0	1
33.62	2.6655	18.05	3	1	0
36.94	2.4332	6.93	2	2	2
40.02	2.2527	6.67	3	2	1
40.02	2.2527	9.75	3	1	2
42.92	2.1072	45.27	4	0	0
45.66	1.9867	8.9	4	1	1
45.66	1.9867	10.67	3	3	0
50.81	1.7971	1.22	3	3	2
53.24	1.7206	3.58	4	2	2
55.6	1.6531	6.39	4	3	1
55.6	1.6531	4.97	4	1	3
55.6	1.6531	1.33	5	0	1
60.13	1.5389	13.25	5	2	1
60.13	1.5389	11.37	5	1	2
62.31	1.4901	20.2	4	4	0
64.46	1.4456	5.75	4	3	3
66.56	1.4048	1.07	6	0	0
68.64	1.3674	2.14	5	3	2
68.64	1.3674	4.39	6	1	1
68.64	1.3674	3.2	5	2	3
72.7	1.3006	1.37	5	4	1
72.7	1.3006	1.04	5	1	4
74.7	1.2707	3.13	6	2	2
76.68	1.2428	1.95	6	3	1
76.68	1.2428	1.54	6	1	3
78.64	1.2166	4.97	4	4	4
80.59	1.192	1.33	7	1	0
84.46	1.147	1.48	6	3	3
84.46	1.147	6.13	5	5	2
88.3	1.1068	1.82	7	3	0
88.3	1.1068	1.57	7	0	3

Cell parameters: $a = 8.429$ Å, $b = 8.429$ Å, $c = 8.429$ Å, $\alpha = 90.0°$, $\beta = 90.0°$, and $\gamma = 90.0°$.

Nb₂O₅

Space Group: $C2/c$ (15)

2Θ (°)	d-Spacing (Å)	Intensity	h k l	Remarks
14.4	6.1523	214	2 0 0	
19.5	4.5386	2	1 1 0	
24.4	3.6425	999	−1 1 1	
26.9	3.314	715	1 1 1	
29.0	3.0761	566	4 0 0	
29.9	2.9811	927	−3 1 1	
32.7	2.7347	17	−2 0 2	
33.3	2.6854	378	0 0 2	
35.8	2.5033	331	3 1 1	
36.8	2.4415	178	0 2 0	
38.3	2.3466	76	−4 0 2	
39.7	2.2693	12	2 2 0	
39.9	2.2561	3	2 0 2	
40.6	2.22	99	−5 1 1	Multiply indexed line
40.6	2.22	99	1 1 2	Multiply indexed line
41.0	2.1976	59	5 1 0	
41.6	2.1659	39	−2 2 1	
44.1	2.0507	55	6 0 0	
44.8	2.0222	1	2 2 1	
46.9	1.9342	4	−5 1 2	
47.5	1.9123	221	4 2 0	
48.2	1.8866	32	5 1 1	
48.3	1.882	53	−6 0 2	
48.7	1.8663	4	3 1 2	
50.0	1.8212	11	−2 2 2	
50.5	1.8065	302	0 2 2	Doubly indexed line
50.5	1.8065	302	4 0 2	Doubly indexed line
53.1	1.7225	118	−1 1 3	
53.3	1.7163	67	4 2 1	
53.6	1.7079	157	−3 1 3	
53.8	1.7016	271	−7 1 1	
54.1	1.6918	242	−4 2 2	
55.4	1.657	6	2 2 2	
55.5	1.6539	4	7 1 0	
57.0	1.6136	71	1 3 0	Doubly indexed line
57.0	1.6136	71	1 1 3	Doubly indexed line
57.7	1.5948	3	−6 2 1	

57.8	1.5922	3	−7 1 2	
58.4	1.5783	8	−5 1 3	
58.7	1.5703	45	6 2 0	
59.2	1.56	47	−1 3 1	
60.1	1.538	42	8 0 0	
60.2	1.5354	26	5 1 2	
60.4	1.5311	52	1 3 1	
61.1	1.5146	36	−8 0 2	
62.1	1.4939	84	−3 3 1	
62.2	1.4905	82	−6 2 2	
62.6	1.4822	112	7 1 1	
63.8	1.4578	38	6 0 2	
64.1	1.4512	91	4 2 2	
64.5	1.4437	8	0 2 3	
64.8	1.4374	75	3 1 3	
65.0	1.4325	40	6 2 1	
65.6	1.4212	127	3 3 1	
66.5	1.4043	3	−1 3 2	
66.9	1.3963	82	−7 1 3	
68.1	1.3755	1	−3 3 2	
68.5	1.3673	42	−4 0 4	
68.8	1.3633	28	−5 3 1	Doubly indexed line
68.8	1.3633	28	1 3 2	Doubly indexed line
69.0	1.3595	6	−9 1 1	
70.0	1.3427	38	0 0 4	
70.5	1.334	4	−3 1 4	Doubly indexed line
70.5	1.334	4	−8 2 1	Doubly indexed line
71.4	1.3194	5	−9 1 2	
71.5	1.3172	7	−6 2 3	Doubly indexed line
71.5	1.3172	7	9 1 0	Doubly indexed line
72.6	1.3013	30	8 2 0	
73.5	1.287	40	−8 2 2	Multiply indexed line
73.5	1.287	40	−6 0 4	Multiply indexed line
74.2	1.2761	1	7 1 2	
74.4	1.2736	4	5 3 1	
74.8	1.2673	1	3 3 2	
75.8	1.2532	5	5 1 3	
75.9	1.2516	8	6 2 2	
76.2	1.2477	23	−10 0 2	
77.5	1.2304	8	10 0 0	
78.2	1.2207	22	0 4 0	
78.7	1.2141	39	−3 3 3	
78.9	1.2119	35	−7 3 1	
79.3	1.2066	14	−2 2 4	Doubly indexed line

(*Continued*)

(*Continued*)

2Θ (°)	*d*-Spacing (Å)	Intensity	*h k l*	Remarks
79.3	1.2066	14	8 0 2	Doubly indexed line
80.0	1.1974	3	2 4 0	
80.4	1.193	32	−4 2 4	
80.6	1.1903	19	0 4 1	
81.4	1.1814	6	−2 4 1	
81.6	1.1788	23	1 3 3	
81.8	1.1765	49	0 2 4	
82.0	1.1733	40	−8 0 4	
82.3	1.1704	13	−7 3 2	
82.8	1.1649	6	−5 3 3	
83.7	1.1545	1	3 1 4	
84.3	1.1473	1	5 3 2	
85.2	1.1381	13	−6 2 4	
85.5	1.1346	40	4 4 0	
85.7	1.1322	20	−4 4 1	
86.1	1.128	20	4 0 4	
86.4	1.1246	52	−11 1 1	Doubly indexed line
86.4	1.1246	52	7 3 1	Doubly indexed line
87.4	1.1147	3	−2 4 2	
87.7	1.1113	49	0 4 2	Doubly indexed line
87.7	1.1113	49	−10 2 2	Doubly indexed line
88.4	1.1046	23	3 3 3	
89.0	1.0988	32	10 2 0	
89.9	1.0896	33	7 1 3	Doubly indexed line
89.9	1.0896	33	11 1 0	Doubly indexed line

Cell parameters: $a = 21{,}74$ Å, $b = 4.883$ Å, $c = 5.561$ Å, $\alpha = 90.0°$, $\beta = 105.02°$, and $\gamma = 90.0°$.

References

[1] Abrahams S C and Marsh P 1986 Defect structure dependence on composition in lithium niobate *Acta Crystallogr.* B **42** 61–8

[2] Lundberg M 1971 The crystal structure of $LiNb_3O_8$ *Acta Chem. Scand.* **25** 3337–46

[3] Grenier J, Martin C and Durif A 1964 Etude cristallographique des orthoniobates et orthotantalates de lithium *Bull. Soc. Fr. Mineral. Cristallogr.* **87** 316–20 (in French)

[4] Ercit T S 1991 Refinement of the structure of ζ-Nb_2O_5 and its relationship to the rutile and thoreaulite structures *Mineral. Petrol.* **43** 217–23